Report of a River Pollution Survey of England and Wales 1970

Volume 2

NOTRE DAME COLLEGE OF EDUCATION
MOUNT PLEASANT
LIVERPOOL L3 5SP

Department of the Environment
The Welsh Office

Report of a River Pollution Survey of England and Wales 1970

Volume 2

Discharges and Forecasts of Improvement

London: Her Majesty's Stationery Office 1972

© Crown copyright 1972

SBN 11 750472 6*

Preface

Contents of Volume 2

1 This Volume contains the survey's results concerning discharges of sewage effluent, crude sewage and industrial effluent; unsatisfactory storm overflows; estimates of the cost of remedial work; and forecasts of the possibilities of river improvement. It also contains the report of the survey undertaken by the Confederation of British Industry on methods of disposal and cost of treatment of industrial effluent, analysed by type of industry.

Details of the scope of the survey and river quality, both chemical and biological, are given in Volume 1 published December 1971.

Acknowledgements

2 The Department are grateful for the whole-hearted co-operation of river authorities and local sewerage authorities who were responsible for giving or collecting data for some 25,000 miles of rivers, canals and estuaries. The Department is also grateful to the Confederation of British Industry for undertaking their useful supplementary survey and making the details available for publication in this Volume.

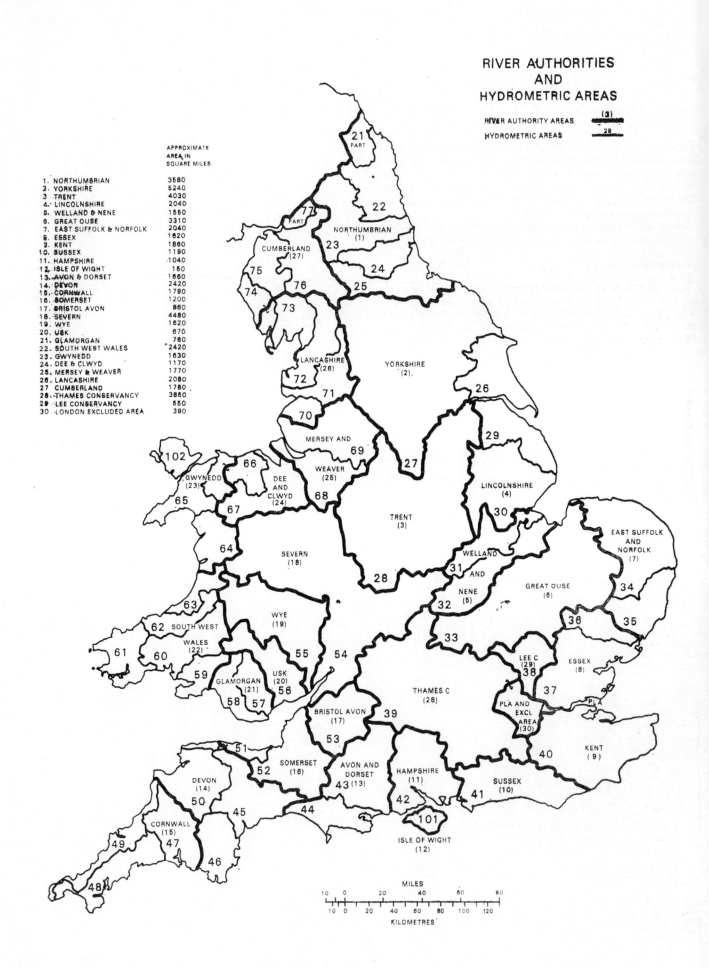

Fig. 1

Note on sequence of presentation

For the purposes of the survey, river authority areas have been numbered in a clockwise sequence round the coast of England and Wales, starting with Northumbrian River Authority in the north east and finishing with Cumberland River Authority in the north west, followed by Thames Conservancy Catchment Board, Lee Conservancy Catchment Board and Port of London Authority (including the London Excluded Area). The numbers are as indicated below and the descriptive text and tables are set out in this order. The authorities are shown in Figure 1.

1. Northumbrian River Authority
2. Yorkshire River Authority
3. Trent River Authority
4. Lincolnshire River Authority
5. Welland and Nene River Authority
6. Great Ouse River Authority
7. East Suffolk and Norfolk River Authority
8. Essex River Authority
9. Kent River Authority
10. Sussex River Authority
11. Hampshire River Authority
12. Isle of Wight River and Water Authority
13. Avon and Dorset River Authority
14. Devon River Authority
15. Cornwall River Authority
16. Somerset River Authority
17. Bristol Avon River Authority
18. Severn River Authority
19. Wye River Authority
20. Usk River Authority
21. Glamorgan River Authority
22. South West Wales River Authority
23. Gwynedd River Authority
24. Dee and Clwyd River Authority
25. Mersey and Weaver River Authority
26. Lancashire River Authority
27. Cumberland River Authority
28. Thames Conservancy Catchment Board
29. Lee Conservancy Catchment Board
30. Port of London Authority (including the London Excluded Area)

Glossary of abbreviations

ARA	Association of River Authorities
BOD	Biochemical Oxygen Demand
BWB	British Waterways Board
CBI	Confederation of British Industry
DOE	Department of the Environment
DWF	Dry Weather Flow
GLC	Greater London Council
ghd	gallons per head per day
gpd	gallons per day
HMSO	Her Majesty's Stationery Office
mgd	million gallons per day
mg/l	milligrams per litre
PLA	Port of London Authority
RA	River Authority
RC	Royal Commission on Sewage Disposal (1898-1915)
SS	Suspended Solids

Contents

	page
Note on sequence of presentation	vii
Glossary of abbreviations	viii

1 Introduction — 1
Main findings — 1
The scope of the survey — 2
Biological classification of rivers — 2

2 Discharges of sewage effluent — 3
Introduction — 3
Details of the survey for England and Wales — 3
River authorities:

Avon and Dorset	16
Bristol Avon	19
Cornwall	17
Cumberland	28
Dee and Clwyd	24
Devon	16
East Suffolk and Norfolk	11
Essex	12
Glamorgan	22
Great Ouse	11
Gwynedd	24
Hampshire	14
Isle of Wight	15
Kent	13
Lancashire	27
Lee Conservancy	30
Lincolnshire	9
Mersey and Weaver	25
Northumbrian	5
Port of London Authority (including the London Excluded Area)	31
Severn	20
Somerset	18
South West Wales	23
Sussex	14
Thames Conservancy	29
Trent	7
Usk	21
Welland and Nene	10
Wye	21
Yorkshire	6

3 Discharges of crude sewage — 33
Introduction — 33
Details of the survey for England and Wales — 33
River authorities:

Avon and Dorset	35
Bristol Avon	35
Cornwall	35
Cumberland	37
Dee and Clwyd	36
Devon	35
East Suffolk and Norfolk	34
Essex	34
Glamorgan	36
Great Ouse	34
Gwynedd	36
Hampshire	35
Isle of Wight	35
Kent	34
Lancashire	37
Lee Conservancy	37
Lincolnshire	34
Mersey and Weaver	36
Northumbrian	34
Port of London Authority (including the London Excluded Area)	37
Severn	36
Somerset	35
South West Wales	36
Sussex	35
Thames Conservancy	37
Trent	34
Usk	36
Welland and Nene	34
Wye	36
Yorkshire	34

4 Discharges of storm sewage from unsatisfactory storm overflows — 39
Introduction — 39
Other intermittent and polluting discharges — 39
Details of the survey for England and Wales — 39
River authorities:

Avon and Dorset	40
Bristol Avon	41
Cornwall	41
Cumberland	42
Dee and Clwyd	41
Devon	40
East Suffolk and Norfolk	40
Essex	40
Glamorgan	41
Great Ouse	40
Gwynedd	41
Hampshire	40
Isle of Wight	40
Kent	40
Lancashire	42
Lee Conservancy	42
Lincolnshire	40
Mersey and Weaver	41
Northumbrian	39
Port of London Authority (including the London Excluded Area)	42
Severn	41
Somerset	41
South West Wales	41
Sussex	40
Thames Conservancy	42
Trent	40
Usk	41
Welland and Nene	40
Wye	41
Yorkshire	40

5 Discharges of industrial effluent — 43
(i) *The Main Survey* — 43
Introduction — 43
Details of the survey for England and Wales — 44
River authorities:

Avon and Dorset	60
Bristol Avon	62
Cornwall	61

Cumberland	67
Dee and Clwyd	65
Devon	61
East Suffolk and Norfolk	58
Essex	59
Glamorgan	64
Great Ouse	58
Gwynedd	65
Hampshire	60
Isle of Wight	60
Kent	59
Lancashire	66
Lee Conservancy	68
Lincolnshire	57
Mersey and Weaver	66
Northumbrian	55
Port of London Authority (including the London Excluded Area)	68
Severn	62
Somerset	62
South West Wales	64
Sussex	59
Thames Conservancy	67
Trent	56
Usk	63
Welland and Nene	57
Wye	63
Yorkshire	56
(ii) *The Confederation of British Industry Survey*	69
Details of the survey for England and Wales	69

6 Expenditure
Introduction	83
Details of the estimates for England and Wales	83
River authorities:	
Avon and Dorset	91
Bristol Avon	93
Cornwall	92
Cumberland	97
Dee and Clwyd	96
Devon	92
East Suffolk and Norfolk	89
Essex	89
Glamorgan	94
Great Ouse	89
Gwynedd	95
Hampshire	91
Isle of Wight	91
Kent	90
Lancashire	96
Lee Conservancy	98
Lincolnshire	88
Mersey and Weaver	96
Northumbrian	87
Port of London Authority (including the London Excluded Area)	98
Severn	93
Somerset	92
South West Wales	95
Sussex	90
Thames Conservancy	97
Trent	87
Usk	94
Welland and Nene	88
Wye	94
Yorkshire	87

7 Forecasts of improvement
Introduction	99
River upgrading. England and Wales	99
Suitability of improved rivers for further public water supplies	101
Details of the survey for England and Wales	101
River authorities:	
Avon and Dorset	106
Bristol Avon	108
Cornwall	107
Cumberland	113
Dee and Clwyd	111
Devon	107
East Suffolk and Norfolk	104
Essex	105
Glamorgan	110
Great Ouse	104
Gwynedd	110
Hampshire	106
Isle of Wight	106
Kent	105
Lancashire	112
Lee Conservancy	113
Lincolnshire	103
Mersey and Weaver	111
Northumbrian	102
Port of London Authority (including the London Excluded Area)	114
Severn	108
Somerset	108
South West Wales	110
Sussex	105
Thames Conservancy	113
Trent	103
Usk	109
Welland and Nene	104
Wye	109
Yorkshire	102

8 River water quality and discharges in England & Wales
	115
Population and River Pollution	115
Discharges and River Pollution	115
(i) Non-tidal rivers	115
(ii) Tidal rivers	116
(iii) Canals	116
Costs of improvements to River Water Quality	116
Summary of the situation in the river authority areas	116
River authorities:	
Avon and Dorset	118
Bristol Avon	120
Cornwall	120
Cumberland	119
Dee and Clwyd	118
Devon	119
East Suffolk and Norfolk	116
Essex	119
Glamorgan	122
Great Ouse	118
Gwynedd	118
Hampshire	119
Isle of Wight	118
Kent	119
Lancashire	122
Lee Conservancy	121
Lincolnshire	118
Mersey and Weaver	122
Northumbrian	120
Port of London Authority (including the London Excluded Area)	123
Severn	120
Somerset	120
South West Wales	121
Sussex	116
Thames Conservancy	119
Trent	122
Usk	121
Welland and Nene	116
Wye	118
Yorkshire	121

Appendix 1 Note on biological classification of rivers — 124
Appendix 2 Questionnaire circulated by the Confederation of British Industries — 125
Additional tables — 129

List of tables

Table No.	Subject	Page
1	Summary of information about sewage treatment works and discharges of sewage effluent. England and Wales	4
2	Satisfactory discharges by population ranges	5
3	Unsatisfactory discharges—inadequate maintenance	5
4–33	Details of sewage treatment works and discharges of sewage effluent	5 to 31

River authorities:

Avon and Dorset	16	Lincolnshire	9
Bristol Avon	19	Mersey and Weaver	26
Cornwall	18	Northumbrian	5
Cumberland	28	Port of London Authority	
Dee and Clwyd	25	(including the London Excluded Area)	31
Devon	17	Severn	20
East Suffolk and Norfolk	12	Somerset	18
Essex	12	South West Wales	23
Glamorgan	23	Sussex	14
Great Ouse	11	Thames Conservancy	29
Gwynedd	24	Trent	7
Hampshire	15	Usk	22
Isle of Wight	15	Welland and Nene	10
Kent	13	Wye	21
Lancashire	27	Yorkshire	6
Lee Conservancy	30		

Table No.	Subject	Page
34	Summary of information about discharges of crude sewage. England and Wales	33
35	Unsatisfactory storm overflows. England and Wales	39
36	Industrial effluent classification code	43
37	Details of discharges of industrial effluent (excluding minewater) to rivers and canals in different chemical classes, with percentage satisfactory. England and Wales	44
38	Number of discharges of industrial effluent and percentage satisfactory	45
39	Number of discharges of industrial effluent excluding discharges of solely cooling water and percentage satisfactory	46
40	Number of discharges of solely cooling water and percentage satisfactory	47
41	Volume of all industrial effluent and percentage satisfactory	48
42	Volume of industrial effluent, excluding discharges of solely cooling water, and percentage satisfactory	49
43	Volume of solely cooling water and percentage satisfactory	50
44	Volume of process water discharged as part of industrial effluent recorded in Table 42	51
45	Volume of cooling water discharged as part of industrial effluent recorded in Table 42	52
46	Total volume of all cooling water consisting of solely cooling water (Table 43) and cooling water, if any, discharged as part of industrial effluent (Table 45)	53
47	Discharges of industrial effluent in each river authority area in order of volume	54
48	Discharges of minewater to rivers and canals	55
49–78	Numbers and volumes of discharges of industrial effluent (excluding minewater discharges)	55 to 69

River authorities:

Avon and Dorset	61	Lincolnshire	57
Bristol Avon	62	Mersey and Weaver	66
Cornwall	61	Northumbrian	55
Cumberland	67	Port of London Authority	
Dee and Clwyd	65	(including the London Excluded Area)	69
Devon	61	Severn	63
East Suffolk and Norfolk	58	Somerset	62
Essex	59	South West Wales	64
Glamorgan	64	Sussex	60
Great Ouse	58	Thames Conservancy	68
Gwynedd	65	Trent	57
Hampshire	60	Usk	63
Isle of Wight	60	Welland and Nene	58
Kent	59	Wye	63
Lancashire	67	Yorkshire	56
Lee Conservancy	68		

79	Comparison of the CBI survey and main survey returns	69
80	CBI survey. Summary of discharges of industrial effluent and costs of disposal	70
81	CBI survey. Information (by industries) about discharges to non-tidal rivers and canals	71
82	CBI survey. Information (by industries) about discharges to controlled tidal rivers and uncontrolled waters	72
83	CBI survey. Information (by industries) about discharges to sewers and about industrial material disposed of by contractors or by incineration	73
84	CBI survey. Summary (by industries) of financial information	74
85	CBI survey. Summary (by industries) of daily volumes of effluent discharged in 1968	76
86	CBI survey. Information (by river authorities) about discharges to non-tidal rivers and canals	77
87	CBI survey. Information (by river authorities) about discharges to controlled tidal rivers and uncontrolled waters	78
88	CBI survey. Information (by river authorities) about discharges to sewers and about industrial material disposed of by contractors or by incineration	79
89	CBI survey. Summary (by river authorities) of financial information	80
90	CBI survey. Summary (by river authorities) of daily volumes of effluent discharged in 1968	82
91	Estimates of costs of remedial works. England and Wales	84
92	Estimates of costs of remedial works on discharges of sewage effluent. England and Wales	84
93	Estimates of costs of remedial works on discharges of crude sewage. England and Wales	85
94	Estimates of costs of remedial works on discharges of industrial effluent. England and Wales	86
95	Estimated costs per head of population for dealing with discharges of sewage effluent and crude sewage. England and Wales	86
96–125	Estimates of costs of remedial works	87 to 98

River authorities:

Avon and Dorset	91		Lincolnshire	88
Bristol Avon	93		Mersey and Weaver	96
Cornwall	92		Northumbrian	87
Cumberland	97		Port of London Authority	
Dee and Clwyd	96		(including the London Excluded Area)	98
Devon	92		Severn	93
East Suffolk and Norfolk	89		Somerset	92
Essex	90		South West Wales	95
Glamorgan	94		Sussex	90
Great Ouse	89		Thames Conservancy	97
Gwynedd	95		Trent	88
Hampshire	91		Usk	94
Isle of Wight	91		Welland and Nene	88
Kent	90		Wye	94
Lancashire	97		Yorkshire	87
Lee Conservancy	98			

126	Upgrading in chemical classification following remedial works on discharges at present unsatisfactory. England and Wales	99
127	Comparison of mileages by chemical classification following remedial works on discharges at present unsatisfactory. Rivers, England and Wales	99
128	Comparison of mileages by chemical classification following remedial works on discharges at present unsatisfactory. Canals, England and Wales	99
129	Estimates of miles of river by chemical classification following remedial works. England and Wales	100
130	Estimates of miles of canal by chemical classification following remedial works. England and Wales	101
131	Suitability for public water supply. Situation after treatment of unsatisfactory discharges. Non-tidal rivers, England and Wales	101
132–166	Comparison of mileages by chemical classification following remedial works on discharges at present unsatisfactory	102 to 114

River authorities:

Avon and Dorset	106		Lincolnshire	103
Bristol Avon	108		Mersey and Weaver	111
Cornwall	107		Northumbrian	102
Cumberland	113		Port of London Authority	
Dee and Clwyd	111		(including the London Excluded Area)	114
Devon	107		Severn	108
East Suffolk and Norfolk	104		Somerset	108
Essex	105		South West Wales	110
Glamorgan	110		Sussex	105
Great Ouse	104		Thames Conservancy	113
Gwynedd	110		Trent	103
Hampshire	106		Usk	109
Isle of Wight	106		Welland and Nene	104
Kent	105		Wye	109
Lancashire	112		Yorkshire	102
Lee Conservancy	113			

167	Summary of selected information about non-tidal rivers. England and Wales	115
168	Summary of selected information about tidal rivers. England and Wales	117
169	Summary of selected information about canals. England and Wales	117

The following additional tables give a further breakdown of details given in Chapters 2 to 6 of the Report, and will be found at the end of the volume.

170	Sewage treatment works. Total number of discharges and percentage considered satisfactory	130
171	Sewage treatment works. Total population served and percentage served by sewage treatment works considered satisfactory	132
172	Sewage treatment works. Total dry weather flow of sewage effluent and percentage discharged from sewage treatment works considered satisfactory	134
173	Sewage treatment works. Total number of discharges required to comply with Royal Commission Standard and percentage considered satisfactory	136
174	Sewage treatment works. Total number of discharges required to comply with more stringent than Royal Commission Standard and percentage considered satisfactory	138
175	Sewage treatment works. Total number of discharges required to comply with less stringent than Royal Commission Standard and percentage considered satisfactory	140
176	Sewage treatment works. Total population served by sewage treatment works required to comply with Royal Commission Standard and percentage served by works considered satisfactory	142
177	Sewage treatment works. Total population served by treatment works required to comply with more stringent than Royal Commission Standard and percentage served by works considered satisfactory	144
178	Sewage treatment works. Total population served by treatment works required with less stringent than Royal Commission Standard and percentage served by works considered satisfactory	146
179	Sewage treatment works. Total dry weather flow of sewage effluent required to comply with Royal Commission Standard and percentage discharged from treatment works considered satisfactory	148
180	Sewage treatment works. Total dry weather flow of sewage effluent required to comply with more stringent than Royal Commission Standard and percentage discharged from treatment works considered satisfactory	150
181	Sewage treatment works. Total dry weather flow of sewage effluent required to comply with less stringent th n Royal Commission Standard and percentage discharged from treatment works considered satisfactory	152
182	Sewage treatment works by population ranges and percentage satisfactory (population up to 50,000)	154
183	Sewage treatment works by population ranges and percentage satisfactory (population over 50,000)	156
184	Sewage treatment works. 1980 situation and changes, 1970–1980	158
185	Discharges of crude sewage. Estimated number, population served, dry weather flow and percentage satisfactory	160
186	Discharges of crude sewage. Estimated number of outlets still remaining in 1980, population served and dry weather flow	162
187	Number of unsatisfactory storm overflows	164
188–217	Main survey. Numbers and volumes of discharges of industrial effluent to rivers and canals and percentage considered satisfactory	166

River authorities:

Avon and Dorset	190		Lincolnshire	172
Bristol Avon	198		Mersey and Weaver	214
Cornwall	194		Northumbrian	166
Cumberland	218		Port of London Authority	
Dee and Clwyd	212		(including the London Excluded Area)	224
Devon	192		Severn	200
East Suffolk and Norfolk	178		Somerset	196
Essex	180		South West Wales	208
Glamorgan	206		Sussex	184
Great Ouse	176		Thames Conservancy	220
Gwynedd	210		Trent	170
Hampshire	186		Usk	204
Isle of Wight	188		Welland and Nene	174
Kent	182		Wye	202
Lancashire	216		Yorkshire	168
Lee Conservancy	222			

218	Estimates of costs of remedial works on discharges of sewage effluent	226
219	Estimates of costs of remedial works on discharges of crude sewage	228
220	Estimates of costs of remedial works on discharges of industrial effluent	230

Figure

1	River authorities and hydrometric areas	vi

Volume 2

Discharges and Forecasts of Improvement

1 Introduction

Main Findings

1 Volume 1 reported the survey results on river quality. This second volume gives the survey results on the main causes of river pollution—the discharges of sewage and industrial effluent. The report of a supplementary survey, by the Confederation of British Industry, of the costs and methods of disposal of trade effluent is contained in Chapter 5.

2 Rivers and canals received the liquid wastes from 88.5 per cent of the population of England and Wales in 1970. Discharges from about 57 per cent of the population were to non-tidal stretches of river, from about 31 per cent to tidal stretches and from about 0.5 per cent to canals.

3 Excluding cooling water, the dry weather flow of effluent to rivers and canals amounted to just over 4,000 million gallons per day (mgd), consisting of sewage effluent, crude sewage, industrial effluent and water raised or drained from underground mines. About 2,600 mgd (64 per cent) was discharged to non-tidal waters, about 1,400 mgd (35 per cent) was discharged to tidal waters and about 50 mgd (1 per cent) was discharged to canals. A flow of 4,000 mgd represents, on average, 83 gallons of liquid waste per day for each person in England and Wales and is nearly 3 times the average flow of the River Thames over Teddington Weir. In addition, the rivers and canals received nearly 15,000 million gallons per day of industrial cooling water. About two-thirds of this was discharged to tidal waters and most of the balance to non-tidal waters, with just over 1 per cent of all the cooling water being discharged to canals. Altogether, all the discharges during a year would fill Lake Windermere 100 times.

4 Excluding cooling water, about half of the effluent came from industry, either in the form of direct discharges from industrial premises to rivers and canals or discharged with treated sewage effluent or crude sewage. The proportions of this industrial effluent discharged to non-tidal waters, tidal waters and canals were very similar to the proportions given in paragraph 3 above for all effluent (excluding cooling water). The greater part of the industrial process water was discharged through some 2,500 outlets direct to rivers and canals rather than to the local authority's sewage treatment works. More than 2,000 of direct discharges were made to non-tidal waters, about 330 discharges were made to tidal waters and 85 were made to canals. About 400 discharges of minewater were made in addition to the above, mostly to non-tidal waters.

5 There were 4,379 discharges of sewage effluent—4,047 to non-tidal waters, 306 to tidal waters and 26 to canals but out of 484 discharges of crude sewage, only 47 were made to non-tidal waters. There were no discharges of crude sewage to canals. The sewage from about 4½ million people was discharged without treatment, mostly to tidal waters. The sewage from about 30 per cent of the population draining to tidal stretches was thus discharged crude. By contrast, the sewage from only about one person in every 700 draining to non-tidal waters was discharged without treatment.

6 A discharge was classified as "satisfactory" if it generally complied with the standard at present required of it by the river authority: it implies no judgment of the standard itself and the discharger might disagree with the assessment. About 37 per cent of all the discharges of sewage effluent were considered to be unsatisfactory, amounting to nearly 60 per cent of the volume of sewage effluent discharged. Almost all the discharges of crude sewage were considered to be unsatisfactory. Excluding discharges consisting solely of cooling water (a high proportion of which were satisfactory) about a half of all the industrial effluent discharges were unsatisfactory and a similar proportion of the volume was not up to standard. In addition more than 2,000 storm overflows, almost all discharging to non-tidal waters, were recorded as needing remedial works. As might be expected, the highest proportions of unsatisfactory discharges were generally associated with the most polluted stretches of river. The survey also showed that a significant improvement in the quality of effluent from some 250 local authority sewage treatment works could be achieved by better maintenance or management. But the main cause of poor quality effluent was overloading. This can only be relieved by additional treatment plant.

7 The expenditure required to bring all discharges of sewage and industrial effluent up to the standards that river authorities expect to impose by 1980 would be about £610 million (at constant 1970 prices), which represents about £12 per head of the present population. About £425 million relates to remedial works for improving discharges of sewage effluent, about £145 million to remedial works for improving discharges of crude sewage, and about £40 million to remedial works for improving discharges of industrial effluent. (This last figure particularly is considered to be lower than the probable figure for reasons which are explained in Chapter 6). The estimates do not include expenditure on sewerage or on the replacement of worn-out equipment and those could well account for as much again. Neither do they include any costs associated with remedial works on storm overflows, because insufficient information was available on which to estimate with any degree of accuracy. A desk cost estimate was however made, based on the application of the available actual cost estimates to the different types of remedial work required. This suggested that remedial works on storm overflows might cost as much as £170 million.

8 Expenditure on remedial works to improve discharges of sewage effluent and crude sewage was divided into expenditure required immediately (effluent already unsatisfactory) and expenditure required in the future (effluent satisfactory at present but likely to become unsatisfactory before 1980 as populations and flows increase). Out of about £570 million likely to be required on those remedial works by 1980, some £410 million was shown to be needed immediately and could be taken as a measure of the back log. This included almost all of the expenditure required on discharges of crude sewage. A proportion of the expenditure on sewage treatment works related to works necessary to achieve higher standards of effluent than those which are considered acceptable at present—usually this would be an improvement from a Royal Commission Standard* at present to a more stringent than Royal Commission Standard before 1980. The survey showed that, by 1980, the sewage effluent from some 20 million people would be subject to higher standards than the corresponding effluent at present.

9 In addition to expenditure associated with the treatment of industrial discharges to controlled waters as indicated above, industry will continue to incur other costs. The CBI survey showed, for example, that between 1960 and 1968, industry spent about £30 million on capital works for pre-treatment of industrial effluent prior to discharge to local authority sewers and were expecting to spend about £20 million on similar works in 1969 and 1970. Furthermore,

---

* The Royal Commission on Sewage Disposal (1889–1915) proposed that with 8 times dilution available in the receiving stream a standard for a discharge of 30 mg/l for suspended solids and 20 mg/l for the biochemical oxygen demand should be satisfactory. It is customary to refer to this as the '30/20' RC standard.

about £5 million was paid to local authorities in 1968 for the conveyance and treatment of industrial effluent.

10 The effect of the overall expenditure on the works required for discharges to comply with the standards expected to be required in 1980 would be to greatly reduce the pollution of rivers but not eliminate it.

11 Expenditure of the order outlined in paragraph 7 would be expected by 1980 to cut the mileage of grossly polluted non-tidal rivers by about 80 per cent, upgrading 753 miles and leaving only 199 miles in class 4, compared with 952 miles in 1970. Over 80 per cent of the length of non-tidal rivers would then be free of pollution (Class 1), 15.3 per cent would be mildly polluted (Class 2), 2.8 per cent badly polluted (Class 3) and 0.9 per cent grossly polluted (Class 4). The mileage of grossly polluted tidal rivers would be cut by about 72 per cent, leaving 59 miles in class 4 compared with 209 miles in 1970. In 1980 there would be 55.8 per cent of the length of tidal rivers in class 1, 28.4 per cent in class 2, 12.5 per cent in class 3 and 3.3 per cent in class 4.

12 A further reduction in pollution of non-tidal rivers could be achieved if discharges from mines were fully controlled. The pollution of tidal rivers could also be reduced if pre-1960 discharges were all brought under control. Ministers have in fact decided that full control should be extended to tidal rivers and estuaries (Circular 10/72 Department of the Environment, 18/72 Welsh Office, on the Report of the Working Party on Sewage Disposal, 8 February 1972).

13 Even with the extended powers of control and consequent additional expenditure required to comply with conditions of consents to discharge, it would not be possible to bring up to class 1 or class 2 all the rivers which receive discharges from densely populated and highly industrialised areas without much higher standards than the river authorities propose to require by 1980. The estimates in the report do not cover the considerable expenditure which would be required for upgrading all rivers to class 1 or class 2 but only the costs of objectives which appear to the river authorities to be attainable in the next decade by the use of their existing powers of control.

The scope of the survey

14 The survey does not cover discharges to septic tanks, cesspools or soakaways. There is no accurate information on the extent of these discharges but it is broadly estimated that rather less than 3 million people are not connected to main drainage. There are also some discharges to underground strata, controlled under section 72 of the Water Resources Act 1963, which are not covered by the survey. Nor are discharges to the sea included, but the Departments are currently collecting information about discharges of sewage to the sea. As stated in Volume 1, some controlled waters were not considered to come within the term "river" as used in the survey—The River Humber, the Wash, the Solent, the controlled parts of the Bristol Channel, the Menai Straits, part of Morecambe Bay and the Solway Firth. There are large discharges into some of these waters, but all discharges to them are excluded from this survey. Information about these discharges is being obtained by the Departments and will be published.

15 The survey is dated 1970, the date of most of the returns, but there have naturally been changes between the time the information was recorded and the date of publication. The information given in the report will therefore already be out of date in some respects. As mentioned in Volume 1, the survey data will be brought up to date at regular intervals, allowing changes to be recorded more promptly in the future. The first updating will be to January 1972.

Biological classification of rivers

16 In Chapter 4 of Volume 1, a summary in the form of a series of tables was given of the biological information provided by the river authorities. These tables showed, amongst other things, the extent of agreement between what were considered to be broadly corresponding classes ie Chemical Class 1 corresponding with Biological Class A, Chemical Class 2 corresponding with Biological Class B and so on. A further note on the subject is at Appendix 1 of this Report where the correlation between the chemical and biological criteria is discussed in greater detail.

2 Discharges of sewage effluent

Introduction

1 This chapter deals with a survey of discharges from sewage treatment works giving full or partial treatment to the sewage. Chapter 3 deals in a similar way with discharges of crude sewage.

2 Before discharge to inland waters most sewage is given treatment at a sewage treatment works and the effluent from the works is normally required to comply with the terms of a consent issued by the river authority. Certain discharges to tidal waters are controlled in the same way.

3 River authorities were asked to record every discharge (to the controlled waters covered by the survey) from treatment works serving or designed to serve populations of 200 or more, specifying the standard of effluent required and whether each discharge was satisfactory or not. There was inevitable difficulty in judging whether a discharge was satisfactory or not in borderline cases. Occasionally a bad sample of effluent might be recorded at a treatment works which normally produces a satisfactory effluent. Such an effluent would not be expected to fall into the "unsatisfactory" category. If the number of bad samples tended to increase, it would be necessary to decide whether this was a symptom of overloading or a symptom of a particular defect which could be remedied by, for example, better maintenance or operational procedures, or, perhaps, stricter trade waste control by the sewerage authority. River authorities were asked to use reasonable discretion in all borderline cases and usually, therefore, where a discharge is described as unsatisfactory, it means that, in the opinion of the river authority, remedial works are required because of significant overloading. It was possible, however, to distinguish in the returns between those discharges which are unsatisfactory because of inadequate capacity at the treatment works and those discharges which are unsatisfactory despite the provision of sufficient treatment capacity. The latter discharges are described in the subsequent tables as unsatisfactory due to "inadequate maintenance or similar", a term appropriate to the great majority of these cases and one which was frequently noted in the returns.

4 It should be noted that a discharge could be described as "satisfactory" and yet still cause some pollution. Such a situation could arise, for example, where a large discharge was made to a relatively small river and the effluent standard was the highest practicable standard that the river authority could impose, or where a higher standard could be achieved only at excessive cost, or where an interim standard had been agreed in order to enable the authority to observe the effect and thus assess the need for a higher standard.

5 The river authorities were also asked to give their views on the date when remedial works would be required and the convention was adopted that, where an existing discharge is unsatisfactory due to overloading, the date '1970' was recorded, even although it would seldom have been possible to carry out the work by that date and even although the river authority might not want to press for remedial works to be carried out as, for example at those works soon to be abandoned. This convention was used in order to obtain some idea of the extent of the backlog of work to be done, but the phasing of these remedial works within a general capital works programme would be a separate matter to be considered independently of the survey. In the majority of '1970' cases, however, the river authority would most probably want early remedial action, but the necessary works would usually have to be spread over several years.

6 It sometimes happens that a discharge is not subject to a formal consent simply because there would be little practical advantage in dealing with the application (under the 1961 Act) until extensions had been carried out. These extensions would be designed and built to meet the standards fixed by the river authority. In completing the returns in such cases, the river authority normally indicated as the "required standard" the standard which they intended to impose, but in a few cases only, the return showed "no standards applicable".

7 River authorities were asked to indicate the probable standards which would be applicable to the discharges in 1980 and in a number of cases these proved to be more restrictive than the present standards. The views of the river authorities in these matters were accepted without question for the purpose of the survey although, in actual practice, and under present legislation, the terms of the consent are a matter to be settled by the river authority, the discharger having a right of appeal to the Secretary of State.

8 The returns showed a few cases where the effluent from a single sewage treatment works discharges to more than one watercourse. Such cases are recorded in the survey and in the tables as though each discharge emanated from one sewage works, with the total population and flow apportioned appropriately between the different rivers or canals.

9 No attempt was made to collect information about sewage treatment works which would be built under schemes of first-time provision of main drainage in rural areas. It was not thought that accurate information about such schemes would be available. Where, therefore, information is given about works to be built in future, this refers mainly to works to be built to serve areas already provided with main drainage or extensions of these areas.

Details of the survey for England and Wales

10 A national summary of the information obtained in the survey is at Table 1. Tables 4–33 give a breakdown of the principal details for each river authority and further details are given in Tables 170–184 at the end of the Report.

11 Some significant points can be noted from Table 1 although it must be remembered that the trends will vary from one river authority area to another as described later in this chapter.

12 The total number of discharges is 4,379, the total population served is 38,931,099* and the dry weather flow is 2,232 million gallons per day which represents about 57.5 gallons per head per day on the average. About 18 per cent of this flow (10 ghd) is recorded as industrial effluent.

13 The total number of discharges is divided in the following ranges:

Under 1,000 population	2,131	(48.7%)
1,001–5,000	1,358	(30.9%)
5,001–10,000	324	(7.4%)
10,001–50,000	435	(10.0%)
50,001–100,000	79	(1.8%)
100,001–500,000	43	(1.0%)
Over 500,000	9	(0.2%)
	4,379	(100%)

Of this total, 2,775 discharges (63 per cent) are considered to be satisfactory. Of the remaining 1,604 unsatisfactory discharges, some

---

* According to the Registrar General's Annual Estimate of Population in England and Wales, the estimated population mid-1970, was 48,987,700.

Table 1   Summary of information about sewage treatment works and discharges of sewage effluent. England and Wales

Description	Rivers Non-tidal	Rivers Tidal	Rivers All rivers	Canals	All rivers and canals
a. Sewage treatment works					
1 Total number of discharges	4,047	306	4,353	26	4,379
2 Number where effluent is satisfactory	2,573	187	2,760	15	2,775
3 Number where effluent is unsatisfactory (capacity)	1,243	106	1,349	10	1,359
4 Number where effluent is unsatisfactory (maintenance)	231	13	244	1	245
b. Populations					
5 Total population served	27,692,263	10,936,594	38,628,857	302,242	38,931,099
6 Total served, effluent satisfactory	13,759,871	3,239,684	16,999,555	246,009	17,245,564
7 Total served, effluent unsatisfactory	13,932,392	7,696,910	21,629,302	56,233	21,685,535
c. Dry weather flow					
8 Present dry weather flow of sewage effluent (x 1,000 gpd)	1,576,093	637,729	2,213,822	18,365	2,232,187
9 Percentage of industrial effluent in DWF	19	14	17	30	18
10 Dry weather flow. Discharges satisfactory (x 1,000 gpd)	723,608	172,817	896.425	14,275	910,700
11 Dry weather flow. Discharges unsatisfactory (x 1,000 gpd)	852,485	464,912	1,317,397	4,090	1,321,487
d. Number of discharges by effluent standards*					
12 Royal Commission Standard					
Number	3,684	186	3,870	26	3,896
Number satisfactory	2,320	121	2,441	15	2,456
Number unsatisfactory	1,364	65	1,429	11	1,440
13 More stringent than Royal Commission Standard					
Number	317	7	324	0	324
Number satisfactory	234	4	238	—	238
Number unsatisfactory	83	3	86	—	86
14 Less stringent than Royal Commission Standard					
Number	45	108	153	0	153
Number satisfactory	18	62	80	—	80
Number unsatisfactory	27	46	73	—	73
e. Population by effluent standards*					
15 Royal Commission Standard					
Total population	22,327,591	5,784,403	28,111,994	302,242	28,414,236
Populations: Satisfactory discharges	10,295,768	814,005	11,109,773	246,009	11,355,782
Unsatisfactory discharges	12,031,823	4,970,398	17,002,221	56,233	17,058,454
16 More stringent than RC Standard					
Total population	4,139,788	3,116,882	7,256,670	0	7,256,670
Satisfactory discharges	2,657,795	1,511,192	4,168,987	0	4,168,987
Unsatisfactory discharges	1,481,993	1,605,690	3,087,683	0	3,087,683
17 Less stringent than RC Standard					
Total population	1,222,884	1,953,979	3,176,863	0	3,176,863
Satisfactory discharges	804,308	912,417	1,716,725	—	1,716,725
Unsatisfactory discharges	418,576	1,041,562	1.460,138	—	1,460,138
f. Dry weather flow by effluent standards*					
18 Royal Commission Standard					
Total DWF (x 1,000 gpd)	1,267,506	351,837	1,619,343	18,365	1,637,708
DWF. Satisfactory discharges (x 1,000 gpd)	537,923	40,435	578,358	14,275	592,633
DWF. Unsatisfactory discharges (x 1,000 gpd)	729,583	311,402	1,040,985	4,090	1,045,075
19 More stringent than RC Standard					
Total DWF (x 1,000 gpd)	235,980	186,667	422,647	0	422,647
DWF. Satisfactory discharges (x 1,000 gpd)	137,710	91,048	228,758	—	228,758
DWF. Unsatisfactory discharges (x 1,000 gpd)	98,270	95,619	193,889	—	193,889
20 Less stringent than RC Standard					
Total DWF (x 1,000 gpd)	72,522	95,149	167,671	0	167,671
DWF. Satisfactory discharges (x 1,000 gpd)	47,890	41,255	89,145	—	89,145
DWF. Unsatisfactory discharges (x 1,000 gpd)	24,632	53,894	78,526	—	78,526
g. Sewage treatment works by population ranges Present population	(%)	(%)	(%)	(%)	(%)
21 1,000 and under Satisfactory	1,391 (69)	74 (67)	1,465 (69)	4 (44)	1,469 (69)
Unsatisfactory	621 (31)	36 (33)	657 (31)	5 (56)	662 (31)
22 1,001 to 5,000 Satisfactory	779 (62)	59 (66)	838 (62)	2 (50)	840 (62)
Unsatisfactory	485 (38)	31 (34)	516 (38)	2 (50)	518 (38)
23 5,001 to 10,000 Satisfactory	154 (53)	19 (61)	173 (54)	2 (50)	175 (54)
Unsatisfactory	135 (47)	12 (39)	147 (46)	2 (50)	149 (46)
24 10,001 to 50,000 Satisfactory	206 (54)	24 (55)	230 (54)	5 (71)	235 (54)
Unsatisfactory	178 (46)	20 (45)	198 (46)	2 (29)	200 (46)
25 50,001 to 100,000 Satisfactory	25 (43)	7 (37)	32 (42)	2 (100)	34 (43)
Unsatisfactory	33 (57)	12 (63)	45 (58)	0	45 (57)
26 100,001 to 500,000 Satisfactory	17 (49)	3 (38)	20 (47)	0	20 (47)
Unsatisfactory	18 (51)	5 (62)	23 (53)	0	23 (53)
27 Over 500,000 Satisfactory	1 (20)	1 (25)	2 (22)	0	2 (22)
Unsatisfactory	4 (80)	3 (75)	7 (78)	0	7 (78)
h. Forecast of 1980 situation					
28 Estimated number of discharges	3,610	291	3,901	20	3,921
29 Estimated population served	34,229,343	13,484,053	47,713,396	334,603	48,047,999
30 Estimated dry weather flow (x 1,000 gpd)	2,201,546	938,112	3,139,658	21,045	3,160,703
j. Changes 1970–1980					
31 No. of existing sewage works expected to close down	528	39	567	6	573
32 No. of new sewage works expected to be built	91	24	115	0	115
k. Improved Standards 1970–1980					
33 No. of works where improved standards are likely to be imposed	481	47	528	5	533
34 Estimated 1980 population served by works where improved standards are likely to be imposed	14,039,713	6,694,845	20,734,558	231,203	20,965,761
35 Estimated 1980 DWF of sewage effluent where improved standards are likely to be imposed (x 1,000 gpd)	985,643	481,523	1,467,166	14,627	1,481,793

*A total of 6 discharges were indicated as having "no standards" and these account for the small differences in the totals against lines 12–14, 15–17 and 18–20 above when compared with the totals against lines 1–4, 5–7 and 8–11 respectively.

15 per cent (or about one out of every seven) are considered unsatisfactory because of inadequate maintenance or similar. A total of 1,359 discharges come from sewage treatment works that are overloaded.

14 Although about two-thirds of the number of discharges are satisfactory only about 44 per cent of the population covered by this survey is served by the sewage treatment works making these satisfactory discharges. Furthermore, only about 41 per cent of all the effluent discharged is to a satisfactory standard. This indicates that it is the larger sewage works which tend to be the more unsatisfactory and this is confirmed from the figures at lines (21) to (27) of Table 1 which are summarised in Table 2. A further breakdown of the figures for unsatisfactory discharges by population ranges is shown in Table 3 which brings out the fact that nearly one-quarter of all the small (under 1,000 population) sewage disposal works considered unsatisfactory are so classified because of inadequate maintenance.

Table 2 Satisfactory discharges by population ranges

Population range	Percentage of works in range considered satisfactory
1,000 and under	69%
1,001— 5,000	62%
5,001— 10,000	54%
10,001— 50,000	54%
50,001—100,000	43%
100,001—500,000	47%
Over 500,000	22%

Table 3 Unsatisfactory discharges—inadequate maintenance

Population range	Number of unsatisfactory discharges in range	Number unsatisfactory due to inadequate maintenance and % of total unsatisfactory in range	
1,000 and under	662	157	(24%)
1,001— 5,000	518	62	(12%)
5,001— 10,000	149	10	(7%)
10,001— 50,000	200	13	(7%)
50,001—100,000	45	3	(7%)
100,001—500,000	23	0	—
Over 500,000	7	0	—
Total	1,604	245	

15 About 89 per cent of all discharges are required to comply with the Royal Commission Standard, about 7 per cent with a better standard and some 4 per cent are permitted a lower standard. The highest proportion of satisfactory discharges is to be found among those which are required to produce a standard more stringent than the Royal Commission's.

16 The figures at line 29 of Table 1 are the sum of all the individual estimates made by sewerage authorities for the 1980 populations draining to sewage treatment works. There are corresponding figures at line 25 of Table 34 (Chapter 3). The estimates are inflated, as the total projection for the population draining to sewers in 1980 amounts to a greater number than the whole population is likely to attain by that date. The population draining to sewers will probably increase by about 950,000 in the period to 1980 through the extension of first-time sewerage to existing settlements in rural areas and when the normal population increase is added, the population draining to sewers in 1980 would be about 46,600,000. Even if the rate of increase of the population on sewers were above normal, as might be expected, since new development is generally connected to sewers, the total of the estimates received (51,919,354) is certainly much too high. The estimates do however broadly represent the populations on which the design of sewage treatment works would be based and are therefore used in the discussion of costs of remedial works in Chapter 6.

17 The returns show that, in keeping with the current trend towards grouping, the river authorities expect nearly 600 of the existing sewage works to be abandoned over the next 10 years and 115 new works to be built. Some of the latter will be required for new towns and major town developments such as Milton Keynes and Peterborough. Also the returns show a significant trend of rising standards, for, at more than 500 existing works, the river authorities are likely to seek better standards of effluent than those applicable at present. The 1980 population at these works according to the estimates will be in the region of 21 million and the corresponding dry weather flow about 1,480 mgd. The estimates suggest that more than 40 per cent of the population served in 1980 will be served by treatment works where higher standards will have been imposed. The proportion of dry weather flow is of the same order.

18 A description of the situation in each river authority area follows.

1 Northumbrian River Authority

Table 4 Details of sewage treatment works and discharges of sewage effluent

Description	Rivers		
	Non-tidal	Tidal	All rivers
a. Sewage treatment works			
1 Total number of discharges	203	9	212
2 Number where effluent is satisfactory	144	6	150
3 Number where effluent is unsatisfactory (capacity)	47	2	49
4 Number where effluent is unsatisfactory (maintenance)	12	1	13
b. Populations			
5 Total population served	760,137	69,116	829,253
6 Total served, effluent satisfactory	389,123	40,769	429,892
7 Total served, effluent unsatisfactory	371,014	28,347	399,361
c. Dry weather flow			
8 Present dry weather flow of sewage effluent (x 1,000 gpd)	32,400	2,811	35,211
9 Percentage of industrial effluent in DWF	14	11	14
10 Dry weather flow. Discharges satisfactory (x 1,000 gpd)	15,365	1,347	16,712
11 Dry weather flow. Discharges unsatisfactory (x 1,000 gpd)	17,035	1,464	18,499
d. Number of discharges by effluent standards			
12 Royal Commission Standard			
Number	196	5	201
Number satisfactory	142	3	145
13 More stringent than RC Standard			
Number	1	0	1
Number satisfactory	1	—	1
14 Less stringent than RC Standard			
Number	6	4	10
Number satisfactory	1	3	4
e. Population by effluent standards			
15 Royal Commission Standard			
Total population	732,169	10,700	742,869
% satisfactory	51	36	51
16 More stringent than RC Standard			
Total population	14,080	0	14,080
% satisfactory	100	—	100
17 Less stringent than RC Standard			
Total population	13,888	58,416	72,304
% satisfactory	1	63	51
f. Dry weather flow by effluent standards			
18 Royal Commission Standard			
Total DWF (x 1,000 gpd)	31,204	355	31,559
% satisfactory	47	44	47
19 More stringent than RC Standard			
Total DWF (x 1,000 gpd)	659	0	659
% satisfactory	100	—	100
20 Less stringent than RC Standard			
Total DWF (x 1,000 gpd)	537	2,456	2,993
% satisfactory	1	48	40
g. Sewage treatment works by population ranges Present population			
21 1,000 and under Satisfactory	83	1	84
Unsatisfactory	26	1	27
22 1,001 to 5,000 Satisfactory	43	3	46
Unsatisfactory	16	0	16
23 5,001 to 10,000 Satisfactory	7	1	8
Unsatisfactory	6	1	7
24 10,001 to 50,000 Satisfactory	11	1	12
Unsatisfactory	10	1	11
25 50,001 to 100,000 Satisfactory	0	0	0
Unsatisfactory	1	0	1

19 The river authority area extends over the counties of Northumberland, Durham and fringe areas of the neighbouring counties of

Cumberland and Westmorland to the west and Yorkshire to the south. It includes the important county boroughs of Teesside, Newcastle-upon-Tyne, Sunderland, South Shields and Gateshead, all with a population over 100,000. The estuaries of the Tees and the Tyne are centres of heavy industry and high populations, but over a large part of the authority's area, development is sparse.

20 The survey shows that the total number of sewage effluent discharges in the area is 212 but although 150 (71 per cent) of these are satisfactory, the treatment works considered satisfactory serve only just over 50 per cent of the total population and discharge just under 50 per cent, of the total effluent. About 20 per cent of the discharges considered unsatisfactory are so classified because of inadequate maintenance. It should be noted that the figures above and following, do not include the many discharges of crude sewage which are of particular importance in the area. These are dealt with in Chapter 3.

21 Notable among the unsatisfactory discharges are those to the Tees basin which receives over 9 million gallons a day of unsatisfactory sewage effluent. Over 7 mgd of this unsatisfactory flow is discharged to the Skerne and its tributaries. The Tyne, the Derwent, the Wear, and the Lumley Park Burn all receive discharges of unsatisfactory effluent in excess of 1 mgd on non-tidal lengths.

22 There are only 7 discharges of more than 1 mgd and only one of more than 5 mgd. More than 80 per cent of all the treatment works serve populations of 5,000 or less and about half of all the treatment works serve populations of less than 1,000.

23 The forecasts for the period up to 1980 show that 15 works are expected to close down and one new works is likely to be built in the period; some degree of centralisation will thus be achieved. The estimated population served is expected to increase to about 1,140,000 and the flow to about 70 mgd. Several schemes to improve unsatisfactory discharges are already under construction.

24 Improved standards are likely to be required at only 8 works. ☐

2 Yorkshire River Authority

Table 5 Details of sewage treatment works and discharges of sewage effluent

Description		Rivers Non-tidal	Rivers Tidal	Rivers All rivers	Canals	All rivers and canals
a. Sewage treatment works						
1 Total number of discharges		413	17	430	2	432
2 Number where effluent is satisfactory		258	15	273	2	275
3 Number where effluent is unsatisfactory (capacity)		153	2	155	0	155
4 Number where effluent is unsatisfactory (maintenance)		2	0	2	0	2
b. Populations						
5 Total population served		3,875,442	177,482	4,052,924	26,874	4,079,798
6 Total served, effluent satisfactory		1,316,545	88,477	1,405,022	26,874	1,431,896
7 Total served, effluent unsatisfactory		2,558,897	89,005	2,647,902	0	2,647,902
c. Dry weather flow						
8 Present dry weather flow of sewage effluent (x 1,000 gpd)		227,260	12,315	239,575	841	240,416
9 Percentage of industrial effluent in DWF		25	32	25	0	25
10 Dry weather flow. Discharges satisfactory (x 1,000 gpd)		59,121	6,851	65,972	841	66,813
11 Dry weather flow. Discharges unsatisfactory (x 1,000 gpd)		168,139	5,464	173,603	0	173,603
d. Number of discharges by effluent standards						
12 Royal Commission Standard						
Number		403	4	407	2	409
Number satisfactory		252	3	255	2	257
13 More stringent than RC Standard						
Number		9	0	9	0	9
Number satisfactory		6	—	6	—	6
14 Less stringent than RC Standard						
Number		1	13	14	0	14
Number satisfactory		0	12	12	—	12
e. Population by effluent standards						
15 Royal Commission Standard						
Total population		3,795,779	94,455	3,890,234	26,874	3,917,108
% satisfactory		33	10	33	100	33
16 More stringent than RC Standard						
Total population		79,438	0	79,438	0	79,438
% satisfactory		69	—	69	—	69
17 Less stringent than RC Standard						
Total population		225	83,027	83,252	0	83,252
% satisfactory		0	95	95	—	95
f. Dry weather flow by effluent standards						
18 Royal Commission Standard						
Total DWF (x 1,000 gpd)		221,585	5,640	227,225	841	228,066
% satisfactory		25	6	25	100	25
19 More stringent than RC Standard						
Total DWF (x 1,000 gpd)		5,666	0	5,666	0	5,666
% satisfactory		56	—	56	—	56
20 Less stringent than RC Standard						
Total DWF (x 1,000 gpd)		9	6,675	6,684	0	6,684
% satisfactory		0	98	98	—	97
g. Sewage treatment works by population ranges Present population						
21 1,000 and under	Satisfactory	129	5	134	0	134
	Unsatisfactory	51	0	51	0	51
22 1,001 to 5,000	Satisfactory	77	5	82	0	82
	Unsatisfactory	45	1	46	0	46
23 5,001 to 10,000	Satisfactory	21	1	22	1	23
	Unsatisfactory	23	0	23	0	23
24 10,001 to 50,000	Satisfactory	25	4	29	1	30
	Unsatisfactory	29	0	29	0	29
25 50,001 to 100,000	Satisfactory	5	0	5	0	5
	Unsatisfactory	2	1	3	0	3
26 100,001 to 500,000	Satisfactory	1	0	1	0	1
	Unsatisfactory	4	0	4	0	4
27 Over 500,000	Satisfactory	0	0	0	0	0
	Unsatisfactory	1	0	1	0	1

25 The area administered by the Yorkshire River Authority is the largest of all river authorities in England and Wales and it covers almost all of the county of Yorkshire and small parts of the counties of Lancaster, Derby and Nottingham. It includes some particularly large county boroughs such as Sheffield, Leeds, Bradford, Kingston-upon-Hull, Huddersfield and York. Apart from Kingston-upon-Hull, they all lie well inland on the upland sections of some of the major rivers in the area, they are important centres of industry in the north of England and form the nucleus of high populations in the West Riding. The remainder of the area is by contrast mainly rural in character with the exception of a number of holiday resorts on the east coast.

26 As to be expected with such a large area, there are many sewage treatment works and the survey records 432 of which 275 (64 per cent) produce satisfactory effluents. In terms of population served and quantity discharged, however, the proportions considered satisfactory are much lower being only 35 per cent and 28 per cent respectively. In all but 2 cases, the treatment works considered unsatisfactory have been so classified because of insufficient capacity. Out of the total effluent discharged of about 240 mgd, some 228 mgd is required to meet Royal Commission standard, but only about 25 per cent of this flow is discharged to a satisfactory standard.

27 The Ouse basin receives most of the discharges and out of 168 mgd discharged in an unsatisfactory condition to non-tidal waters, 166 mgd is discharged in this basin. Conspicuous in this category are the Don (33 mgd unsatisfactory effluent), the Aire (34 mgd unsatisfactory), the Calder (35 mgd unsatisfactory) and the Wyke Beck (33 mgd unsatisfactory), and to a lesser extent the Spen Beck (6.4 mgd unsatisfactory), the Mill Shaw Beck (2.3 mgd unsatisfactory), the Wharfe (1.9 mgd unsatisfactory) and the Nidd (2.3 mgd unsatisfactory). Most of the unsatisfactory effluent to tidal water is discharged to the Don.

28 There are some 30 discharges of 1 mgd or more and these account for about three-quarters of the total flow of effluent. Three of the discharges exceed 20 mgd and 2 exceed 30 mgd. There are 6 discharges serving populations of 100,000 or more and 5 of these are considered unsatisfactory. At the other end of the scale, nearly three-quarters of all the discharges come from sewage treatment works serving populations of 5,000 or less.

29 The forecasts for the period up to 1980 show that some 40 existing works are likely to be abandoned and 1 new works will be built, and that the population served will rise to something over 4,800,000 and the dry weather flow to about 327 mgd.

30 The river authority will be seeking, in the future, better standards of effluent at 192 existing treatment works. The 1980 population which will be served by these works will be about 4 million and the dry weather flow about 275 mgd, about 85 per cent of the total 1980 flow. Of special significance in this context are many of the discharges to the Ouse, the Don, the Went, the Dearne, the Rother, the Aire and the Calder. Most of the discharges to these rivers are likely to be required to meet considerably higher standards than at present—in many cases as high as 10 mg/l SS and 10 mg/l BOD. The same will apply to many of their tributaries.

31 A number of schemes to group discharges are envisaged and many schemes to improve unsatisfactory discharges are already under construction. □

3 Trent River Authority

Table 6   Details of sewage treatment works and discharges of sewage effluent

Description	Rivers			Canals	All rivers and canals
	Non-tidal	Tidal	All rivers		
a. Sewage treatment works					
1 Total number of discharges	458	12	470	13	483
2 Number where effluent is satisfactory	252	6	258	6	264
3 Number where effluent is unsatisfactory (capacity)	184	5	189	6	195
4 Number where effluent is unsatisfactory (maintenance)	22	1	23	1	24
b. Populations					
5 Total population served	5,426,439	14,395	5,440,834	247,638	5,688,472
6 Total served, effluent satisfactory	3,397,530	6,555	3,404,085	201,260	3,605,345
7 Total served, effluent unsatisfactory	2,028,909	7,840	2,036,749	46,378	2,083,127
c. Dry weather flow					
8 Present dry weather flow of sewage effluent (x 1,000 gpd)	327,499	591	328,090	16,272	344,362
9 Percentage of industrial effluent in DWF	27	4	27	34	27
10 Dry weather flow. Discharges satisfactory (x 1,000 gpd)	209,292	292	209,584	12,654	222,238
11 Dry weather flow. Discharges unsatisfactory (x 1,000 gpd)	118,207	299	118,506	3,618	122,124
d. Number of discharges by effluent standards					
12 Royal Commission Standard					
Number	444	12	456	13	469
Number satisfactory	248	6	254	6	260
13 More stringent than RC Standard					
Number	2	0	2	0	2
Number satisfactory	1	—	1	—	1
14 Less stringent than RC Standard					
Number	12	0	12	0	12
Number satisfactory	3	—	3	—	3
e. Population by effluent standards					
15 Royal Commission Standard					
Total population	4,341,787	14,395	4,356,182	247,638	4,603,820
% satisfactory	66	45	66	81	67
16 More stringent than RC Standard					
Total population	196,990	0	196,990	0	196,990
% satisfactory	3	—	3	—	3
17 Less stringent than RC Standard					
Total population	887,662	0	887,662	0	887,662
% satisfactory	58	—	58	—	58

continued

(3 Trent River Authority—Table 6 continued)

Description	Rivers			Canals	All rivers and canals
	Non-tidal	Tidal	All rivers		
f. Dry weather flow by effluent standards					
18 Royal Commission Standard					
Total DWF (x 1,000 gpd)	248,686	591	249,277	16,272	265,549
% satisfactory	70	49	70	78	70
19 More stringent than RC Standard					
Total DWF (x 1,000 gpd)	21,219	0	21,219	0	21,219
% satisfactory	1	—	1	—	1
20 Less stringent than RC Standard					
Total DWF (x 1,000 gpd)	57,594	0	57,594	0	57,594
% satisfactory	61	—	61	—	61
g. Sewage treatment works by population ranges					
Present population					
21 1,000 and under Satisfactory	94	4	98	0	98
Unsatisfactory	67	4	71	2	73
22 1,001 to 5,000 Satisfactory	85	2	87	0	87
Unsatisfactory	84	2	86	2	88
23 5,001 to 10,000 Satisfactory	28	0	28	1	29
Unsatisfactory	18	0	18	1	19
24 10,001 to 50,000 Satisfactory	35	0	35	3	38
Unsatisfactory	27	0	27	2	29
25 50,001 to 100,000 Satisfactory	6	0	6	2	8
Unsatisfactory	6	0	6	0	6
26 100,001 to 500,000 Satisfactory	3	0	3	0	3
Unsatisfactory	4	0	4	0	4
27 Over 500,000 Satisfactory	1	0	1	0	1
Unsatisfactory	0	0	0	0	0

Notes: A number of sewage treatment works in the area have more than one discharge and these discharges are to different rivers or canals, eg Stoke-on-Trent (Burslem)—2 discharges; Stoke-on-Trent (Hanley)—3 discharges; Stoke-on-Trent (Tunstall)—2 discharges; Wolverhampton (Barnhurst)—2 discharges;
These are recorded throughout the Tables as discharges rather than treatment works, ie as 9 discharges in the appropriate lines and columns. The populations and flows have been apportioned between the various discharge points in accordance with the information supplied.
Four of the 12 discharges recorded to tidal waters discharge to small and un-coded drains near the mouth of the River Trent. For the purpose of the survey they have been regarded as discharges to the tidal section of the Trent.

32 The Trent River Authority covers a large area in England being third in size to the Yorkshire and Severn authorities. It covers most of the counties of Nottingham, Derby, Stafford, and Leicester, the northern part of Warwickshire and parts of Lincolnshire and Yorkshire. It includes the large industrial and population complex of the West Midlands including Birmingham, Walsall, West Bromwich and part of Wolverhampton together with the county boroughs of Nottingham, Leicester, Stoke, Derby, Burton-on-Trent and part of Doncaster. These large towns are mostly located well inland and are sited on the upper stretches of the Trent and its tributaries.

33 The survey records 483 discharges of sewage effluent—the largest number in any river authority—and 264 (55 per cent) of those are satisfactory. The volume discharged to a satisfactory standard represents about 65 per cent of the total volume of sewage effluent discharged in the area. Of the treatment works which produce unsatisfactory effluents 195 are considered unsatisfactory because of insufficient capacity and 24 because of poor maintenance.

34 The River Trent receives the largest number of discharges (37) with a total of about 59 mgd of effluent, but the River Tame, one of its tributaries, receives the largest quantity of sewage effluent, about 103 mgd, or 30 per cent of all the effluent discharged in the authority's area; 70 per cent of this quantity comes from the works, at Minworth, which is the largest of the 36 treatment plants operated by the Upper Tame Main Drainage Authority. Tributaries of the Tame also receive a further 43 mgd. Other rivers which receive particularly large quantities of effluent are the Soar (23 mgd) and the Derwent (24 mgd) and large discharges are also made to several canals—notably the Trent and Mersey (2.4 mgd), Staffordshire and Worcestershire (6 mgd), Shropshire Union (4.5 mgd) and the Nottingham Canal (2 mgd).

35 The Derwent receives the greatest quantity of unsatisfactory effluent (about 22 mgd and almost all required to comply with a more stringent than Royal Commission standard) and big volumes of unsatisfactory effluent are also discharged to the Tame (15.4 mgd), the Trent (13.4 mgd), the Cole (10.1 mgd), the Erewash (6.6 mgd) and the Wolverhampton Arm of the Tame (6.4 mgd). There are also a number of rivers where the unsatisfactory discharges, although less in volume, represent a very high proportion of the total discharged to the river.

36 Downstream of the Minworth discharge, about 90 per cent of the total flow in the River Tame in summer is sewage effluent. In the Maun, the summer flow in the upper reaches below Mansfield consists of around 80 per cent of sewage effluent.

37 By size, the treatment works vary widely. Minworth, serving a population of close on one million, produces the biggest single discharge to non-tidal waters in the country. There are a further 21 works serving populations of 50,000 or more, but about 70 per cent of all the treatment works serve populations of 5,000 or less and about a half of those serve populations of 1,000 or less. Some 50 or so discharges exceed 1 mgd and, in total they discharge about 280 mgd or about 81 per cent of all the effluent discharged in the area.

38 Considerable changes are likely to take place over the period to 1980, the forecast being that 99 works will be abandoned and 9 new works will be built, thus reducing the number of discharges from 483 to 393. The population to be served in 1980 is estimated at about 6,700,000 with a dry weather flow of about 456 mgd.

39 The survey shows that higher standards are likely to be called for in 30 cases. The estimated 1980 population affected is about 4,400,000 and the dry weather flow about 334 mgd. Thus more than 70 per cent of the 1980 flow will be required to meet higher standards than those for the corresponding discharges at present. Rivers which are prominent among those where higher standards will be required are the Trent, the Derwent and the Tame. On many other rivers, single large discharges will be required to meet more stringent standards.

40 Many schemes involving large expenditure are already under construction in the area.

4 Lincolnshire River Authority

Table 7 Details of sewage treatment works and discharges of sewage effluent

Description	Rivers Non-tidal	Rivers Tidal	Rivers All rivers	Canals	All rivers and canals
a. *Sewage treatment works*					
1 Total number of discharges	116	2	118	1	119
2 Number where effluent is satisfactory	97	1	98	1	99
3 Number where effluent is unsatisfactory (capacity)	19	0	19	0	19
4 Number where effluent is unsatisfactory (maintenance)	0	1	1	0	1
b. *Populations*					
5 Total population served	405,636	1,088	406,724	770	407,494
6 Total served, effluent satisfactory	350,401	200	350,601	770	351,371
7 Total served, effluent unsatisfactory	55,235	888	56,123	0	56,123
c. *Dry weather flow*					
8 Present dry weather flow of sewage effluent (x 1,000 gpd)	15,719	31	15,750	23	15,773
9 Percentage of industrial effluent in DWF	13	0	13	0	13
10 Dry weather flow. Discharges satisfactory (x 1,000 gpd)	13,713	15	13,728	23	13,751
11 Dry weather flow. Discharges unsatisfactory (x 1,000 gpd)	2,006	16	2,022	0	2,022
d. *Number of discharges by effluent standards**					
12 Royal Commission Standard					
Number	98	1	99	1	100
Number satisfactory	82	1	83	1	84
13 More stringent than RC Standard					
Number	17	0	17	0	17
Number satisfactory	14	—	14	—	14
14 Less stringent than RC Standard					
Number	0	1	1	0	1
Number satisfactory	0	0	0	0	0
e. *Population by effluent standards**					
15 Royal Commission Standard					
Total population	250,194	200	250,394	770	251,164
% satisfactory	80	100	80	100	80
16 More stringent than RC Standard					
Total population	153,442	0	153,442	0	153,442
% satisfactory	97	—	97	—	97
17 Less stringent than RC Standard					
Total population	0	888	888	0	888
% satisfactory	—	0	0	—	0
f. *Dry weather flow by effluent standards**					
18 Royal Commission Standard					
Total DWF (x 1,000 gpd)	7,911	15	7,926	23	7,949
% satisfactory	77	100	77	100	77
19 More stringent than RC Standard					
Total DWF (x 1,000 gpd)	7,723	0	7,723	0	7,723
% satisfactory	98	—	98	—	98
20 Less stringent than RC Standard					
Total DWF (x 1,000 gpd)	0	16	16	0	16
% satisfactory	—	0	0	—	0
g. *Sewage treatment works by population ranges*					
Present population					
21 1,000 and under Satisfactory	48	1	49	1	50
Unsatisfactory	5	1	6	0	6
22 1,001 to 5,000 Satisfactory	39	0	39	0	39
Unsatisfactory	11	0	11	0	11
23 5,001 to 10,000 Satisfactory	3	0	3	0	3
Unsatisfactory	2	0	2	0	2
24 10,001 to 50,000 Satisfactory	6	0	6	0	6
Unsatisfactory	1	0	1	0	1
25 50,001 to 100,000 Satisfactory	1	0	1	0	1
Unsatisfactory	0	0	0	0	0

*One discharge is indicated in the survey as having "no standards". This accounts for the small differences in the totals against lines 12–14, 15–17 and 18–20 when compared with the totals against lines 1–4, 5–7, and 8–11 respectively.

41 The Lincolnshire River Authority lies mainly within the county of Lincoln extending from the Humber to the Wash and westwards just into the counties of Nottingham and Leicester; it includes the county boroughs of Grimsby and Lincoln. Most of the land is used for agriculture and population densities are low.

42 The number of treatment works where the sewage effluent is considered satisfactory represents some 83 per cent of the total number and the proportions of population served by satisfactory treatment works, and of effluent discharged to a satisfactory standard, are slightly higher. Only one out of 20 works considered unsatisfactory is so classified because of poor maintenance.

43 The river receiving the greatest quantity of sewage effluent is the Witham—the survey records the discharge of about 6.7 mgd from some 137,000 population to this river, almost all on the non-tidal section, and all of this effluent is satisfactory.

44 There are very few treatment works discharging large quantities of effluent. The biggest treatment works is at Lincoln, serving about 80,000 population. All but 13 of the works serve populations of less than 5,000 and almost half of all the treatment works serve populations of less than 1,000. All of those with a flow greater than 1 mgd are satisfactory.

45 Some changes are forecast by 1980. The number of works is likely to decrease to 116 by the abandonment of 3, the estimated population served in 1980 is forecasted to increase to about 490,000 and the volume discharged to about 21.7 mgd. The survey also shows that improved effluent standards will be called for at 4 works. The population affected (1980) will be about 39,000 and the flow just over 0.5 mgd.

46 A number of schemes of remedial works are under construction.

5 Welland and Nene River Authority

Table 8 Details of sewage treatment works and discharges of sewage effluent

Description	Rivers			Canals*	All rivers and canals
	Non-tidal	Tidal	All rivers		
a. Sewage treatment works					
1 Total number of discharges	153	2	155	2	157
2 Number where effluent is satisfactory	68	1	69	1	70
3 Number where effluent is unsatisfactory (capacity)	50	1	51	1	52
4 Number where effluent is unsatisfactory (maintenance)	35	0	35	0	35
b. Populations					
5 Total population served	484,642	91,450	576,092	1,071	577,163
6 Total served, effluent satisfactory	304,852	75,250	380,102	297	380,399
7 Total served, effluent unsatisfactory	179,790	16,200	195,990	774	196,764
c. Dry weather flow					
8 Present dry weather flow of sewage effluent (x 1,000 gpd)	25,061	5,046	30,107	96	30,203
9 Percentage of industrial effluent in DWF	14	11	13	2	13
10 Dry weather flow. Discharges satisfactory (x 1,000 gpd)	16,170	4,473	20,643	4	20,647
11 Dry weather flow. Discharges unsatisfactory (x 1,000 gpd)	8,891	573	9,464	92	9,556
d. Number of discharges by effluent standards					
12 Royal Commission Standard					
Number	137	1	138	2	140
Number satisfactory	59	1	60	1	61
13 More stringent than RC Standard					
Number	6	0	6	0	6
Number satisfactory	5	—	5	—	5
14 Less stringent than RC Standard					
Number	10	1	11	0	11
Number satisfactory	4	0	4	—	4
e. Population by effluent standards					
15 Royal Commission Standard					
Total population	466,754	75,250	542,004	1,071	543,075
% satisfactory	62	100	67	28	67
16 More stringent than RC Standard					
Total population	12,991	0	12,991	0	12,991
% satisfactory	93	—	93	—	93
17 Less stringent than RC Standard					
Total population	4,897	16,200	21,097	0	21,097
% satisfactory	43	0	10	—	10
f. Dry weather flow by effluent standards					
18 Royal Commission Standard					
Total DWF (x 1000 gpd)	24,399	4,473	28,872	96	28,968
% satisfactory	64	100	70	4	70
19 More stringent than RC Standard					
Total DWF (x 1,000 gpd)	481	0	481	0	481
% satisfactory	91	—	91	—	91
20 Less stringent than RC Standard					
Total DWF (x 1,000 gpd)	181	573	754	0	754
% satisfactory	45	0	11	—	11
g. Sewage treatment works by population ranges Present population					
21 1,000 and under — Satisfactory	39	0	39	1	40
Unsatisfactory	54	0	54	1	55
22 1,001 to 5,000 — Satisfactory	23	0	23	0	23
Unsatisfactory	24	0	24	0	24
23 5,001 to 10,000 — Satisfactory	2	0	2	0	2
Unsatisfactory	4	0	4	0	4
24 10,001 to 50,000 — Satisfactory	3	0	3	0	3
Unsatisfactory	3	1	4	0	4
25 50,001 to 100,000 — Satisfactory	0	1	1	0	1
Unsatisfactory	0	0	0	0	0
26 100,001 to 500,000 — Satisfactory	1	0	1	0	1
Unsatisfactory	0	0	0	0	0

*Data relates to one discharge to the Grand Union Canal and one discharge to the Norton Stream, a feeder of the Grand Union Canal.

47 This area extends over several counties, covering parts of Lincolnshire, Leicestershire, Northamptonshire, Rutland, the county of Huntingdon and Peterborough, and Cambridge and Isle of Ely. The principal rivers are the Welland and the Nene both draining through fenland to The Wash. Northampton and Peterborough, both lying on the River Nene, are the principal centres of population and industrial activity in the area. There are several smaller centres of population but the rest of the area is predominantly agricultural.

48 Only 70 out of 157 sewage treatment works (45 per cent) are considered satisfactory and 35 out of the 87 considered unsatisfactory are so classified because of poor or inadequate maintenance. Almost all of the unsatisfactory discharges are to non-tidal waters and the volume discharged to an unsatisfactory standard represents about one third of all the sewage effluent. The greater volumes of unsatisfactory effluent are discharged to the Welland (about 1.25 mgd of which about 0.6 mgd discharges to the tidal section), the Nene (about 1.2 mgd), the Rushden Brook (about 1 mgd) and the River Ise (about 3 mgd).

49 The larger discharges of sewage effluent are all within the Nene catchment and about half of the total effluent discharged in the area is discharged to the Nene itself. About 70 per cent of the effluent discharged in the Nene basin is discharged to a satisfactory standard.

50 There is a very high proportion of small treatment plants in the area, all but 15 out of 157 serving populations of 5,000 or less and 95 works serve populations of 1,000 or less. Of the 35 works considered unsatisfactory for poor maintenance, 27 are in this latter group.

51 By 1980 the number of works is likely to decrease to 151 by the abandonment of 14 existing works and the construction of 8 new. It is forecast that the population served will increase to about one million and that the quantity of sewage effluent discharged will increase to

over 65 mgd. Much of the increase will be attributable to the expansion of Peterborough and Northampton. Higher standards of effluent are likely to be called for at 37 treatment works, the estimated 1980 flow from which will be about 31 mgd.

52 Schemes to improve unsatisfactory discharges are under construction at a number of the existing sewage treatment works.

6 Great Ouse River Authority

Table 9 Details of sewage treatment works and discharges of sewage effluent

Description	Rivers		
	Non-tidal	Tidal	All rivers
a. Sewage treatment works			
1 Total number of discharges	254	9	263
2 Number where effluent is satisfactory	178	9	187
3 Number where effluent is unsatisfactory (capacity)	63	0	63
4 Number where effluent is unsatisfactory (maintenance)	13	0	13
b. Populations			
5 Total population served	949,232	16,497	965,729
6 Total served, effluent satisfactory	494,052	16,497	510,549
7 Total served, effluent unsatisfactory	455,180	0	455,180
c. Dry weather flow			
8 Present dry weather flow of sewage effluent (x 1,000 gpd)	47,844	822	48,666
9 Percentage of industrial effluent in DWF	13	0	13
10 Dry weather flow. Discharges satisfactory (x 1,000 gpd)	22,106	822	22,928
11 Dry weather flow. Discharges unsatisfactory (x 1,000 gpd)	25,738	0	25,738
d. Number of discharges by effluent standards			
12 Royal Commission Standard			
Number	230	9	239
Number satisfactory	157	9	166
13 More stringent than RC Standard			
Number	24	0	24
Number satisfactory	21	—	21
14 Less stringent than RC Standard			
Number	0	0	0
Number satisfactory	—	—	—
e. Population by effluent standards			
15 Royal Commission Standard			
Total population	770,615	16,497	787,112
% satisfactory	43	100	44
16 More stringent than RC Standard			
Total population	178,617	0	178,617
% satisfactory	92	—	92
17 Less stringent than RC Standard			
Total population	0	0	0
% satisfactory	—	—	—
f. Dry weather flow by effluent standards			
18 Royal Commission Standard.			
Total DWF (x 1,000 gpd)	40,001	822	40,823
% satisfactory	37	100	39
19 More stringent than RC Standard			
Total DWF (x 1,000 gpd)	7,843	0	7,843
% satisfactory	91	—	91
20 Less stringent than RC Standard			
Total DWF (x 1,000 gpd)	0	0	0
% satisfactory	—	—	—
g. Sewage treatment works by population ranges			
Present population			
21 1,000 and under Satisfactory	93	4	97
Unsatisfactory	32	0	32
22 1,001 to 5,000 Satisfactory	69	4	73
Unsatisfactory	28	0	28
23 5,001 to 10,000 Satisfactory	8	1	9
Unsatisfactory	7	0	7
24 10,001 to 50,000 Satisfactory	7	0	7
Unsatisfactory	7	0	7
25 50,001 to 100,000 Satisfactory	1	0	1
Unsatisfactory	1	0	1
26 100,001 to 500,000 Satisfactory	0	0	0
Unsatisfactory	1	0	1

Note: Discharges to the Ouse Relief Channel are included in the totals for discharges to tidal waters.

53 The Great Ouse Authority area extends from The Wash through the adjacent fenlands to the Chilterns in the south and to the Northampton Uplands in the west, and it includes nearly the whole of the counties of Bedford, Cambridge and Isle of Ely, and, Huntingdon and Peterborough, and parts of the counties of Norfolk, Suffolk, Essex, Hertford, Buckingham and Northampton. The principal towns are Cambridge and Bedford, both municipal boroughs in an area which is predominantly agricultural and where the overall population density is low. The main river basin is that of the Great Ouse which flows into The Wash.

54 The total number of sewage effluent discharges in the area is 263 and the majority of these (187 or 71 per cent) are satisfactory. The treatment works considered satisfactory serve just over 50 per cent of the total population and discharge some 47 per cent of the total effluent; 13 out of the 63 discharges which are considered unsatisfactory are from works where maintenance is poor.

55 The rivers receiving the most sewage effluent are the Great Ouse (34 discharges, some 11.3 mgd of effluent, and about 9 mgd of this unsatisfactory), the Cam (6 discharges, some 8 mgd of effluent, 7.6 mgd unsatisfactory) and the Ouzel (11 discharges, some 5.2 mgd of effluent, and about 1.5 mgd unsatisfactory). These 3 rivers receive about half of all the effluent discharged in the area. Other watercourses receiving substantial discharges of unsatisfactory effluent are those in the Middle Level Drain and Twenty Foot River System, the River Ivel and its tributaries.

56 Throughout the area there are only 9 discharges of more than 1 mgd and only 2 of these exceed 5 mgd but together they represent nearly 55 per cent of all the effluent discharged in the area. Nearly 90 per cent of all treatment works serve populations of 5,000 and under and over a half of these serve populations of 1,000 and less.

57 All but 24 of the treatment works are required to discharge effluents to Royal Commission standard and nearly 70 per cent of these discharges are considered satisfactory but only about 40 per cent of the volume discharged meets Royal Commission standard. All but 3 of the works required to comply with more stringent standards do so.

58 By 1980, it is expected that 24 works will have closed down and 6 new works opened so that the number of works remaining in 1980 will be 245. The estimated population served is expected to increase to just over one and a half million and the flow discharged to about 100 mgd, a very considerable increase on the corresponding 1970 figures. A big proportion of the increase is attributable to developments at the new town of Milton Keynes. There is also a major scheme already under construction at Kings Lynn for diverting the existing crude sewage outfalls to a new treatment works. Several other schemes to improve unsatisfactory discharges are also under construction.

59 Improved standards are likely to be required at 34 works all in the Ouse basin. The corresponding 1980 flow will be about 20 mgd.

7 East Suffolk and Norfolk River Authority
(see Table 10 over page)

60 This river authority covers mainly the eastern part of the county of Norfolk and most of East Suffolk and includes the coastline from Hunstanton round to Harwich. There are 3 county boroughs in the area, Ipswich, Norwich and Great Yarmouth, all located at or below the tidal limits of the Yare and Orwell. The countryside is the centre of a flourishing agricultural community and the holiday facilities of the Norfolk Broads and the surrounding area, centred on the River Bure, are popular. The area is mostly flat and the tidal limits of the principal rivers are generally well inland. Because of this, a very big proportion (about 80 per cent) of the sewage effluent is discharged to tidal waters, although this emanates from only 32 out of 115 sewage treatment works. The majority of the smaller works discharge to the non-tidal waters.

61 Only 20 discharges are considered unsatisfactory, but these works serve about 60 per cent of the population and account for about 70 per cent of the total sewage effluent in the area. The biggest

(7 East Suffolk and Norfolk River Authority continued)

Table 10 Details of sewage treatment works and discharges of sewage effluent

Description	Rivers		
	Non-tidal	Tidal	All rivers
a. *Sewage treatment works*			
1 Total number of discharges	83	32	115
2 Number where effluent is satisfactory	68	27	95
3 Number where effluent is unsatisfactory (capacity)	12	5	17
4 Number where effluent is unsatisfactory (maintenance)	3	0	3
b. *Populations*			
5 Total population served	126,369	389,579	515,948
6 Total served, effluent satisfactory	92,895	107,819	200,714
7 Total served, effluent unsatisfactory	33,474	281,760	315,234
c. *Dry weather flow*			
8 Present dry weather flow of sewage effluent (x 1,000 gpd)	4,355	17,854	22,209
9 Percentage of industrial effluent in DWF	12	12	12
10 Dry weather flow. Discharges satisfactory (x 1,000 gpd)	3,159	3,288	6,447
11 Dry weather flow. Discharges unsatisfactory (x 1,000 gpd)	1,196	14,566	15,762
d. *Number of discharges by effluent standards*			
12 Royal Commission Standard			
Number	71	16	87
Number satisfactory	58	14	72
13 More stringent than RC Standard			
Number	12	3	15
Number satisfactory	10	2	12
14 Less stringent than RC Standard			
Number	0	13	13
Number satisfactory	—	11	11
e. *Population by effluent standards*			
15 Royal Commission Standard			
Total population	104,038	49,273	153,311
% satisfactory	72	88	77
16 More stringent than RC Standard			
Total population	22,331	4,162	26,493
% satisfactory	78	81	79
17 Less stringent than RC Standard			
Total population	0	336,144	336,144
% satisfactory	—	18	18
f. *Dry weather flow by effluent standards*			
18 Royal Commission Standard			
Total DWF (x 1,000 gpd)	3,456	1,584	5,040
% satisfactory	73	89	78
19 More stringent than RC Standard			
Total DWF (x 1,000 gpd)	899	124	1,023
% satisfactory	72	81	73
20 Less stringent than RC Standard			
Total DWF (x 1,000 gpd)	0	16,146	16,146
% satisfactory	—	11	11
g. *Sewage treatment works by population ranges* Present population			
21 1,000 and under Satisfactory	37	11	48
Unsatisfactory	8	2	10
22 1,001 to 5,000 Satisfactory	28	10	38
Unsatisfactory	5	0	5
23 5,001 to 10,000 Satisfactory	3	3	6
Unsatisfactory	2	1	3
24 10,001 to 50,000 Satisfactory	0	3	3
Unsatisfactory	0	0	0
25 50,001 to 100,000 Satisfactory	0	0	0
Unsatisfactory	0	0	0
26 100,001 to 500,000 Satisfactory	0	0	0
Unsatisfactory	0	2	2

quantities of unsatisfactory effluent are discharged to the tidal length of the Yare (9.5 mgd unsatisfactory) and the tidal length of the Orwell (5 mgd unsatisfactory). All but one of the discharges to the Bure and the Ant—the principal Broadland rivers—are satisfactory.

62 There are only 2 large discharges of sewage effluent, both over 5 mgd and both discharging to tidal waters. The effluent standard at both is less stringent than Royal Commission and together they account for most of the effluent considered unsatisfactory. The distribution of works by population ranges shows that 88 per cent of the total number serve populations of 5,000 and less and a little over a half of these serve populations of 1,000 and less.

63 The forecasts for 1980 show that the number of treatment works is not expected to change, but the population served should increase to about 720,000 and the effluent discharged to nearly 37 mgd, the increases being well spread without any particular areas being significant. Remedial works are in hand to improve a number of the unsatisfactory discharges. Improved standards will be required at 3 works by 1980.

64 It should be noted that none of the figures include the many discharges of crude sewage at Great Yarmouth. These are discussed in Chapter 3. □

8 Essex River Authority

Table 11 Details of sewage treatment works and discharges of sewage effluent

Description	Rivers		
	Non-tidal	Tidal*	All rivers
a. *Sewage treatment works*			
1 Total number of discharges	131	27	158
2 Number where effluent is satisfactory	98	18	116
3 Number where effluent is unsatisfactory (capacity)	29	7	36
4 Number where effluent is unsatisfactory (maintenance)	4	2	6
b. *Populations*			
5 Total population served	499,738	292,341	792,079
6 Total served, effluent satisfactory	342,883	129,412	472,295
7 Total served, effluent unsatisfactory	156,855	162,929	319,784
c. *Dry weather flow*			
8 Present dry weather flow of sewage effluent (x 1,000 gpd)	19,936	14,666	34,602
9 Percentage of industrial effluent in DWF	5	22	12
10 Dry weather flow. Discharges satisfactory (x 1,000 gpd)	13,966	6,243	20,209
11 Dry weather flow. Discharges unsatisfactory (x 1,000 gpd)	5,970	8,423	14,393
d. *Number of discharges by effluent standards*			
12 Royal Commission Standard			
Number	92	23	115
Number satisfactory	65	15	80
13 More stringent than RC Standard			
Number	39	0	39
Number satisfactory	33	—	33
14 Less stringent than RC Standard			
Number	0	4	4
Number satisfactory	—	3	3
e *Population by effluent standards*			
15 Royal Commission Standard			
Total population	197,942	267,263	465,205
% satisfactory	59	42	49
16 More stringent than RC Standard			
Total population	301,796	0	301,796
% satisfactory	75	—	75
17 Less stringent than RC Standard			
Total population	0	25,078	25,078
% satisfactory	—	72	72
f. *Dry weather flow by effluent standards*			
18 Royal Commission Standard			
Total DWF (x 1,000 gpd)	7,001	13,371	20,372
% satisfactory	65	39	48
19 More stringent than RC Standard			
Total DWF (x 1,000 gpd)	12,935	0	12,935
% satisfactory	73	—	73
20 Less stringent than RC Standard			
Total DWF (x 1,000 gpd)	0	1,295	1,295
% satisfactory	—	80	80
g. *Sewage treatment works by population ranges* Present population			
21 1,000 and under Satisfactory	49	7	56
Unsatisfactory	11	2	13
22 1,001 to 5,000 Satisfactory	31	6	37
Unsatisfactory	13	2	15
23 5,001 to 10,000 Satisfactory	9	2	11
Unsatisfactory	4	1	5
24 10,001 to 50,000 Satisfactory	8	2	10
Unsatisfactory	5	3	8
25 50,001 to 100,000 Satisfactory	1	1	2
Unsatisfactory	0	1	1

*The above figures do not include discharges to tidal waters of rivers discharging to the River Thames, which are controlled by the Port of London Authority and are recorded in the section of the report dealing with that authority.

65 The area, which borders on the Thames estuary and the North Sea, extends over most of the county of Essex and includes also small parts of Suffolk and Cambridgeshire and Isle of Ely and a part of the Greater London area. The major centres of population are the London Boroughs of Barking, Havering, Redbridge and parts of Newham. Some built-up and industrialised areas lie in the strip between London and Southend-on-Sea and include Thurrock, Basildon and Brentwood. Other main centres of population are at Colchester and Chelmsford, but over much of the area population densities are low and land is generally devoted to agriculture.

66 In all there are 158 discharges of sewage effluent in the area, 3 of the treatment works being owned by GLC. Nearly 75 per cent of the discharges are satisfactory and the population served by the satisfactory treatment works, and the effluent discharged to a satisfactory standard represent about 60 per cent of the total.

67 So far as non-tidal waters are concerned, the Roding receives the greatest amount of effluent—about 5.6 mgd and about a third of this is unsatisfactory. The Blackwater receives the greatest amount discharged to tidal waters—6.3 mgd of which about 5.5 mgd is considered unsatisfactory.

68 There are 7 treatment works in the area where the discharge exceeds 1 mgd and one of these exceeds 5 mgd. Together they account for nearly a half of all the sewage effluent discharged in the area, but over 120 out of the total of 158 treatment works serve populations of less than 5,000 and more than half of these serve 1,000 and less.

69 Most works are required to discharge effluents to Royal Commission standard and about half of the total volume of effluent from these is satisfactory, but a fairly big proportion of all the effluent (nearly 40 per cent) is required to meet standards more stringent than Royal Commission, and 73 per cent of this is satisfactory.

70 By 1980, 24 works are expected to close down and 4 to open so that by then only 138 works should be in operation. The estimated population to be served in 1980 is just over a million with an expected volume of sewage effluent of about 52 mgd. Improved standards are envisaged at 12 works all discharging to non-tidal waters, the corresponding 1980 flow being about 4 mgd.

71 A number of schemes are in hand in the area to improve unsatisfactory discharges. □

9 Kent River Authority
(see Table 12)

72 The area extends over the entire county of Kent, parts of Sussex and Surrey and parts of the London boroughs of Bexley and Bromley. There are several centres of population along and close to the Thames estuary at such places as Dartford, Gravesend, Rochester, Chatham and Gillingham and also at Canterbury, Maidstone and Royal Tunbridge Wells. Population densities tend to be high in these areas and in the neighbourhood of the many coastal resorts, but there is also a considerable amount of land devoted to market gardening and the fruit-growing industry.

73 There are 156 sewage treatment works discharging in the area of which around three-quarters produce satisfactory effluents by number, population served and volume discharged. Most discharges are required to meet Royal Commission standard and about 70 per cent of these are satisfactory. Of the 47 treatment works considered unsatisfactory 10 are so classified because of poor maintenance.

74 The total volume of effluent is divided about equally between non-tidal and tidal waters. The biggest discharges to non-tidal waters are to the Medway (1.9 mgd, 94 per cent satisfactory) the Great Stour (5.3 mgd, 50 per cent satisfactory) and the Grom (1.6 mgd, all unsatisfactory). By far the largest discharges to tidal waters are to the Medway (13.9 mgd, 97 per cent satisfactory).

Table 12   Details of sewage treatment works and discharges of sewage effluent

Description	Rivers Non-tidal	Tidal	All rivers
a. Sewage treatment works			
1 Total number of discharges	134	22	156
2 Number where effluent is satisfactory	95	14	109
3 Number where effluent is unsatisfactory (capacity)	29	8	37
4 Number where effluent is unsatisfactory (maintenance)	10	0	10
b. Populations			
5 Total population served	414,264	451,694	865,958
6 Total served, effluent satisfactory	292,786	407,307	700,093
7 Total served, effluent unsatisfactory	121,478	44,387	165,865
c. Dry weather flow			
8 Present dry weather flow of sewage effluent (x 1,000 gpd)	19,504	19,206	38,710
9 Percentage of industrial effluent in DWF	7	15	11
10 Dry weather flow. Discharges satisfactory (x 1,000 gpd)	12,667	16,571	29,238
11 Dry weather flow. Discharges unsatisfactory (x 1,000 gpd)	6,837	2,635	9,472
d. Number of discharges by effluent standards			
12 Royal Commission Standard Number	107	16	123
Number satisfactory	72	10	82
13 More stringent than RC Standard Number	27	0	27
Number satisfactory	23	—	23
14 Less stringent than RC Standard Number	0	6	6
Number satisfactory	—	4	4
e. Population by effluent standards			
15 Royal Commission Standard Total population	273,649	170,617	444,266
% satisfactory	67	78	71
16 More stringent than RC Standard Total population	140,615	0	140,615
% satisfactory	78	—	78
17 Less stringent than RC Standard Total population	0	281,077	281,077
% satisfactory	—	97	97
f. Dry weather flow by effluent standards			
18 Royal Commission Standard Total DWF (x 1,000 gpd)	12,629	8,671	21,300
% satisfactory	60	73	65
19 More stringent than RC Standard Total DWF (x 1,000 gpd)	6,875	0	6,875
% satisfactory	73	—	73
20 Less stringent than RC Standard Total DWF (x 1,000 gpd)	0	10,535	10,535
% satisfactory	—	98	98
g. Sewage treatment works by population ranges Present population			
21 1,000 and under Satisfactory	43	2	45
Unsatisfactory	20	1	21
22 1,001 to 5,000 Satisfactory	37	3	40
Unsatisfactory	14	4	18
23 5,001 to 10,000 Satisfactory	11	3	14
Unsatisfactory	2	2	4
24 10,001 to 50,000 Satisfactory	4	4	8
Unsatisfactory	3	1	4
25 50,001 to 100,000 Satisfactory	0	1	1
Unsatisfactory	0	0	0
26 100,001 to 500,000 Satisfactory	0	1	1
Unsatisfactory	0	0	0

75 There are 10 treatment works discharging over 1 mgd with one of these discharging over 5 mgd and they account for nearly 65 per cent of the total sewage discharged in the area with the majority being satisfactory. One of these larger discharges which is unsatisfactory is of note because it is to the River Grom where the dilution afforded by the natural flow in the river is especially low. Nearly 80 per cent of the works serve populations of less than 5,000 and more than half of these serve populations of 1,000 and less.

76 The forecasts up to 1980 show that 23 works are likely to be abandoned and 8 new works opened thus reducing the number to 141, whilst the population served is likely to rise to just over 1,200,000

and the sewage effluent discharged to nearly 65 mgd. Improved standards of effluent are likely to be called for at 9 works, the corresponding 1980 flow being about 7 mgd.

77 A number of schemes of remedial works are at present under construction in the area. ☐

10 Sussex River Authority

Table 13 Details of sewage treatment works and discharges of sewage effluent

Description	Rivers		
	Non-tidal	Tidal	All rivers
a. Sewage treatment works			
1 Total number of discharges	104	14	118
2 Number where effluent is satisfactory	70	7	77
3 Number where effluent is unsatisfactory (capacity)	25	5	30
4 Number where effluent is unsatisfactory (maintenance)	9	2	11
b. Populations			
5 Total population served	307,821	54,229	362,050
6 Total served, effluent satisfactory	181,074	14,333	195,407
7 Total served, effluent unsatisfactory	126,747	39,896	166,643
c. Dry weather flow			
8 Present dry weather flow of sewage effluent (x 1,000 gpd)	12,337	3,556	15,893
9 Percentage of industrial effluent in DWF	7	3	6
10 Dry weather flow. Discharges satisfactory (x 1,000 gpd)	7,074	496	7,570
11 Dry weather flow. Discharges unsatisfactory (x 1,000 gpd)	5,263	3,060	8,323
d. Number of discharges by effluent standards			
12 Royal Commission Standard			
Number	95	11	106
Number satisfactory	61	5	66
13 More stringent than RC Standard			
Number	9	0	9
Number satisfactory	9	—	9
14 Less stringent than RC Standard			
Number	0	3	3
Number satisfactory	—	2	2
e Population by effluent standards			
15 Royal Commission Standard			
Total population	247,186	31,413	278,599
% satisfactory	49	15	45
16 More stringent than RC Standard			
Total population	60,635	0	60,635
% satisfactory	100	—	100
17 Less stringent than RC Standard.			
Total population	0	22,816	22,816
% satisfactory	—	43	43
f. Dry weather flow by effluent standards			
18 Royal Commission Standard			
Total DWF (x 1,000 gpd)	10,641	2,240	12,881
% satisfactory	51	6	43
19 More stringent than RC Standard			
Total DWF (x 1,000 gpd)	1,696	0	1,696
% satisfactory	100	—	100
20 Less stringent than RC Standard			
Total DWF (x 1,000 gpd)	0	1,316	1,316
% satisfactory	—	28	28
g. Sewage treatment works by population ranges			
Present population			
21 1,000 and under Satisfactory	42	4	46
Unsatisfactory	17	4	21
22 1,001 to 5,000 Satisfactory	20	2	22
Unsatisfactory	11	1	12
23 5,001 to 10,000 Satisfactory	2	1	3
Unsatisfactory	2	0	2
24 10,001 to 50,000 Satisfactory	6	0	6
Unsatisfactory	4	2	6

78 The area includes the major parts of the counties of East and West Sussex and small parts of Surrey and Hampshire. The principal towns are the seaside resorts of Brighton, Hastings and Eastbourne and a very big proportion of the total population is concentrated along the coast where most of the sewage is discharged to the sea and therefore outside the scope of this survey. Land use in the area is largely for agriculture, with a thriving holiday industry also of importance.

79 The total number of sewage effluent discharges recorded is 118 and 65 per cent of these are satisfactory. Of the total population served by inland sewage treatment works, just over a half are served by works considered satisfactory and just under a half of the total volume discharged is considered satisfactory. About one-third of the works discharging unsatisfactory effluents do so as a result of poor maintenance. The greatest quantities of unsatisfactory effluent are discharged to the Arun (1.7 mgd to non-tidal waters), the Ouse (about 1 mgd to non-tidal and 1 mgd to tidal) and the Ham Brook (2 mgd to tidal).

80 Most discharges (about 90 per cent) are required to meet Royal Commission standard although only about two-thirds are satisfactory. All works requiring effluent standards more stringent than Royal Commission do comply and two out of the three where standards less stringent than Royal Commission apply, are satisfactory.

81 There are 3 works which treat flows in excess of 1 mgd and they account for nearly one-third of the total flow. Two are unsatisfactory and together they account for nearly half of all the unsatisfactory effluent in the area. All but 17 of the treatment works serve populations of 5,000 or less whilst nearly 60 per cent of all the treatment works serve populations of 1,000 or less.

82 The forecasts for the period up to 1980 show that 13 works are expected to close and 3 new works are likely to be built in the period, thus reducing the total number to 108. The estimated population served is expected to increase to just over half a million and the flow to about 25 mgd. Improved standards are likely to be required at 4 works, the corresponding 1980 flow being about 2.5 mgd.

83 A number of schemes to improve unsatisfactory discharges are under construction in the area. ☐

11 Hampshire River Authority
(see Table 14)

84 The area of the river authority lies almost entirely within the county of Hampshire but includes small fringe areas in the counties of Berkshire and Wiltshire. Most of the population and industry are located around Southampton and Portsmouth and adjacent areas such as Fareham and Gosport on the coast and much of the remaining area is taken up for agriculture. The major river basins discharge into The Solent and about three-quarters of all the sewage effluent recorded is discharged to tidal waters. Discharges to The Solent itself are not included in the survey.

85 There are 47 discharges of sewage effluent of which about 65 per cent are satisfactory but only about 30 per cent of the population s served by treatment works considered satisfactory and only about 30 per cent of the total effluent discharged is considered satisfactory. About 80 per cent of all the unsatisfactory effluent is discharged to tidal waters. The Itchen stands out as the river receiving the greatest amount of unsatisfactory effluent on its non-tidal length (3.2 mgd unsatisfactory) whereas on tidal lengths, the Wallington receives 3.4 mgd unsatisfactory, the Test 7.5 mgd unsatisfactory and the Itchen 4.5 mgd unsatisfactory.

86 There are 7 discharges which exceed 1 mgd and 2 of these exceed 5 mgd. These 7 discharges account for some 26 mgd of effluent which is nearly 90 per cent of the total volume discharged in the area. Five of these discharges are to tidal waters, 4 are unsatisfactory and they account for a big proportion of the total unsatisfactory effluent in the area. The distribution by population shows that 33 works out of the total of 47 serve populations of 5,000 or less and a little over half of these serve populations of 1,000 or less. Most of these works discharge to non-tidal waters.

Table 14 Details of sewage treatment works and discharges of sewage effluent

Description	Rivers		
	Non-tidal	Tidal	All rivers
a. *Sewage treatment works*			
1 Total number of discharges	33	14	47
2 Number where effluent is satisfactory	24	6	30
3 Number where effluent is unsatisfactory (capacity)	8	8	16
4 Number where effluent is unsatisfactory (maintenance)	1	0	1
b. *Populations*			
5 Total population served	126,848	434,938	561,786
6 Total served, effluent satisfactory	62,714	109,588	172,302
7 Total served, effluent unsatisfactory	64,134	325,350	389,484
c. *Dry weather flow*			
8 Present dry weather flow of sewage effluent (x 1,000 gpd)	7,316	23,070	30,386
9 Percentage of industrial effluent in DWF	7	23	19
10 Dry weather flow. Discharges satisfactory (x 1,000 gpd)	3,042	6,613	9,655
11 Dry weather flow. Discharges unsatisfactory (x 1,000 gpd)	4,274	16,457	20,731
d. *Number of discharges by effluent standards**			
12 Royal Commission Standard			
Number	30	7	37
Number satisfactory	23	3	26
13 More stringent than RC Standard			
Number	3	0	3
Number satisfactory	1	—	1
14 Less stringent than RC Standard			
Number	0	2	2
Number satisfactory	—	1	1
e. *Population by effluent standards**			
15 Royal Commission Standard			
Total population	120,547	282,808	403,355
% satisfactory	51	32	38
16 More stringent than RC Standard			
Total population	6,301	0	6,301
% satisfactory	29	—	29
17 Less stringent than RC Standard			
Total population	0	70,800	70,800
% satisfactory	—	23	23
f. *Dry weather flow by effluent standards**			
18 Royal Commission Standard			
Total DWF (x 1,000 gpd)	6,947	16,600	23,547
% satisfactory	41	36	38
19 More stringent than RC Standard			
Total DWF (x 1,000 gpd)	369	0	369
% satisfactory	53	—	53
20 Less stringent than RC Standard			
Total DWF (x 1,000 gpd)	0	2,394	2,394
% satisfactory	—	21	21
g. *Sewage treatment works by population ranges*			
Present population			
21 1,000 and under Satisfactory	13	2	15
Unsatisfactory	4	0	4
22 1,001 to 5,000 Satisfactory	8	2	10
Unsatisfactory	3	1	4
23 5,001 to 10,000 Satisfactory	2	0	2
Unsatisfactory	1	1	2
24 10,001 to 50,000 Satisfactory	1	1	2
Unsatisfactory	1	2	3
25 50,001 to 100,000 Satisfactory	0	1	1
Unsatisfactory	0	3	3
26 100,001 to 500,000 Satisfactory	0	0	0
Unsatisfactory	0	1	1

* A total of 5 discharges were indicated as having "no standards" and these account for the small differences in the totals against lines 12–14, 15–17 and 18–20 above when compared with the totals against lines 1–4, 5–7 and 8–11 respectively. These are pre-1960 discharges to tidal waters and are not subject to control. They relate to a population of about 81,000 and a flow of 4 mgd, virtually the whole of which is classed as unsatisfactory.

87 The future of the South Hampshire area is somewhat uncertain and this made it difficult for the river authority to forecast the 1980 situation, but the information they were able to provide indicated that 10 works were expected to close and 2 new were likely to open leaving 39 in operation at the end of the period. The estimated population served in 1980 would be about 840,000 and the total volume of effluent discharged about 56.5 mgd, but these figures could be subject to considerable variation. Improved standards are likely to be called for at 5 works, 4 of these being works where no standards are at present applicable.

88 Some schemes for remedial works have been approved or are under construction at present. ☐

12 Isle of Wight River and Water Authority

Table 15 Details of sewage treatment works and discharges of sewage effluent

Description	Rivers		
	Non-tidal	Tidal	All rivers
a. *Sewage treatment works*			
1 Total number of discharges	11	2	13
2 Number where effluent is satisfactory	6	1	7
3 Number where effluent is unsatisfactory (capacity)	3	1	4
4 Number where effluent is unsatisfactory (maintenance)	2	0	2
b. *Populations*			
5 Total population served	8,319	22,083	30,402
6 Total served, effluent satisfactory	3,989	743	4,732
7 Total served, effluent unsatisfactory	4,330	21,340	25,670
c. *Dry weather flow*			
8 Present dry weather flow of sewage effluent (x 1,000 gpd)	487	891	1,378
9 Percentage of industrial effluent in DWF	0	4	3
10 Dry weather flow. Discharges satisfactory (x 1,000 gpd)	205	110	315
11 Dry weather flow. Discharges unsatisfactory (x 1,000 gpd)	282	781	1,063
d. *Number of discharges by effluent standards*			
12 Royal Commission Standard			
Number	10	2	12
Number satisfactory	5	1	6
13 More stringent than RC Standard			
Number	1	0	1
Number satisfactory	1	—	1
14 Less stringent than RC Standard			
Number	0	0	0
Number satisfactory	—	—	—
e. *Population by effluent standards*			
15 Royal Commission Standard			
Total population	8,169	22,083	30,252
% satisfactory	47	3	15
16 More stringent than RC Standard			
Total population	150	0	150
% satisfactory	100	—	100
17 Less stringent than RC Standard			
Total population	0	0	0
% satisfactory	—	—	—
f. *Dry weather flow by effluent standards*			
18 Royal Commission Standard			
Total DWF (x 1,000 gpd)	483	891	1,374
% satisfactory	42	12	23
19 More stringent than RC Standard			
Total DWF (x 1,000 gpd)	4	0	4
% satisfactory	100	—	100
20 Less stringent than RC Standard			
Total DWF (x 1,000 gpd)	0	0	0
% satisfactory	—	—	—
g. *Sewage treatment works by population ranges*			
Present population			
21 1,000 and under Satisfactory	4	1	5
Unsatisfactory	2	0	2
22 1,001 to 5,000 Satisfactory	2	0	2
Unsatisfactory	3	0	3
23 5,001 to 10,000 Satisfactory	0	0	0
Unsatisfactory	0	0	0
24 10,001 to 50,000 Satisfactory	0	0	0
Unsatisfactory	0	1	1

89 The island is noted mainly as a holiday resort with agriculture, horticulture, and forestry in the rural areas and engineering and light industry near the urban areas. The major towns are Newport, Cowes, Ryde, Sandown, Shanklin and Ventnor. There are a large number of

river basins around the island, the principal basins terminating at the controlled waters of The Solent. A number of small rivers also discharge to the English Channel between The Needles in the west and Foreland and Bembridge in the east. Nearly two-thirds of all the effluent recorded is discharged to tidal waters. Direct discharges to The Solent are not included in the survey.

90 There are 13 sewage effluent discharges and about half are satisfactory, but in terms of population and volume the satisfactory discharges represent only some 16 per cent of the population served and about 23 per cent of the volume discharged. All of the unsatisfactory effluent discharged to tidal waters is discharged to the Medina. Unsatisfactory effluent to non-tidal rivers is divided mainly between the Carpenters Dyke, the Wroxall Stream and the Brightstone Brook. Most effluents are required to comply with Royal Commission standard but again the percentages of those that are satisfactory in terms of population served and volume discharged are low. Apart from Newport, with a treatment works serving just over 21,000 people, all the works serve populations of 5,000 and less and 7 of them serve less than 1,000.

91 By 1980, 2 new works are expected to open, increasing the number to 15. The population served is expected to rise to nearly 40,000 and the flow discharged to about 2.5 mgd. Improved standards are envisaged at only one works.

92 A number of schemes for improving existing discharges are in preparation. ☐

13 Avon and Dorset River Authority
(see Table 16)

93 The area extends over most of Dorset, the southern half of Wiltshire and small parts of Hampshire and Somerset. The main centres of population are on the coast near Bournemouth, Christchurch, Poole and Weymouth, all well-known holiday resorts. Inland, the principal town is Salisbury, and the area is mainly devoted to agriculture.

94 There are 85 sewage treatment works recorded in the survey and 73 of them (about 85 per cent) are satisfactory. The proportions of population served by satisfactory treatment works and volume discharged to a satisfactory standard are of the same order. About one-third of all the sewage effluent is discharged to tidal waters and all of that effluent is considered to be satisfactory. Most of the discharges in the area are required to meet Royal Commission standard.

95 The Stour receives by far the greatest amount of unsatisfactory effluent (about 2.1 mgd). The balance of the unsatisfactory effluent is divided in fairly small quantities between a number of rivers, notably the Avon, the Bourne, the Bibberne Brook, the Bride and the Char.

96 Five of the discharges are greater than 1 mgd and the total volume of these is about 60 per cent of the total sewage discharged. Four of these discharges are satisfactory and the remaining one represents a big proportion of the unsatisfactory volume discharged in the area. All but 13 of the discharges are from treatment works serving populations less than 5,000 and nearly half of all the discharges are from treatment works serving less than 1,000 population.

97 The forecasts up to 1980 show that only 1 works is likely to close down and that the population is likely to increase to about 720,000 and the flow to about 32 mgd. Big expected increases in the Bournemouth and Poole areas are responsible for a large proportion of the additional population and flow. Improved standards are likely to be required at 2 works affecting some 30,000 people and about 1.5 mgd of effluent by 1980.

98 There is not a great deal to be done in the area, but a number of schemes to improve unsatisfactory discharges have recently been commissioned and others are in hand. ☐

Table 16 Details of sewage treatment works and discharges of sewage effluent

Description	Rivers Non-tidal	Tidal	All rivers
a. *Sewage treatment works*			
1 Total number of discharges	79	6	85
2 Number where effluent is satisfactory	67	6	73
3 Number where effluent is unsatisfactory (capacity)	11	0	11
4 Number where effluent is unsatisfactory (maintenance)	1	0	1
b. *Populations*			
5 Total population served	299,579	126,545	426,124
6 Total served, effluent satisfactory	233,925	126,545	360,470
7 Total served, effluent unsatisfactory	65,654	0	65,654
c. *Dry weather flow*			
8 Present dry weather flow of sewage effluent (x 1,000 gpd)	12,644	5,766	18,410
9 Percentage of industrial effluent in DWF	7	10	8
10 Dry weather flow. Discharges satisfactory (x 1,000 gpd)	9,972	5,766	15,738
11 Dry weather flow. Discharges unsatisfactory (x 1,000 gpd)	2,672	0	2,672
d. *Number of discharges by effluent standards*			
12 Royal Commission Standard Number	72	6	78
Number satisfactory	62	6	68
13 More stringent than RC Standard Number	6	0	6
Number satisfactory	5	—	5
14 Less stringent than RC Standard Number	1	0	1
Number satisfactory	0	—	0
e. *Population by effluent standards*			
15 Royal Commission Standard Total population	291,150	126,545	417,695
% satisfactory	79	100	85
16 More stringent than RC Standard Total population	6,429	0	6,429
% satisfactory	77	—	77
17 Less stringent than RC Standard Total population	2,000	0	2,000
% satisfactory	0	—	0
f. *Dry weather flow by effluent standards*			
18 Royal Commission Standard Total DWF (x 1,000 gpd)	12,337	5,766	18,103
% satisfactory	80	100	86
19 More stringent than RC Standard Total DWF (x 1,000 gpd)	241	0	241
% satisfactory	55	—	55
20 Less stringent than RC Standard Total DWF (x 1,000 gpd)	66	0	66
% satisfactory	0	—	0
g. *Sewage treatment works by population ranges* Present population			
21 1,000 and under Satisfactory	32	2	34
Unsatisfactory	4	0	4
22 1,001 to 5,000 Satisfactory	25	2	27
Unsatisfactory	7	0	7
23 5,001 to 10,000 Satisfactory	3	0	3
Unsatisfactory	0	0	0
24 10,001 to 50,000 Satisfactory	7	1	8
Unsatisfactory	1	0	1
25 50,001 to 100,000 Satisfactory	0	1	1
Unsatisfactory	0	0	0

14 Devon River Authority
(see Table 17)

99 The Devon River Authority area extends from the English Channel across to the Bristol Channel and covers most of the county of Devon and fringe areas of Somerset and Dorset. The main centres of population are at Torbay and Exeter, at the several residential and holiday resorts along the north and south coast and at the inland towns of Newton Abbot, Tiverton and Barnstaple. Much of the area is devoted to agriculture and to a thriving holiday industry.

Table 17 Details of sewage treatment works and discharges of sewage effluent

Description	Rivers		
	Non-tidal	Tidal	All rivers
a. Sewage treatment works			
1 Total number of discharges	123	13	136
2 Number where effluent is satisfactory	78	4	82
3 Number where effluent is unsatisfactory (capacity)	41	9	50
4 Number where effluent is unsatisfactory (maintenance)	4	0	4
b. Populations			
5 Total population served	128,393	157,370	285,763
6 Total served, effluent satisfactory	73,516	15,622	89,138
7 Total served, effluent unsatisfactory	54,877	141,748	196,625
c. Dry weather flow			
8 Present dry weather flow of sewage effluent (x 1,000 gpd)	6,899	7,341	14,240
9 Percentage of industrial effluent in DWF	16	16	16
10 Dry weather flow. Discharges satisfactory (x 1,000 gpd)	3,744	625	4,369
11 Dry weather flow. Discharges unsatisfactory (x 1,000 gpd)	3,155	6,716	9,871
d. Number of discharges by effluent standards			
12 Royal Commission Standard			
Number	112	6	118
Number satisfactory	75	1	76
13 More stringent than RC Standard			
Number	11	1	12
Number satisfactory	3	0	3
14 Less stringent than RC Standard			
Number	0	6	6
Number satisfactory	—	3	3
e. Population by effluent standards			
15 Royal Commission Standard			
Total population	116,352	123,787	240,139
% satisfactory	62	0*	30
16 More stringent than RC Standard			
Total population	12,041	4,920	16,961
% satisfactory	13	0	9
17 Less stringent than RC Standard			
Total population	0	28,663	28,663
% satisfactory	—	53	53
f. Dry weather flow by effluent standards			
18 Royal Commission Standard			
Total DWF (x 1,000 gpd)	6,395	5,308	11,703
% satisfactory	58	0*	32
19 More stringent than RC Standard			
Total DWF (x 1,000 gpd)	504	196	700
% satisfactory	10	0	7
20 Less stringent than RC Standard			
Total DWF (x 1,000 gpd)	0	1,837	1,837
% satisfactory	—	33	33
g. Sewage treatment works by population ranges Present population			
21 1,000 and under Satisfactory	59	2	61
Unsatisfactory	26	2	28
22 1,001 to 5,000 Satisfactory	18	0	18
Unsatisfactory	17	3	20
23 5,001 to 10,000 Satisfactory	0	2	2
Unsatisfactory	2	2	4
24 10,001 to 50,000 Satisfactory	1	0	1
Unsatisfactory	0	1	1
25 50,001 to 100,000 Satisfactory	0	0	0
Unsatisfactory	0	1	1

* One discharge is satisfactory but the population and flow represent only about 0.3% of the totals in the "Tidal" column at lines 15 and 18.

100 There are 136 sewage treatment works in the area, 13 of which discharge to tidal waters. The latter discharges account for more than half of all the effluent, and for about 70 per cent of all the unsatisfactory effluent, nearly all of that discharged to tidal waters being considered unsatisfactory. Most of the effluent to tidal waters that is satisfactory is required to comply with a standard less stringent than Royal Commission. Nearly two-thirds of the treatment works discharging to non-tidal waters are considered satisfactory, the corresponding proportions of population served and flow discharged being in the region of 55 per cent. Nearly half of all the effluent discharged is to the River Exe and its tributaries.

101 On tidal lengths the greatest quantities of unsatisfactory effluent are discharged to the Exe (4.1 mgd), the Teign (1.3 mgd) and the Dart (1 mgd). None of the effluents discharged to these rivers on their tidal reaches is considered satisfactory. The unsatisfactory discharges to the non-tidal rivers are spread over a wide area, mostly in relatively small quantities, the largest ones being to the Axe (0.3 mgd unsatisfactory), the Otter (0.3 mgd unsatisfactory), the Dart (0.7 mgd unsatisfactory) and the Okement (0.3 mgd unsatisfactory).

102 There are only 3 discharges of 1 mgd and over but they account for more than 40 per cent of the total flow of effluent and for about 60 per cent of all the unsatisfactory effluent. Most works are small with all but 9 serving populations of 5,000 and less and 89 out of a total of 136 serving 1,000 and less.

103 The survey shows that, in the period up to 1980, 10 existing works are likely to be abandoned and 3 new works built thus reducing the numbers to 129. The population to be served in 1980 is estimated to be nearly 370,000 with a dry weather flow of about 23 mgd. Improved standards are envisaged at one small treatment works only.

104 Several schemes are already under construction in the area to improve unsatisfactory discharges and other schemes are under preparation including one at Newton Abbot where major re-organisation is planned. □

15 Cornwall River Authority
(see Table 18 over page)

105 The Cornwall River Authority covers all of the county of Cornwall and also an adjacent strip of the county of Devon which includes the county borough of Plymouth. The three main centres of population are at Plymouth, St. Austell with Fowey, and Camborne-Redruth. The area is noted for its tourist industry and for industrial activity associated with the mining of china clay and other minerals. Much of the area is of a rural character and apart from a few centres, the density of resident population is low.

106 There are 101 sewage treatment works recorded, and two-thirds are satisfactory. All but 2 are required to comply with the Royal Commission standard and of the 36 unsatisfactory treatment works, 17 are so classified because of poor maintenance. The data also shows that almost three-quarters of the population are served by the treatment works considered satisfactory and about the same proportion of the total effluent is discharged to a satisfactory standard.

107 Over half of the sewage effluent in the area is discharged to the basin of the River Tamar and some 5 mgd is discharged to 2 rivers in the Tamar basin—the Tavy and the Camel's Head Stream. It should be noted that there are many discharges of crude sewage in the Tamar estuary and elsewhere which are not included in these figures. They are described in Chapter 3.

108 The biggest quantities of unsatisfactory effluent to non-tidal rivers are to the Tamar (0.25 mgd) and to 2 small rivers, the Cober (0.4 mgd) and the Crantock Stream (0.4 mgd). The balance is distributed mostly in small amounts over many rivers. Most of the unsatisfactory effluent to tidal waters is discharged to the Kenwyn.

109 Only 2 sewage treatment works discharge a flow greater than 1 mgd, both are satisfactory and they account for a big proportion of all the effluent. All but 8 of the works serve populations of 5,000 and less and 66 out of the total of 101 in the survey serve populations of under 1,000.

110 The returns show that It is proposed to abandon 8 works and to commission 2 new works, so that by 1980 the number would reduce to 95. The population should rise to a little over 260,000 and the dry weather flow to just over 14 mgd. No changes in effluent standards are likely to be required.

111 Several schemes to improve unsatisfactory discharges are imminent in the area. □

(15 Cornwall River Authority continued)

Table 18 Details of sewage treatment works and discharges of sewage effluent

Description	Rivers Non-tidal	Tidal	All rivers
a. Sewage treatment works			
1 Total number of discharges	92	9	101
2 Number where effluent is satisfactory	60	5	65
3 Number where effluent is unsatisfactory (capacity)	16	3	19
4 Number where effluent is unsatisfactory (maintenance)	16	1	17
b. Populations			
5 Total population served	182,128	31,404	213,532
6 Total served, effluent satisfactory	143,114	12,693	155,807
7 Total served, effluent unsatisfactory	39,014	18,711	57,725
c. Dry weather flow			
8 Present dry weather flow of sewage effluent (x 1,000 gpd)	9,966	1,569	11,535
9 Percentage of industrial effluent in DWF	4	0	3
10 Dry weather flow. Discharges satisfactory (x 1,000 gpd)	7,948	570	8,518
11 Dry weather flow. Discharges unsatisfactory (x 1,000 gpd)	2,018	999	3,017
d. Number of discharges by effluent standards			
12 Royal Commission Standard Number	90	9	99
Number satisfactory	58	5	63
13 More stringent than RC Standard Number	2	0	2
Number satisfactory	2	—	2
14 Less stringent than RC Standard Number	0	0	0
Number satisfactory	—	—	—
e. Population by effluent standards			
15 Royal Commission Standard Total population	178,063	31,404	209,467
% satisfactory	78	40	72
16 More stringent than RC Standard Total population	4,065	0	4,065
% satisfactory	100	—	100
17 Less stringent than RC Standard Total population	0	0	0
% satisfactory	—	—	—
f. Dry weather flow by effluent standards			
18 Royal Commission Standard Total DWF (x 1,000 gpd)	9,827	1,569	11,396
% satisfactory	79	36	73
19 More stringent than RC Standard Total DWF (x 1,000 gpd)	139	0	139
% satisfactory	100	—	100
20 Less stringent than RC Standard Total DWF (x 1,000 gpd)	0	0	0
% satisfactory	—	—	—
g. Sewage treatment works by population ranges Present population			
21 1,000 and under — Satisfactory	40	3	43
Unsatisfactory	21	2	23
22 1,001 to 5,000 — Satisfactory	15	1	16
Unsatisfactory	10	1	11
23 5,001 to 10,000 — Satisfactory	2	0	2
Unsatisfactory	1	0	1
24 10,001 to 50,000 — Satisfactory	2	1	3
Unsatisfactory	0	1	1
25 50,001 to 100,000 — Satisfactory	1	0	1
Unsatisfactory	0	0	0

16 Somerset River Authority
(see Table 19)

112 The area lies almost entirely within the county of Somerset, apart from small areas to the south within the county of Dorset and to the west in the county of Devon. The largest town is the holiday resort of Weston-super-Mare, and the other main centres are at Taunton, Yeovil and Bridgwater. The sewage of Weston-super-Mare is discharged to the Severn estuary and it is not therefore included in the survey. The population density throughout the area is generally low and most of the area is given over to farming, with the holiday and tourist industry also prominent.

Table 19 Details of sewage treatment works and discharges of sewage effluent

Description	Rivers Non-tidal	Tidal	All rivers
a. Sewage treatment works			
1 Total number of discharges	104	0	104
2 Number where effluent is satisfactory	64	0	64
3 Number where effluent is unsatisfactory (capacity)	37	0	37
4 Number where effluent is unsatisfactory (maintenance)	3	0	3
b. Populations			
5 Total population served	222,789	0	222,789
6 Total served, effluent satisfactory	67,987	0	67,987
7 Total served, effluent unsatisfactory	154,802	0	154,802
c. Dry weather flow			
8 Present dry weather flow of sewage effluent (x 1,000 gpd)	15,352	0	15,352
9 Percentage of industrial effluent in DWF	14	0	14
10 Dry weather flow. Discharges satisfactory (x 1,000 gpd)	3,514	0	3,514
11 Dry weather flow. Discharges unsatisfactory (x 1,000 gpd)	11,838	0	11,838
d. Number of discharges by effluent standards			
12 Royal Commission Standard Number	60	0	60
Number satisfactory	40	—	40
13 More stringent than RC Standard Number	44	0	44
Number satisfactory	24	—	24
14 Less stringent than RC Standard Number	0	0	0
Number satisfactory	—	—	—
e. Population by effluent standards			
15 Royal Commission Standard Total population	63,194	0	63,194
% satisfactory	55	—	55
16 More stringent than RC Standard Total population	159,595	0	159,595
% satisfactory	21	—	21
17 Less stringent than RC Standard Total population	0	0	0
% satisfactory	—	—	—
f. Dry weather flow by effluent standards			
18 Royal Commission Standard Total DWF (x 1,000 gpd)	3,082	0	3,082
% satisfactory	48	—	48
19 More stringent than RC Standard Total DWF (x 1,000 gpd)	12,270	0	12,270
% satisfactory	17	—	17
20 Less stringent than RC Standard Total DWF (x 1,000 gpd)	0	0	0
% satisfactory	—	—	—
g. Sewage treatment works by population ranges Present population			
21 1,000 and under — Satisfactory	46	0	46
Unsatisfactory	18	0	18
22 1,001 to 5,000 — Satisfactory	16	0	16
Unsatisfactory	14	0	14
23 5,001 to 10,000 — Satisfactory	2	0	2
Unsatisfactory	6	0	6
24 10,001 to 50,000 — Satisfactory	0	0	0
Unsatisfactory	2	0	2

Note:
One discharge is not included in the table although it is subject to the consent of the authority. The discharge is to a diverted watercourse which flows to rivers in the Bristol Avon area and it is included in the tables for that authority.

113 A total of 64 out of the 104 sewage treatment works are satisfactory but about 70 per cent of the total population is served by treatment works considered unsatisfactory and nearly 80 per cent of the total volume of effluent is discharged to an unsatisfactory standard. There are no discharges of sewage effluent to tidal waters although there are several discharges of crude sewage which are described in Chapter 3.

114 The rivers receiving the greatest volumes of unsatisfactory effluent are the Tone (about 4 mgd) where all discharges received are unsatisfactory and the Yeovil Yeo (about 3 mgd unsatisfactory) where all but the 2 smallest out of 6 discharges are unsatisfactory. Both rivers are tributaries of the River Parrett.

115 Although nearly 60 per cent of the works are required to discharge effluents to Royal Commission standard, about 80 per cent of the volume of effluent is from treatment works where the standard imposed is more stringent than Royal Commission. These higher standards mainly affect discharges in the River Parrett basin which receives the greatest volume of sewage effluent in the area.

116 Two large discharges, both greater than 2 mgd, account for more than a third of all the sewage effluent. Both are unsatisfactory and together they account for nearly one half of all the unsatisfactory effluent discharged. The number of small treatment works is relatively high, all but 10 serve populations of less than 5,000 and about two-thirds serve 1,000 and less. All but 2 of the 10 larger works are considered to be unsatisfactory.

117 By 1980 the number of works is likely to be reduced by 14 with the abandonment of 16 and the construction of 2 new works. It is forecast that the population served will increase to about 370,000 and the dry weather flow to nearly 25 mgd. Improved standards are envisaged at 27 works, 21 of which discharge to rivers in the Parrett basin. The corresponding 1980 volume of effluent for these 27 works is estimated at about 10 mgd. With 44 treatment works already required to discharge to more stringent than Royal Commission standard, the number in the future will constitute a big proportion of all the discharges in the area.

118 Several schemes to improve unsatisfactory discharges are under construction including a major regional scheme which will lead to the closing down of a number of existing works. Other schemes, some of a regional nature, are under consideration.

17 Bristol Avon River Authority
(see Table 20)

119 The Bristol Avon River Authority area extends over parts of the counties of Gloucester, Wiltshire and Somerset and includes the cities of Bristol and Bath and several inland centres of population notably Keynsham, Chippenham and Trowbridge. Otherwise the population in the area is not particularly high with the land generally taken up for farming.

120 The total number of discharges of sewage effluent in the area is 87 and about two-thirds of these are satisfactory, but the proportions of population served by satisfactory treatment works and volume discharged to a satisfactory standard are just over 40 per cent. Of the 31 works which are unsatisfactory, 10 are so classified because of poor maintenance. All except one of the discharges are to non-tidal waters and most discharges are required to comply with the Royal Commission standard. It should be noted that the table does not include any discharges from Bristol as the discharges from Bristol to the Avon are of crude sewage and are described in Chapter 3. The effluent from the Bristol sewage treatment works is controlled by the Severn River Authority.

121 The Bristol Avon itself receives the largest number of discharges and the greatest quantity of unsatisfactory effluent—some 5.7 mgd. Other rivers which receive large quantities of unsatisfactory effluent are the Biss (1.8 mgd), the Byde Mill Brook (0.5 mgd), the Pudding Brook (0.7 mgd) and the River Marden (0.7 mgd). A number of rivers and streams are significant because of their small natural flow in relation to the quantity of effluent they carry, the Biss being a particularly good example.

122 There are 5 treatment works in the area each discharging over 1 mgd of effluent and together they account for more than half of all the effluent discharged in the area. Three of the discharges are unsatisfactory. There is a preponderance of small works in the area with 70 out of 87 serving populations of 5,000 and less and 40 serving 1,000 and less.

Table 20 Details of sewage treatment works and discharges of sewage effluent

Description	Rivers Non-tidal	Tidal	All rivers
a. Sewage treatment works			
1 Total number of discharges	86	1	87
2 Number where effluent is satisfactory	56	0	56
3 Number where effluent is unsatisfactory (capacity)	20	1	21
4 Number where effluent is unsatisfactory (maintenance)	10	0	10
b. Populations			
5 Total population served	351,081	1,199	352,280
6 Total served, effluent satisfactory	153,154	0	153,154
7 Total served, effluent unsatisfactory	197,927	1,199	199,126
c. Dry weather flow			
8 Present dry weather flow of sewage effluent (x 1,000 gpd)	17,920	59	17,979
9 Percentage of industrial effluent in DWF	11	0	11
10 Dry weather flow. Discharges satisfactory (x 1,000 gpd)	7,585	0	7,585
11 Dry weather flow. Discharges unsatisfactory (x 1,000 gpd)	10,335	59	10,394
d. Number of discharges by effluent standards			
12 Royal Commission Standard Number	80	0	80
Number satisfactory	52	—	52
13 More stringent than RC Standard Number	6	0	6
Number satisfactory	4	—	4
14 Less stringent than RC Standard Number	0	1	1
Number satisfactory	—	0	0
e. Population by effluent standards			
15 Royal Commission Standard Total population	343,303	0	343,303
% satisfactory	43	—	43
16 More stringent than RC Standard Total population	7,778	0	7,778
% satisfactory	81	—	81
17 Less stringent than RC Standard Total population	0	1,199	1,199
% satisfactory	—	0	0
f. Dry weather flow by effluent standards			
18 Royal Commission Standard Total DWF (x 1,000 gpd)	17,613	0	17,613
% satisfactory	42	—	42
19 More stringent than RC Standard Total DWF (x 1,000 gpd)	307	0	307
% satisfactory	86	—	86
20 Less stringent than RC Standard Total DWF (x 1,000 gpd)	0	59	59
% satisfactory	—	0	0
g. Sewage treatment works by population ranges Present population			
21 1,000 and under Satisfactory	28	0	28
Unsatisfactory	12	0	12
22 1,001 to 5,000 Satisfactory	18	0	18
Unsatisfactory	11	1	12
23 5,001 to 10,000 Satisfactory	7	0	7
Unsatisfactory	4	0	4
24 10,001 to 50,000 Satisfactory	3	0	3
Unsatisfactory	2	0	2
25 50,001 to 100,000 Satisfactory	0	0	0
Unsatisfactory	1	0	1

Note:
See footnote on Table 19.

123 The forecast for the period to 1980 shows that the number of treatment works is likely to be reduced to 81, the population served will be about 475,000 and the dry weather flow nearly 27 mgd. It is also estimated that, by 1980, improved effluent standards will be required at 33 treatment works, or more than a third of the total number. The population which will be served by these works in 1980 is estimated to be some 203,000 and the dry weather flow nearly 13 mgd.

124 A number of schemes to improve unsatisfactory discharges are under construction in the area.

18 Severn River Authority

Table 21 Details of sewage treatment works and discharges of sewage effluent

Description	Rivers			Canals	All rivers and canals
	Non-tidal	Tidal	All rivers		
a. Sewage treatment works					
1 Total number of discharges	342	4	346	1	347
2 Number where effluent is satisfactory	221	2	223	1	224
3 Number where effluent is unsatisfactory (capacity)	106	2	108	0	108
4 Number where effluent is unsatisfactory (maintenance)	15	0	15	0	15
b. Populations					
5 Total population served	2,256,229	74,218	2,330,447	2,100	2,332,547
6 Total served, effluent satisfactory	1,296,921	1,942	1,298,863	2,100	1,300,963
7 Total served, effluent unsatisfactory	959,308	72,276	1,031,584	0	1,031,584
c. Dry weather flow					
8 Present dry weather flow of sewage effluent (x 1,000 gpd)	122,188	5,209	127,397	240	127,637
9 Percentage of industrial effluent in DWF	16	10	16	0	16
10 Dry weather flow. Discharges satisfactory (x 1,000 gpd)	69,570	128	69,698	240	69,938
11 Dry weather flow. Discharges unsatisfactory (x 1,000 gpd)	52,618	5,081	57,699	0	57,699
d. Number of discharges by effluent standards					
12 Royal Commission Standard					
Number	307	2	309	1	310
Number satisfactory	198	1	199	1	200
13 More stringent than RC Standard					
Number	34	0	34	0	34
Number satisfactory	22	—	22	—	22
14 Less stringent than RC Standard					
Number	1	2	3	0	3
Number satisfactory	1	1	2	—	2
e. Population by effluent standards					
15 Royal Commission Standard					
Total population	1,470,552	2,740	1,473,292	2,100	1,475,392
% satisfactory	46	32	46	100	46
16 More stringent than RC Standard					
Total population	509,177	0	509,177	0	509,177
% satisfactory	68	—	68	—	68
17 Less stringent than RC Standard					
Total population	276,500	71,478	347,978	0	347,978
% satisfactory	100	2	80	—	80
f. Dry weather flow by effluent standards					
18 Royal Commission Standard					
Total DWF (x 1,000 gpd)	79,389	128	79,517	240	79,757
% satisfactory	46	27	46	100	46
19 More stringent than RC Standard					
Total DWF (x 1,000 gpd)	30,619	0	30,619	0	30,619
% satisfactory	69	—	69	—	69
20 Less stringent than RC Standard					
Total DWF (x 1,000 gpd)	12,180	5,081	17,261	0	17,261
% satisfactory	100	2	71	—	71
g. Sewage treatment works by population ranges					
Present population					
21 1,000 and under Satisfactory	141	1	142	0	142
Unsatisfactory	47	0	47	0	47
22 1,001 to 5,000 Satisfactory	53	1	54	1	55
Unsatisfactory	41	1	42	0	42
23 5,001 to 10,000 Satisfactory	7	0	7	0	7
Unsatisfactory	14	0	14	0	14
24 10,001 to 50,000 Satisfactory	15	0	15	0	15
Unsatisfactory	13	0	13	0	13
25 50,001 to 100,000 Satisfactory	2	0	2	0	2
Unsatisfactory	5	1	6	0	6
26 100,001 to 500,000 Satisfactory	3	0	3	0	3
Unsatisfactory	1	0	1	0	1

125 The area administered by the Severn River Authority is second in size to the largest authority in England, the Yorkshire River Authority, but covers only one river basin, that of the Severn. Parts of 13 different counties are included in the area, the principal ones being the counties of Salop, Montgomery, Worcester, Gloucester, Hereford and Warwick. The greater part of the area lies in England but a small part at the head of the basin lies in Wales. Most of the area is utilised for agriculture, the main exceptions being the fringes of the West Midlands conurbation at Wolverhampton, the areas around Coventry, Worcester, Gloucester and Rugby, where industrial activity is intense, and the several well populated but less industrialised towns. Many of the towns from which the biggest quantities of sewage effluent are discharged are situated on the rivers in the upper parts of the catchment. The city of Bristol, though lying mostly in the area of the Bristol Avon River Authority, discharges much of its sewage effluent into a watercourse controlled by the Severn River Authority.

126 The survey shows that there are 347 sewage treatment works in the area nearly two-thirds of which are satisfactory. Some 55 per cent of the population is served by the satisfactory works and about the same proportion of the total volume of sewage effluent is discharged to a satisfactory standard. Only 15 out of the 123 works regarded as unsatisfactory are so classed as a result of poor maintenance.

127 Most of the works (310) are required to discharge effluents complying with the Royal Commission standard but only 200 of these and less than half of the total flows from them are satisfactory. Where standards more stringent than Royal Commission apply, about 70 per cent of the volume is satisfactory.

128 The greatest numbers of discharges are made to the Severn which receives 23, and to the Avon (Warwickshire) which receives

20, but the River Sowe, a tributary of the Avon in its upper length, receives the largest quantity of effluent (24.5 mgd) all of which is satisfactory. The Severn receives 17.2 mgd of which 11.7 mgd is unsatisfactory, the Stour (Worcestershire), 16.1 mgd of which 11.1 mgd is unsatisfactory and the Avon (Warwickshire), 11.4 mgd of which 6.3 mgd is unsatisfactory. Another watercourse, the Holesmouth Pill which is comparatively very short, receives some 12.2 mgd of effluent (all satisfactory) and discharges it to the Severn estuary. Several smaller rivers receive significant discharges which are 100 per cent unsatisfactory, for example the Frome (3.06 mgd), the Chelt (5.5 mgd) and the Smestow Brook (1.85 mgd). Substantial discharges of unsatisfactory effluent are also made to the Tach Brook (4.4 mgd), the Hatherly Brook (1.8 mgd) and the Arrow (1.9 mgd). The Sowe, the Worcestershire Stour, the Avon and the Chelt stand out as rivers where a particularly high proportion of the total flow in dry weather is sewage effluent.

129 Some 26 discharges exceed 1 mgd, 5 of these exceed 5 mgd and 3 of them exceed 10 mgd. In total, they account for about 98 mgd or nearly 80 per cent of all the effluent discharged in the area. At the other end of the scale, 286 out of the 347 treatment works serve populations of 5,000 and less and nearly two-thirds of these serve populations of 1,000 and less.

130 Considerable changes are likely to take place over the period to 1980 with the forecast that the number of discharges will be reduced to 311. The population to be served by these discharges in 1980 is estimated at almost 3,270,000 and the dry weather flow about 210 mgd. Improved standards are likely to be imposed at 28 treatment works which, by 1980, will serve a population of some 840,000 with a dry weather flow of about 55 mgd.

131 A number of schemes to improve existing discharges are under construction and the forecasts for the period to 1980 indicate that many major schemes will be constructed and others will involve the closing of small works and the treatment of the sewage on a more regional basis.

19 Wye River Authority

(see Table 22)

132 The Wye Authority area, one of several draining to the Severn estuary, includes the major parts of the counties of Hereford and Radnor, together with small fringe areas of other neighbouring counties. About half the area lies in England and half in Wales and it is very rural in general character and sparsely populated. The City of Hereford is the largest centre of population.

133 All but 8 of the 54 discharges are considered satisfactory and the 8 works that are unsatisfactory are so classified because of lack of capacity. These unsatisfactory works serve just over half of the population but discharge about three-quarters of all unsatisfactory effluent in the area. All but 3 of the 54 discharges are required to meet Royal Commission standard.

134 The Wye is by far the longest river in the area and receives the highest number of discharges. It also receives the greatest quantity of effluent (7.4 mgd) of which about 90 per cent is classed as unsatisfactory. The small balance of unsatisfactory effluent is divided among several tributaries. About 6.5 mgd out of the total of 6.8 mgd classed as unsatisfactory comes from 2 treatment works discharging to the Wye.

135 Only 5 treatment works serve populations of over 5,000 people and over two-thirds of all the treatment works serve populations of 1,000 and less.

136 Few changes are forecast in the period up to 1980 with only one works likely to be abandoned. The population served is expected to increase to around 165,000 and the dry weather flow to just over 12 mgd. Improved standards will be required at two works where the flow in 1980 is estimated at under 1 mgd.

137 Only a small amount of remedial works are in hand at present but a number of others, including one major scheme, are in preparation.

Table 22 Details of sewage treatment works and discharges of sewage effluent

Description	Rivers Non-tidal	Tidal	All rivers
a. Sewage treatment works			
1 Total number of discharges	52	2	54
2 Number where effluent is satisfactory	44	2	46
3 Number where effluent is unsatisfactory (capacity)	8	0	8
4 Number where effluent is unsatisfactory (maintenance)	0	0	0
b. Populations			
5 Total population served	110,988	2,905	113,893
6 Total served, effluent satisfactory	50,313	2,905	53,218
7 Total served, effluent unsatisfactory	60,675	0	60,675
c. Dry weather flow			
8 Present dry weather flow of sewage effluent (x 1,000 gpd)	8,978	107	9,085
9 Percentage of industrial effluent in DWF	25	12	25
10 Dry weather flow. Discharges satisfactory (x 1,000 gpd)	2,154	107	2,261
11 Dry weather flow. Discharges unsatisfactory (x 1,000 gpd)	6,824	0	6,824
d. Number of discharges by effluent standards			
12 Royal Commission Standard Number	50	1	51
Number satisfactory	43	1	44
13 More stringent than RC Standard Number	2	0	2
Number satisfactory	1	—	1
14 Less stringent than RC Standard Number	0	1	1
Number satisfactory	—	1	1
e. Population by effluent standards			
15 Royal Commission Standard Total population	109,344	2,380	111,724
% satisfactory	46	100	47
16 More stringent than RC Standard Total population	1,644	0	1,644
% satisfactory	31	—	31
17 Less stringent than RC Standard Total population	0	525	525
% satisfactory	—	100	100
f. Dry weather flow by effluent standards			
18 Royal Commission Standard Total DWF (x 1,000 gpd)	8,924	83	9,007
% satisfactory	24	100	25
19 More stringent than RC Standard Total DWF (x 1,000 gpd)	54	0	54
% satisfactory	28	—	28
20 Less stringent than RC Standard Total DWF (x 1,000 gpd)	0	24	24
% satisfactory	—	100	100
g. Sewage treatment works by population ranges Present population			
21 1,000 and under Satisfactory	33	1	34
Unsatisfactory	3	0	3
22 1,001 to 5,000 Satisfactory	8	1	9
Unsatisfactory	3	0	3
23 5,001 to 10,000 Satisfactory	3	0	3
Unsatisfactory	1	0	1
24 10,001 to 50,000 Satisfactory	0	0	0
Unsatisfactory	1	0	1

20 Usk River Authority

(see Table 23 over page)

138 This area, whose rivers discharge into the Severn estuary, lies wholly in Wales and extends over most of the county of Monmouth, a major part of the county of Brecon and small fringe areas of Carmarthen and Glamorgan. The River Usk flows down the valley between the Black Mountains and the Brecon Beacons and much of the land in the county of Brecon is mountainous and sparsely populated. Brecon is the principal town and centre of population in this upper area. Towards the coast in the county of Monmouth lies the eastern part of the Welsh industrial and mining belt formed by Pontypool, Cwmbran, Ebbw Vale and other smaller urban areas and

(20 Usk River Authority continued)

Table 23 Details of sewage treatment works and discharges of sewage effluent

Description	Rivers		
	Non-tidal	Tidal	All rivers
a. Sewage treatment works			
1 Total number of discharges	27	4	31
2 Number where effluent is satisfactory	22	2	24
3 Number where effluent is unsatisfactory (capacity)	4	2	6
4 Number where effluent is unsatisfactory (maintenance)	1	0	1
b. Populations			
5 Total population served	47,638	115,173	162,811
6 Total served, effluent satisfactory	38,558	5,675	44,233
7 Total served, effluent unsatisfactory	9,080	109,498	118,578
c. Dry weather flow			
8 Present dry weather flow of sewage effluent (x 1,000 gpd)	2,531	4,692	7,223
9 Percentage of industrial effluent in DWF	4	12	9
10 Dry weather flow. Discharges satisfactory (x 1,000 gpd)	1,994	171	2,165
11 Dry weather flow. Discharges unsatisfactory (x 1,000 gpd)	537	4,521	5,058
d. Number of discharges by effluent standards			
12 Royal Commission Standard			
Number	25	1	26
Number satisfactory	21	1	22
13 More stringent than RC Standard			
Number	1	0	1
Number satisfactory	0	—	0
14 Less stringent than RC Standard			
Number	1	3	4
Number satisfactory	1	1	2
e. Population by effluent standards			
15 Royal Commission Standard			
Total population	41,358	200	41,558
% satisfactory	93	100	93
16 More stringent than RC Standard			
Total population	6,000	0	6,000
% satisfactory	0	—	0
17 Less stringent than RC Standard			
Total population	280	114,973	115,253
% satisfactory	100	5	5
f. Dry weather flow by effluent standards			
18 Royal Commission Standard			
Total DWF (x 1,000 gpd)	2,184	6	2,190
% satisfactory	91	100	91
19 More stringent than RC Standard			
Total DWF (x 1,000 gpd)	340	0	340
% satisfactory	0	—	0
20 Less stringent than RC Standard			
Total DWF (x 1,000 gpd)	7	4,686	4,693
% satisfactory	100	4	4
g. Sewage treatment works by population ranges			
Present population			
21 1,000 and under Satisfactory	14	1	15
Unsatisfactory	3	0	3
22 1,001 to 5,000 Satisfactory	6	0	6
Unsatisfactory	1	0	1
23 5,001 to 10,000 Satisfactory	1	1	2
Unsatisfactory	1	0	1
24 10,001 to 50,000 Satisfactory	1	0	1
Unsatisfactory	0	1	1
25 50,001 to 100,000 Satisfactory	0	0	0
Unsatisfactory	0	1	1

Note:
   The accompanying text contains a special reference relating to unsatisfactory discharges to the tidal waters of the River Usk.

this lower part of the river authority area is therefore heavily populated. Newport county borough is by far the largest town in the area. It lies on the coast but it discharges most of its sewage without treatment to the Usk estuary and these discharges are described later in the report. Direct discharges to the Severn estuary which include those from the Western Valleys Sewerage Board's system as far inland as Tredegar, are outside the scope of the report.

139 There are 31 sewage treatment works in the area, all but 7 of which are satisfactory, but in terms of population served and volume of effluent discharged only some 30 per cent are satisfactory. About 90 per cent of all the unsatisfactory effluent is discharged to tidal waters.

140 The Usk receives altogether some 4 mgd out of the area total of about 5 mgd of unsatisfactory effluent, and almost all of this from a single discharge on its tidal length. (Although this discharge is "unsatisfactory" in terms of the river authority's intended consent, it complies with a temporary standard imposed by the Secretary of State for Wales). Other rivers receiving significant quantities of unsatisfactory effluent are the Liswerry Pill (0.6 mgd unsatisfactory) and the Clydach (0.3 mgd unsatisfactory). Most discharges are required to meet Royal Commission standard. The previously mentioned discharge which is responsible for most of the unsatisfactory effluent to the Usk is subject to standards less stringent than Royal Commission.

141 Only one discharge exceeds 1 mgd and all but 6 discharges are from populations of 5,000 and less; over half of all discharges are from populations of 1,000 and less.

142 By 1980, 3 of the existing works are likely to have been abandoned and one new works should become operational leaving 29 works in operation. The estimated population to be served in 1980 is nearly 300,000 and the dry weather flow about 16 mgd. A big proportion of this additional population and flow will be from the treatment works to be provided for Newport. Improved standards will be required at one works where the 1980 population is estimated to be about 126,000 and the dry weather flow nearly 7 mgd.

143 A major scheme is under consideration for the improvement of conditions in the Newport area and other schemes are in preparation for early construction.   □

21 Glamorgan River Authority
(see Table 24)

144 The authority administers an area extending over a large part of the county of Glamorgan and small areas of the neighbouring counties of Brecon and Monmouth in which is located the larger part of the principal industrial and mining region of South Wales. Within the area is Cardiff, the capital city of Wales, the county borough of Merthyr Tydfil and several municipal boroughs of which Rhondda and Port Talbot are the largest. Population densities throughout the area are generally high. Development of the area to accommodate industry, mine workings and its associated processes covers most of the inland area with a coastline including busy docks and seaside resorts.

145 Most of the sewage from the principal coastal towns from Cardiff in the east round to Neath in the west as well as sewage from the Rhymney and Rhondda valleys is discharged to the Bristol Channel and these important discharges are not included in this survey.

146 There are no sewage effluent discharges to either tidal waters or canals and of the 40 discharges to non-tidal rivers about one-third are satisfactory. All the unsatisfactory works are so classified because of lack of suitable capacity for treatment and the satisfactory discharges relate only to some 17 per cent of the total population served and to about 8 per cent of the total volume of effluent discharged. Most treatment works are required to discharge effluents to a Royal Commission standard but under a third comply and they only discharge some 7 per cent of the total effluent volume required to meet this standard.

147 The three rivers receiving the most sewage effluent are the Taff (about 20 mgd), the Ogmore (about 2 mgd) and the Ely (about 1 mgd). Most of this is from 5 large discharges, all of it is unsatisfactory and the total accounts for more than 90 per cent of all the unsatisfactory effluent discharged from sewage treatment works to rivers in the area.

Table 24 Details of sewage treatment works and discharges of sewage effluent

Description	Rivers		
	Non-tidal	Tidal	All rivers
a. Sewage treatment works			
1 Total number of discharges	40	0	40
2 Number where effluent is satisfactory	13	0	13
3 Number where effluent is unsatisfactory (capacity)	27	0	27
4 Number where effluent is unsatisfactory (maintenance)	0	0	0
b. Populations			
5 Total population served	278,963	0	278,963
6 Total served, effluent satisfactory	46,575	0	46,575
7 Total served, effluent unsatisfactory	232,388	0	232,388
c. Dry weather flow			
8 Present dry weather flow of sewage effluent (x 1,000 gpd)	26,435	0	26,435
9 Percentage of industrial effluent in DWF	8	—	8
10 Dry weather flow. Discharges satisfactory (x 1,000 gpd)	1,995	0	1,995
11 Dry weather flow. Discharges unsatisfactory (x 1,000 gpd)	24,440	0	24,440
d. Number of discharges by effluent standards			
12 Royal Commission Standard			
Number	38	0	38
Number satisfactory	12	—	12
13 More stringent than RC Standard			
Number	1	0	1
Number satisfactory	0	—	0
14 Less stringent than RC Standard			
Number	1	0	1
Number satisfactory	1	—	1
e. Population by effluent standards			
15 Royal Commission Standard			
Total population	262,143	0	262,143
% satisfactory	16	—	16
16 More stringent than RC Standard			
Total population	13,300	0	13,300
% satisfactory	0	—	0
17 Less stringent than RC Standard			
Total population	3,520	0	3,520
% satisfactory	100	—	100
f. Dry weather flow by effluent standards			
18 Royal Commission Standard			
Total DWF (x 1,000 gpd)	25,873	0	25,873
% satisfactory	7	—	7
19 More stringent than RC Standard			
Total DWF (x 1,000 gpd)	400	0	400
% satisfactory	0	—	0
20 Less stringent than RC Standard			
Total DWF (x 1,000 gpd)	162	0	162
% satisfactory	100	—	100
g. Sewage treatment works by population ranges			
Present population			
21 1,000 and under Satisfactory	6	0	6
Unsatisfactory	7	0	7
22 1,001 to 5,000 Satisfactory	4	0	4
Unsatisfactory	10	0	10
23 5,001 to 10,000 Satisfactory	2	0	2
Unsatisfactory	2	0	2
24 10,001 to 50,000 Satisfactory	1	0	1
Unsatisfactory	7	0	7
25 50,001 to 100,000 Satisfactory	0	0	0
Unsatisfactory	1	0	1

148 The sizes of works by population ranges are almost evenly divided between works serving populations of from 5,000 to 100,000 (13), 1,000 to 5,000 (14) and 1,000 and under (13).

149 By 1980, 22 works—more than half of these existing at present—are expected to close down and 4 new works should open. The estimated population to be served in 1980 will be nearly 500,000 and the volume of effluent discharged nearly 40 mgd. Improved standards are envisaged at 3 works with an estimated 1980 flow of about 250,000 gpd.

150 A large programme of sewerage and sewage treatment is in hand. Major regional schemes in the Taff and Ogmore valleys are already under construction and others, including one for the Ely valley, are planned for the immediate future.

22 South West Wales River Authority

Table 25 Details of sewage treatment works and discharges of sewage effluent

Description	Rivers		
	Non-tidal	Tidal	All rivers
a. Sewage treatment works			
1 Total number of discharges	94	28	122
2 Number where effluent is satisfactory	61	22	83
3 Number where effluent is unsatisfactory (capacity)	22	4	26
4 Number where effluent is unsatisfactory (maintenance)	11	2	13
b. Populations			
5 Total population served	130,916	102,273	233,189
6 Total served, effluent satisfactory	97,142	70,816	167,958
7 Total served, effluent unsatisfactory	33,774	31,457	65,231
c. Dry weather flow			
8 Present dry weather flow of sewage effluent (x 1,000 gpd)	6,769	5,149	11,918
9 Percentage of industrial effluent in DWF	3	20	10
10 Dry weather flow. Discharges satisfactory (x 1,000 gpd)	5,262	3,134	8,396
11 Dry weather flow. Discharges unsatisfactory (x 1,000 gpd)	1,507	2,015	3,522
d. Number of discharges by effluent standards			
12 Royal Commission Standard			
Number	89	15	104
Number satisfactory	58	14	72
13 More stringent than RC Standard			
Number	4	0	4
Number satisfactory	3	—	3
14 Less stringent than RC Standard			
Number	1	13	14
Number satisfactory	0	8	8
e. Population by effluent standards			
15 Royal Commission Standard			
Total population	128,215	12,624	140,839
% satisfactory	74	91	76
16 More stringent than RC Standard			
Total population	2,373	0	2,373
% satisfactory	69	—	69
17 Less stringent than RC Standard			
Total population	328	89,649	89,977
% satisfactory	0	66	66
f. Dry weather flow by effluent standards			
18 Royal Commission Standard			
Total DWF (x 1,000 gpd)	6,644	474	7,118
% satisfactory	78	89	79
19 More stringent than RC Standard			
Total DWF (x 1,000 gpd)	104	0	104
% satisfactory	66	—	66
20 Less stringent than RC Standard			
Total DWF (x 1,000 gpd)	21	4,675	4,696
% satisfactory	0	58	58
g. Sewage treatment works by population ranges			
Present population			
21 1,000 and under Satisfactory	49	12	61
Unsatisfactory	21	2	23
22 1,001 to 5,000 Satisfactory	8	5	13
Unsatisfactory	12	3	15
23 5,001 to 10,000 Satisfactory	1	3	4
Unsatisfactory	0	0	0
24 10,001 to 50,000 Satisfactory	3	2	5
Unsatisfactory	0	1	1

151 The South West Wales River Authority is the largest authority in Wales and covers the counties of Carmarthen and Pembroke and most of the county of Cardigan, as well as small parts of the counties of Glamorgan, Brecon and Montgomery. The area is mainly agricultural but the south-east corner, which includes Swansea, is one of the largest centres for industry and commerce on the Welsh coast. There is now also substantial industrialisation around the shores of Milford Haven which is the second largest oil port in the United Kingdom. On the whole the area is not heavily populated.

152 There are 122 discharges of sewage effluent of which about two-thirds are satisfactory and in terms of population served and

volume of effluent discharged approximately 70 per cent are satisfactory; 13 of the works are unsatisfactory because of poor maintenance. Although most of the discharges are to non-tidal waters, they represent only about 55 per cent of all the effluent and nearly 80 per cent is satisfactory. Discharges to tidal waters, although they number only about 20 per cent of the total in the area, account for some 45 per cent of the effluent volume and about 60 per cent of the effluent is satisfactory. Most works are required to discharge effluents to Royal Commission standard and nearly 80 per cent of the effluent discharged from these is satisfactory. On tidal waters by far the greater volume is required to discharge to less stringent than Royal Commission standard and nearly 60 per cent of this is satisfactory.

153 The Loughor receives the most sewage effluent (3.3 mgd all satisfactory), the Tawe receives 1.9 mgd (all satisfactory) and the Lliw 1.7 mgd (mostly unsatisfactory).

154 There are only 4 sewage effluent discharges over 1 mgd to rivers in the area, two to the Loughor and one each to the Tawe and the Lliw. These details do not include direct discharges to the Bristol Channel, nor the many crude sewage discharges which are described later in the report.

155 The majority of sewage treatment works are small, over 90 per cent serving 5,000 population and less and three-quarters of these serving 1,000 and less.

156 By 1980, 5 treatment works are likely to be abandoned and 5 new ones constructed, but the estimated population to be served in 1980 will rise to nearly 370,000 and the dry weather flow to over 19 mgd. Improved standards are envisaged at 5 works, 4 of which are to tidal waters, the corresponding 1980 flow being about 1.4 mgd. ☐

23 Gwynedd River Authority
(see Table 26)

157 The Gwynedd River Authority area covers the north-west corner of Wales from near Aberystwyth on the west coast to Colwyn Bay on the north coast, extending over the whole of Caernarvonshire and parts of the counties of Merioneth, Montgomery, Cardigan and Denbigh. It also includes Anglesey. The major towns are Colwyn Bay, Llandudno, Bangor and Conway all in the north. Most of the northern half of the area comprises the Snowdonia National Park where there is very little population and the rest of the area is largely agricultural; apart from the north coastal strip the average population density is low. There is an influx of summer visitors, however, particularly in the coastal areas.

158 There are 43 sewage treatment works in the area and nearly two-thirds are satisfactory. Most of the effluent is discharged to non-tidal rivers where more than half is satisfactory, but of that discharged to tidal rivers, about 80 per cent is unsatisfactory. Direct discharges to the Menai Straits have not been included in the survey.

159 The largest discharges of unsatisfactory effluent are to the non-tidal rivers Bowydd (0.1 mgd), Ogwen (0.2 mgd) and Machno (0.1 mgd) and to the tidal part of the river Glaslyn (0.24 mgd) where, in each case, the quantity represents practically all the effluent discharged to the river.

160 All the treatment works are small, and serve populations of less than 5,000 and most are required to discharge effluent complying with Royal Commission standard.

161 By 1980, the number of works is likely to decrease to 39 by the abandonment of 7 existing works and the commissioning of 3 new works. The estimated population served in 1980 will be nearly 75,000 and the dry weather flow nearly 3.5 mgd. Improved effluent standards will be called for at 3 works and the population involved (1980) will be about 9,000 and the flow about 0.8 mgd. ☐

Table 26 Details of sewage treatment works and discharges of sewage effluent

Description	Rivers		
	Non-tidal	Tidal	All rivers
a. Sewage treatment works			
1 Total number of discharges	35	8	43
2 Number where effluent is satisfactory	22	5	27
3 Number where effluent is unsatisfactory (capacity)	11	3	14
4 Number where effluent is unsatisfactory (maintenance)	2	0	2
b. Populations			
5 Total population served	41,032	7,496	48,528
6 Total served, effluent satisfactory	23,653	3,056	26,709
7 Total served, effluent unsatisfactory	17,379	4,440	21,819
c. Dry weather flow			
8 Present dry weather flow of sewage effluent (x 1,000 gpd)	1,547	388	1,935
9 Percentage of industrial effluent in DWF	9	6	8
10 Dry weather flow. Discharges satisfactory (x 1,000 gpd)	859	78	937
11 Dry weather flow. Discharges unsatisfactory (x 1,000 gpd)	688	310	998
d. Number of discharges by effluent standards			
12 Royal Commission Standard. Number	34	4	38
Number satisfactory	21	3	24
13 More stringent than RC Standard Number	0	0	0
Number satisfactory	—	—	—
14 Less stringent than RC Standard Number	1	4	5
Number satisfactory	1	2	3
e. Population by effluent standards			
15 Royal Commission Standard Total population	38,482	3,306	41,788
% satisfactory	55	78	57
16 More stringent than RC Standard Total population	0	0	0
% satisfactory	—	—	—
17 Less stringent than RC Standard Total population	2,550	4,190	6,740
% satisfactory	100	11	45
f. Dry weather flow by effluent standards			
18 Royal Commission Standard Total DWF (x 1,000 gpd)	1,429	87	1,516
% satisfactory	52	75	53
19 More stringent than RC Standard Total DWF (x 1,000 gpd)	0	0	0
% satisfactory	—	—	—
20 Less stringent than RC Standard Total DWF (x 1,000 gpd)	118	301	419
% satisfactory	100	4	31
g. Sewage treatment works by population ranges Present population			
21 1,000 and under Satisfactory	15	4	19
Unsatisfactory	8	2	10
22 1,001 to 5,000 Satisfactory	7	1	8
Unsatisfactory	5	1	6

24 Dee and Clwyd River Authority
(see Table 27)

162 The Dee and the Clwyd River Authority area lies mostly in Wales and covers the greater parts of the counties of Denbigh and Flint, and small areas of Merioneth, Cheshire and Salop. Chester and Wrexham are the only large towns; there are smaller communities at Rhyl, Prestatyn, Connah's Quay and Flint, but the population density in the area is generally low. There is an influx of summer visitors on the north coast particularly at Prestatyn and Rhyl.

163 There are 109 discharges of sewage effluent in the area and about two-thirds are satisfactory. Most of the discharges are to non-tidal waters which receive about two-thirds of all the sewage effluent

Table 27 Details of sewage treatment works and discharges of sewage effluent

Description	Rivers Non-tidal	Tidal	All rivers
a. *Sewage treatment works*			
1 Total number of discharges	100	9	109
2 Number where effluent is satisfactory	68	3	71
3 Number where effluent is unsatisfactory (capacity)	29	6	35
4 Number where effluent is unsatisfactory (maintenance)	3	0	3
b. *Populations*			
5 Total population served	252,694	115,723	368,417
6 Total served, effluent satisfactory	199,029	91,669	290,698
7 Total served, effluent unsatisfactory	53,665	24,054	77,719
c. *Dry weather flow*			
8 Present dry weather flow of sewage effluent (x 1,000 gpd)	9,755	5,564	15,319
9 Percentage of industrial effluent in DWF	8	5	7
10 Dry weather flow. Discharges satisfactory (x 1,000 gpd)	7,824	4,519	12,343
11 Dry weather flow. Discharges unsatisfactory (x 1,000 gpd)	1,931	1,045	2,976
d. *Number of discharges by effluent standards*			
12 Royal Commission Standard Number	94	5	99
Number satisfactory	65	2	67
13 More stringent than RC Standard Number	6	0	6
Number satisfactory	3	—	3
14 Less stringent than RC Standard Number	0	4	4
Number satisfactory	—	1	1
e. *Population by effluent standards*			
15 Royal Commission Standard Total population	217,467	101,953	319,420
% satisfactory	85	87	85
16 More stringent than RC Standard Total population	35,227	0	35,227
% satisfactory	43	—	43
17 Less stringent than RC Standard Total population	0	13,770	13,770
% satisfactory	—	23	23
f. *Dry weather flow by effluent standards*			
18 Royal Commission Standard Total DWF (x 1,000 gpd)	8,598	4,999	13,597
% satisfactory	86	89	87
19 More stringent than RC Standard Total DWF (x 1,000 gpd)	1,157	0	1,157
% satisfactory	40	—	40
20 Less stringent than RC Standard Total DWF (x 1,000 gpd)	0	565	565
% satisfactory	—	16	16
g. *Sewage treatment works by population ranges* Present population			
21 1,000 and under Satisfactory	38	0	38
Unsatisfactory	22	1	23
22 1,001 to 5,000 Satisfactory	19	1	20
Unsatisfactory	7	3	10
23 5,001 to 10,000 Satisfactory	5	0	5
Unsatisfactory	2	2	4
24 10,001 to 50,000 Satisfactory	6	1	7
Unsatisfactory	1	0	1
25 50,001 to 100,000 Satisfactory	0	1	1
Unsatisfactory	0	0	0

in the area and some 80 per cent of this is satisfactory. The balance of the effluent, discharged to tidal waters, also includes some 80 per cent which is satisfactory. Over 90 per cent of the discharges are required to be to Royal Commission standard and two-thirds of these are satisfactory. There are no discharges to non-tidal rivers where the standard is less stringent than Royal Commission.

164 The greater part of all the sewage effluent is discharged in the Dee basin and the River Dee itself receives the largest quantity, 0.6 mgd to non-tidal waters and 5.1 mgd to tidal, and the non-tidal River Clywedog receives 2.2 mgd. Most of this effluent is satisfactory. There are no large discharges of unsatisfactory sewage effluent in the area although the non-tidal Worthenbury Brook receives 0.4 mgd, the River Gwenfro and the Black Brook, a tributary of the Alyn, 0.3 mgd each and the Pulford Brook 0.2 mgd, whilst the tidal stretches of the Dee receive some 0.9 mgd. In a few rivers the returns indicate that the natural flow is almost nil during summer so there is little dilution of effluent.

165 Only 2 works have discharges exceeding 2 mgd, one to a non-tidal river and one to a tidal river. All the remaining discharges are under 1 mgd. Most works are small in size, 91 out of the total of 109 serving populations of less than 5,000 and two-thirds of these serving populations of 1,000 and less.

166 By 1980, 23 works are expected to close down and 2 to commence operation thus reducing the number to 88. The population to be served by these is estimated to be just over half a million and the dry weather flow nearly 26 mgd. Improved standards are likely to be called for at 3 works representing some 100,000 people and a dry weather flow of about 8 mgd.

167 As indicated by the big reduction in numbers of discharges, the river authority visualise the diversion of sewage flows from many works to larger works or, in some cases, to new works or to the sea. A number of schemes to this end are already under construction or in preparation. □

25 Mersey and Weaver River Authority
(see Table 28 over page)

168 The Mersey and Weaver River Authority covers the greater part of Cheshire, a smaller part of the neighbouring county of Lancaster to the north and fringe areas of the counties of Salop, Stafford, Derby and the West Riding to the south and east. In the northern half of the area lie the two major conurbations of Merseyside and South-East Lancashire which together contain a population approaching 4 million resulting in very high population densities in areas of considerable industrial activity. The South-East Lancashire conurbation, which includes Manchester, extends well inland towards the Pennines where the sources of many of the rivers in the Mersey basin are to be found. To the south in Cheshire, and providing a contrast with the northern part, the area is predominantly rural and devoted to farming, the exceptions being the Crewe, Northwich and Macclesfield areas.

169 There are 189 discharges of sewage effluent in the area and about half of these discharges are unsatisfactory, 7 being so classed because of poor maintenance. The population served by these unsatisfactory treatment works and the effluent discharged from them represent about two-thirds of the respective totals in the area. The majority of all the discharges are to non-tidal waters with one discharge being to a canal and in all but 5 cases, the effluent is required to comply with the Royal Commission standard.

170 It should be noted that the many discharges of crude sewage to the Mersey estuary are not included in the above details. They are dealt with in Chapter 3.

171 The River Weaver receives the largest number of discharges (11) representing a dry weather flow of some 7 mgd but this is very small compared with the flow from the 9 discharges to the Manchester Ship Canal which amounts to 83 mgd. About 80 mgd of this flow is derived from 2 works, the Davyhulme works of Manchester and the Weaste works of Salford. Other rivers which receive particularly large quantities of effluent are the Irwell (32 mgd), the Mersey (22 mgd), the Tame (12 mgd), the Roch (11 mgd) and the Alt (10 mgd); several others receive quantities of between 5 and 10 mgd. The Manchester Ship Canal also receives the greatest quantity of unsatisfactory effluent (about 81 mgd) but some large volumes are also discharged to the Mersey (about 15 mgd), the Roch (11.2 mgd), the Alt (9.1 mgd), the Tame (8.6 mgd), the Sankey and Hardshaw Brooks (8.2 mgd) and the River Bollin (6.9 mgd). Several other rivers and brooks receive unsatisfactory discharges of between 1 and 3 mgd representing either all or most of the total discharges received into them.

172 With an area of such variety in character, the sizes of the treatment works also vary widely. At the upper end of the range, the

(25 Mersey and Weaver River Authority continued)

Table 28 Details of sewage treatment works and discharges of sewage effluent

Description	Rivers Non-tidal	Rivers Tidal	Rivers All rivers	Canals*	All rivers and canals
a. *Sewage treatment works*					
1 Total number of discharges	178	10	188	1	189
2 Number where effluent is satisfactory	91	7	98	0	98
3 Number where effluent is unsatisfactory (capacity)	80	3	83	1	84
4 Number where effluent is unsatisfactory (maintenance)	7	0	7	0	7
b. *Populations*					
5 Total population served	3,819,912	185,086	4,004,998	730	4,005,728
6 Total served, effluent satisfactory	1,335,530	141,555	1,477,085	0	1,477,085
7 Total served, effluent unsatisfactory	2,484,382	43,531	2,527,913	730	2,528,643
c. *Dry weather flow*					
8 Present dry weather flow of sewage effluent (x 1,000 gpd)	248,779	8,992	257,771	25	257,796
9 Percentage of industrial effluent in DWF	23	32	23	0	23
10 Dry weather flow. Discharges satisfactory (x 1,000 gpd)	77,438	7,534	84,972	0	84,972
11 Dry weather flow. Discharges unsatisfactory (x 1,000 gpd)	171,341	1,458	172,799	25	172,824
d. *Number of discharges by effluent standards*					
12 Royal Commission Standard Number	175	8	183	1	184
Number satisfactory	89	5	94	0	94
13 More stringent than RC Standard Number	2	0	2	0	2
Number satisfactory	2	—	2	—	2
14 Less stringent than RC Standard Number	1	2	3	0	3
Number satisfactory	0	2	0	—	2
e. *Population by effluent standards*					
15 Royal Commission Standard Total population	3,730,792	60,146	3,790,938	730	3,791,668
% satisfactory	34	28	34	0	34
16 More stringent than RC Standard Total population	67,120	0	67,120	0	67,120
% satisfactory	100	—	100	—	100
17 Less stringent than RC Standard Total population	22,000	124,940	146,940	0	146,940
% satisfactory	0	100	85	—	85
f. *Dry weather flow by effluent standards*					
18 Royal Commission Standard Total DWF (x 1,000 gpd)	242,882	2,155	245,037	25	245,062
% satisfactory	30	32	30	0	30
19 More stringent than RC Standard Total DWF (x 1,000 gpd)	4,605	0	4,605	0	4,605
% satisfactory	100	—	100	—	100
20 Less stringent than RC Standard Total DWF (x 1,000 gpd)	1,292	6,837	8,129	0	8,129
% satisfactory	0	100	84	—	84
g. *Sewage treatment works by population ranges* Present population					
21 1,000 and under Satisfactory	30	1	31	0	31
Unsatisfactory	15	1	16	1	17
22 1,001 to 5,000 Satisfactory	24	3	27	0	27
Unsatisfactory	22	0	22	0	22
23 5,001 to 10,000 Satisfactory	8	1	9	0	9
Unsatisfactory	13	0	13	0	13
24 10,001 to 50,000 Satisfactory	22	1	23	0	23
Unsatisfactory	27	2	29	0	29
25 50,001 to 100,000 Satisfactory	5	0	5	0	5
Unsatisfactory	5	0	5	0	5
26 100,001 to 500,000 Satisfactory	2	1	3	0	3
Unsatisfactory	4	0	4	0	4
27 Over 500,000 Satisfactory	0	0	0	0	0
Unsatisfactory	1	0	1	0	1

* For the purpose of the survey, the Manchester Ship Canal is considered to be a part of the river systems of the Mersey and the Irwell and details relating to discharge to it are recorded under the details for rivers.

Davyhulme treatment works of the Manchester Corporation serves over 800,000 people and constitutes the largest discharge in the area and there are 18 works serving populations greater than 50,000. At the lower end of the range about half of all the works serve populations of 5,000 and less and a quarter serve 1,000 and less. There are altogether 44 discharges of sewage effluent of 1 mgd and over, 11 of these exceed 5 mgd, 3 of them exceed 10 mgd and altogether they account for some 85 per cent of all the sewage effluent discharged in the area. Half of these 1 mgd-and-above discharges are unsatisfactory and they account for some 85 per cent of all the unsatisfactory effluent discharged.

173 The forecasts for the period up to 1980 show that 41 existing treatment works (including the one works discharging to the Weaver Navigation Canal) are likely to be abandoned and 7 new works will be built, thus reducing the number of discharges to rivers to 155. It is estimated that the population served will rise to a little over 4,600,000 and the dry weather flow to about 360 mgd by 1980. The river authority are likely to be seeking improved standards at 5 works by 1980 affecting nearly 500,000 people with a corresponding dry weather flow of about 38 mgd.

174 Several large schemes involving extensions to treatment works are envisaged during the period up to 1980 mainly in the basin of the River Mersey and schemes to improve unsatisfactory discharges are already under construction. A Steering Committee has been set up to study and make recommendations on the course of action to be taken to reduce the pollution of the Mersey estuary.

26 Lancashire River Authority

Table 29 Details of sewage treatment works and discharges of sewage effluent

Description	Rivers Non-tidal	Rivers Tidal	Rivers All rivers	Canals	All rivers and canals
a. Sewage treatment works					
1 Total number of discharges	130	26	156	1	157
2 Number where effluent is satisfactory	44	10	54	0	54
3 Number where effluent is unsatisfactory (capacity)	70	14	84	1	85
4 Number where effluent is unsatisfactory (maintenance)	16	2	18	0	18
b. Populations					
5 Total population served	841,285	276,000	1,117,285	351	1,117,636
6 Total served, effluent satisfactory	152,085	36,665	188,750	0	188,750
7 Total served, effluent unsatisfactory	689,200	239,335	928,535	351	928,886
c. Dry weather flow					
8 Present dry weather flow of sewage effluent (x 1,000 gpd)	52,921	17,191	70,112	15	70,127
9 Percentage of industrial effluent in DWF	21	23	22	0	22
10 Dry weather flow. Discharges satisfactory (x 1,000 gpd)	8,348	2,542	10,890	0	10,890
11 Dry weather flow. Discharges unsatisfactory (x 1,000 gpd)	44,573	14,649	59,222	15	59,237
d. Number of discharges by effluent standards					
12 Royal Commission Standard Number	119	8	127	1	128
Number satisfactory	37	7	44	0	44
13 More stringent than RC Standard Number	3	0	3	0	3
Number satisfactory	2	—	2	—	2
14 Less stringent than RC Standard Number	8	18	26	0	26
Number satisfactory	5	3	8	—	8
e. Population by effluent standards					
15 Royal Commission Standard Total population	718,569	33,361	751,930	351	752,281
% satisfactory	20	99	23	—	23
16 More stringent than RC Standard Total population	118,299	0	118,299	0	118,299
% satisfactory	7	—	7	—	7
17 Less stringent than RC Standard Total population	4,417	242,639	247,056	0	247,056
% satisfactory	58	1	2	—	2
f. Dry weather flow by effluent standards					
18 Royal Commission Standard Total DWF (x 1,000 gpd)	43,824	2,411	46,235	15	46,250
% satisfactory	18	100	22	0	22
19 More stringent than RC Standard Total DWF (x 1,000 gpd)	8,924	0	8,924	0	8,924
% satisfactory	5	—	5	—	5
20 Less stringent than RC Standard Total DWF (x 1,000 gpd)	173	14,780	14,953	0	14,953
% satisfactory	54	1	2	—	2
g. Sewage treatment works by population ranges Present population					
21 1,000 and under Satisfactory	27	4	31	0	31
Unsatisfactory	36	6	42	1	43
22 1,001 to 5,000 Satisfactory	10	5	15	0	15
Unsatisfactory	24	7	31	0	31
23 5,001 to 10,000 Satisfactory	2	0	2	0	2
Unsatisfactory	9	1	10	0	10
24 10,001 to 50,000 Satisfactory	5	1	6	0	6
Unsatisfactory	14	0	14	0	14
25 50,001 to 100,000 Satisfactory	0	0	0	0	0
Unsatisfactory	1	1	2	0	2
26 100,001 to 500,000 Satisfactory	0	0	0	0	0
Unsatisfactory	2	1	3	0	3

175 The Lancashire River Authority area covers that part of the county of Lancashire north of a line passing south of Southport, Wigan and Burnley, that part of the West Riding on the western slopes of the Pennines and the southern part of Westmorland and the Cumbrian mountains including Lake Windermere and the Lune valley. The major centres of population occur in the south at Preston, Blackburn, Wigan and Burnley along the northern fringe of the industrial conurbation of South-East Lancashire and in the coastal towns of Southport, Blackpool and, to the north of Morecambe Bay, Barrow-in-Furness. In the north population densities are very low in the mountainous and moorland areas, increasing towards the industrial areas near the southern boundary.

176 There are 157 discharges of sewage effluent from treatment works in the area but only about one-third are satisfactory and of those that are unsatisfactory 18 are so classed because of poor maintenance. The total population served by unsatisfactory treatment works and the quantity of effluent discharged by them represent about 85 per cent of the respective totals in the area. A high proportion of the effluent discharged to tidal waters is required to comply with a standard less stringent than Royal Commission only.

177 The greatest numbers of discharges are made to the Douglas (9), the Ribble (8), the Lune (7) and the Calder (6) but the Ribble receives the greatest quantity of effluent—13.1 mgd. The Hole Brook receives

8.5 mgd of effluent, the Calder 8.1 mgd and the Douglas 7.2 mgd. A number of other rivers receive flows of between 1 mgd and 5 mgd. The greatest quantities of unsatisfactory effluent are discharged to the Ribble (10.6 mgd, mostly from a single discharge), the Hole Brook (8.5 mgd, all from a single discharge), the Calder (7.9 mgd) and the Douglas (6.7 mgd). Several other rivers receive significant discharges of unsatisfactory effluent, notably in the Douglas and Ribble basins, Crossens Pool, the Yarrow, the Darwen, the Hyndburn and Colne Water being prominent; also, in the north, the Kent.

178 There are 21 discharges of 1 mgd or above and 3 of these exceed 5 mgd. In total they discharge about 80 per cent of all the effluent discharged, 16 of them are unsatisfactory and they account for about 80 per cent of all unsatisfactory effluent discharged in the area. There is a high proportion of small treatment works and the survey shows that 120 out of 157 serve populations less than 5,000 and just over 60 per cent of these serve 1,000 and less. There are only 5 works serving populations of 50,000 and over.

179 Considerable changes are likely to take place over the period to 1980, the forecast being that 50 works will be abandoned (including the works discharging to the Leeds and Liverpool Canal) and 7 new works are likely to be built thus reducing the number of discharges to 114. Many of these changes are envisaged in the Ribble basin and involve reorganisation of treatment facilities. The population to be served in 1980 is estimated at nearly 1.5 million and the sewage dry weather flow at about 117 mgd. The survey shows that higher standards are likely to be called for in 26 cases. The estimated 1980 population affected is a little over 900,000 and the dry weather flow about 76 mgd representing about 65 per cent of the 1980 flow. The improved standards will apply mainly to discharges in the Douglas and Ribble basins.

180 A number of remedial schemes are already under construction in the area. ☐

27 Cumberland River Authority
(see Table 30)

181 The Cumberland River Authority area is bordered on the south by the Cumbrian mountains, on the east by the Pennines, on the north by the Solway Firth and the Scottish border, and by the Irish sea on the west. It covers virtually all of Cumberland, the northern part of Westmorland and small fringe areas in the county of Northumberland. Much of the area is mountainous with farming in the valleys, and population densities are generally low except in and around the main towns such as Carlisle and the industrial centres of Workington and Whitehaven.

182 The survey records 94 discharges of sewage effluent in this area mostly to non-tidal waters, with only 18 (19 per cent) regarded as satisfactory, the lowest proportion in the entire survey. Many of these discharges are from treatment works where maintenance is not considered to be up to standard. The unsatisfactory discharges occur mainly in the basins of the Derwent where there are 21, the Eden where there are 31 and the Ellen where there are 10. Some 80 per cent of the population is served by unsatisfactory treatment works and these works discharge about 70 per cent of all the effluent. All but one of the discharges are required to comply with the Royal Commission standard.

183 The River Eden receives more than half of all the effluent, nearly 6.4 mgd, and 6.3 mgd of this is unsatisfactory. Other rivers receiving substantial quantities of effluent are the Derwent which receives about 2.1 mgd (about 0.5 mgd unsatisfactory) and the Keekle which receives about 1.3 mgd, mostly satisfactory.

184 There are only 3 sewage effluent discharges exceeding 1 mgd and one of these discharges is about 5 mgd. One is unsatisfactory and accounts not only for a high proportion of all the unsatisfactory effluent but also for almost all of the unsatisfactory effluent discharged to the River Eden. The survey shows that there are only 4 works where the population exceeds 5,000 and three-quarters of all the treatment works serve populations of under 1,000.

185 Few changes are envisaged for the period up to 1980. The forecast is that only one works is likely to close down so that 93 will remain and they will serve a little over 210,000 people and discharge just over 13 mgd. An improved standard will be sought at one works affecting a 1980 population of nearly 6,000 and a dry weather flow of about 0.25 mgd.

186 Because of the very large numbers of small treatment works, the remedial measures will mainly be spread in relatively small schemes throughout the area. Some remedial works are already in hand. ☐

Table 30 Details of sewage treatment works and discharges of sewage effluent

Description		Rivers		
		Non-tidal	Tidal	All rivers
a.	*Sewage treatment works*			
1	Total number of discharges	90	4	94
2	Number where effluent is satisfactory	18	0	18
3	Number where effluent is unsatisfactory (capacity)	48	4	52
4	Number where effluent is unsatisfactory (maintenance)	24	0	24
b.	*Populations*			
5	Total population served	189,979	1,797	191,776
6	Total served, effluent satisfactory	34,282	0	34,282
7	Total served, effluent unsatisfactory	155,697	1,797	157,494
c.	*Dry weather flow*			
8	Present dry weather flow of sewage effluent (x 1,000 gpd)	12,239	59	12,298
9	Percentage of industrial effluent in DWF	18	6	18
10	Dry weather flow. Discharges satisfactory (x 1,000 gpd)	3,448	0	3,448
11	Dry weather flow. Discharges unsatisfactory (x 1,000 gpd)	8,791	59	8,850
d.	*Number of discharges by effluent standards*			
12	Royal Commission Standard Number	89	4	93
	Number satisfactory	17	0	17
13	More stringent than RC Standard Number	0	0	0
	Number satisfactory	—	—	—
14	Less stringent than RC Standard Number	1	0	1
	Number satisfactory	1	—	1
e.	*Population by effluent standards*			
15	Royal Commission Standard Total population	185,362	1,797	187,159
	% satisfactory	16	0	16
16	More stringent than RC Standard Total population	0	0	0
	% satisfactory	—	—	—
17	Less stringent than RC Standard Total population	4,617	0	4,617
	% satisfactory	100	—	100
f.	*Dry weather flow by effluent standards*			
18	Royal Commission Standard Total DWF (x 1,000 gpd)	12,057	59	12,116
	% satisfactory	27	0	27
19	More stringent than RC Standard Total DWF (x 1,000 gpd)	0	0	0
	% satisfactory	—	—	—
20	Less stringent than RC Standard Total DWF (x 1,000 gpd)	182	0	182
	% satisfactory	100	—	100
g.	*Sewage treatment works by population ranges Present population*			
21	1,000 and under Satisfactory	12	0	12
	Unsatisfactory	56	4	60
22	1,001 to 5,000 Satisfactory	4	0	4
	Unsatisfactory	14	0	14
23	5,001 to 10,000 Satisfactory	1	0	1
	Unsatisfactory	0	0	0
24	10,001 to 50,000 Satisfactory	1	0	1
	Unsatisfactory	1	0	1
25	50,001 to 100,000 Satisfactory	0	0	0
	Unsatisfactory	1	0	1

28 Thames Conservancy Catchment Board

Table 31 Details of sewage treatment works and discharges of sewage effluent

Description	Rivers			Canals	All rivers and canals
	Non-tidal	Tidal	All rivers		
a. Sewage treatment works					
1 Total number of discharges	345	0	345	5	350
2 Number where effluent is satisfactory	257	—	257	4	261
3 Number where effluent is unsatisfactory (capacity)	83	—	83	1	84
4 Number where effluent is unsatisfactory (maintenance)	5	—	5	0	5
b. Populations					
5 Total population served	3,247,588	0	3,247,588	22,708	3,270,296
6 Total served, effluent satisfactory	1,742,011	—	1,742,011	14,708	1,756,719
7 Total served, effluent unsatisfactory	1,505,577	—	1,505 577	8,000	1,513,577
c. Dry weather flow					
8 Present dry weather flow of sewage effluent (x 1,000 gpd)	174,472	0	174,472	853	175,325
9 Percentage of industrial effluent in DWF	13	—	13	6	13
10 Dry weather flow. Discharges satisfactory (x 1,000 gpd)	90,785	—	90,785	513	91,298
11 Dry weather flow. Discharges unsatisfactory (x 1,000 gpd)	83,687	—	83,687	340	84,027
d. Number of discharges by effluent standards					
12 Royal Commission Standard					
Number	320	0	320	5	325
Number satisfactory	237	—	237	4	241
13 More stringent than RC Standard					
Number	25	0	25	0	25
Number satisfactory	20	—	20	—	20
14 Less stringent than RC Standard					
Number	0	0	0	0	0
Number satisfactory	—	—	—	—	—
e. Population by effluent standards					
15 Royal Commission Standard					
Total population	2,020,359	0	2,020,359	22,708	2,043,067
% satisfactory	58	—	58	65	58
16 More stringent than RC Standard					
Total population	1,227,229	0	1,227,229	0	1,227,229
% satisfactory	47	—	47	—	47
17 Less stringent than RC Standard					
Total population	0	0	0	0	0
% satisfactory	—	—	—	—	—
f. Dry weather flow by effluent standards					
18 Royal Commission Standard					
Total DWF (x 1,000 gpd)	108,505	0	108,505	853	109,358
% satisfactory	56	—	56	60	56
19 More stringent than RC Standard					
Total DWF (x 1,000 gpd)	65,967	0	65,967	0	65,967
% satisfactory	45	—	45	—	45
20 Less stringent than RC standard					
Total DWF (x 1,000 gpd)	0	0	0	0	0
% satisfactory	—	—	—	—	—
g. Sewage treatment works by population ranges					
Present population					
21 1,000 and under Satisfactory	134	0	134	2	136
Unsatisfactory	25	0	25	0	25
22 1,001 to 5,000 Satisfactory	72	0	72	1	73
Unsatisfactory	29	0	29	0	29
23 5,001 to 10,000 Satisfactory	14	0	14	0	14
Unsatisfactory	9	0	9	1	10
24 10,001 to 50,000 Satisfactory	30	0	30	1	31
Unsatisfactory	17	0	17	0	17
25 50,001 to 100,000 Satisfactory	3	0	3	0	3
Unsatisfactory	6	0	6	0	6
26 100,001 to 500,000 Satisfactory	4	0	4	0	4
Unsatisfactory	1	0	1	0	1
27 Over 500,000 Satisfactory	0	0	0	0	0
Unsatisfactory	1	0	1	0	1

187 The Thames Conservancy administers the catchment area of the non-tidal length of the River Thames above the tidal limit at Teddington Weir. This area stretches from the Cotswolds in Gloucestershire to London and extends over most of Surrey, Berkshire, Oxfordshire, Buckinghamshire, parts of Gloucestershire, Wiltshire and Hampshire and also small fringe areas of several other neighbouring counties. The population densities are highest in that part of the area surrounding Greater London, but otherwise population density is generally low except in and around the main towns which are widely spaced throughout the area. In most towns commercial and industrial interests are well established but elsewhere the area is noted for farming and market gardening.

188 There are 350 sewage treatment works recorded in the survey and about three-quarters of these discharge satisfactory effluents. Most of the works which are unsatisfactory are so classed because of inadequate capacity, only 5 being considered unsatisfactory because of poor maintenance. Around 53 per cent of the population are served by treatment works considered satisfactory and a similar proportion of effluent is discharged to a satisfactory standard. Some two-thirds of the effluent is required to comply with the Royal Commission standard but only 56 per cent does, and the remaining one-third is required to comply with standards more stringent than Royal Commission, with 45 per cent meeting the necessary standards. Five of the works discharge to 3 canals in the area.

189 The Thames receives the largest number of discharges (22) followed by the Cherwell (19), the Thame (16), the Wey (12) and the Evenlode (11). Several other rivers receive up to 10 discharges each. The largest quantities of effluent, however, are discharged to the Colne—30.2 mgd, virtually all from one works classed as unsatisfactory. Other rivers receiving large quantities of effluent are the Wey (12 mgd of which about 8 mgd is unsatisfactory), the Thames (about 12 mgd of which 3 mgd is unsatisfactory), the Foudry Brook, Reading (10 mgd mostly satisfactory), the Roundmoor Ditch, Slough (10 mgd of which 9.2 mgd is unsatisfactory), the Hogsmill River (9.4 mgd of which 1.5 mgd is unsatisfactory), the Mole (9.1 mgd of which 2.7 mgd is unsatisfactory), the Northfield Brook, Oxford (8.1 mgd mostly satisfactory) and the Blackwater (8 mgd of which 5.5 mgd is unsatisfactory). Large quantities of unsatisfactory effluent (representing 100 per cent of the discharges) are also discharged to the Earlswood Brook and Gatwick Stream in Surrey and to the River Wye in Buckinghamshire. The flows in a few of the rivers receiving large quantities of effluent, and several others receiving smaller amounts, consist almost wholly of the sewage effluent discharged to them, the more prominent being the Hogsmill River, the Gatwick Stream, the River Bourne, the Roundmoor Ditch and the upper reaches of the River Blackwater.

190 There are 38 works in the area out of the 350 recorded where the effluent discharged is greater than 1 mgd, 6 of these discharges are between 5 mgd and 10 mgd and one of them amounts to 30 mgd. The total flow discharged from these 38 works represents just over three-quarters of all the effluent discharged in the area. About half of them are unsatisfactory and they account for nearly 90 per cent of all the unsatisfactory effluent. There are many smaller works and about three-quarters of all the works serve populations of 5,000 and less and about 45 per cent serve populations of 1,000 and less.

191 In the period up to 1980, 43 works are likely to close down and 7 to commence operation leaving a total of 314 operating. The population to be served in 1980 is estimated to be just over 4 million and the flow nearly 250 mgd. The survey shows too that more stringent standards of effluent from 36 works are likely to be required by 1980, the estimated population affected being about 1.5 million and the flow 94 mgd.

192 Many schemes involving large expenditure are planned and some are already under construction. □

29 Lee Conservancy Catchment Board
(see Table 32)

193 The Lee Conservancy area extends over most of Hertfordshire and parts of Essex and Bedfordshire. It also includes all or parts of the London boroughs of Enfield, Barnet, Haringey, Waltham Forest, Hackney, Newham, Tower Hamlets and Redbridge. The sewage from the whole or parts of 5 of these boroughs—Enfield, Barnet, Haringey, Waltham Forest and Redbridge—drains to treatment works whose discharges are controlled by the Board. The other London boroughs are served by sewage treatment works discharging to the tidal Thames and controlled by the Port of London Authority. Apart from the London boroughs, principal towns are Luton, Stevenage, Harlow, Welwyn Garden City and Bishop's Stortford. The area as a whole has the highest population density of any river authority in the country, but is of a rural character in the northern parts. All of the effluent in the area ultimately reaches the River Lee before it flows into the tidal River Thames.

194 The survey shows that 27 out of the 32 treatment works are satisfactory but less than 40 per cent of the population are served by treatment works considered satisfactory and less than 40 per cent of the effluent is discharged to a satisfactory standard. The 5 unsatisfactory works are so classed because of insufficient treatment

Table 32 Details of sewage treatment works and discharges of sewage effluent

Description	Rivers Non-tidal	Tidal	All rivers
a. Sewage treatment works			
1 Total number of discharges	32	0	32
2 Number where effluent is satisfactory	27	0	27
3 Number where effluent is unsatisfactory (capacity)	5	0	5
4 Number where effluent is unsatisfactory (maintenance)	0	0	0
b. Populations			
5 Total population served	1,208,272	0	1,208,272
6 Total served, effluent satisfactory	456,782	0	456,782
7 Total served, effluent unsatisfactory	751,490	0	751,490
c. Dry weather flow			
8 Present dry weather flow of sewage effluent (x 1,000 gpd)	70,005	0	70,005
9 Percentage of industrial effluent in DWF	13	—	13
10 Dry weather flow. Discharges satisfactory (x 1,000 gpd)	27,316	0	27,316
11 Dry weather flow. Discharges unsatisfactory (x 1,000 gpd)	42,689	0	42,689
d. Number of discharges by effluent standards			
12 Royal Commission Standard Number	13	0	13
Number satisfactory	10	—	10
13 More stringent than RC Standard Number	19	0	19
Number satisfactory	17	—	17
14 Less stringent than RC Standard Number	0	0	0
Number satisfactory	—	—	—
e. Population by effluent standards			
15 Royal Commission Standard Total population	745,597	0	745,597
% satisfactory	2	—	2
16 More stringent than RC Standard Total population	462,675	0	462,675
% satisfactory	95	—	95
17 Less stringent than RC Standard Total population	0	0	0
% satisfactory	—	—	—
f. Dry weather flow by effluent standards			
18 Royal Commission Standard Total DWF (x 1,000 gpd)	41,997	0	41,997
% satisfactory	1	—	1
19 More stringent than RC Standard Total DWF (x 1,000 gpd)	28,008	0	28,008
% satisfactory	96	—	96
20 Less stringent than RC Standard Total DWF (x 1,000 gpd)	0	0	0
% satisfactory	—	—	—
g. Sewage treatment works by population ranges Present population			
21 1,000 and under Satisfactory	13	0	13
Unsatisfactory	0	0	0
22 1,001 to 5,000 Satisfactory	10	0	10
Unsatisfactory	1	0	1
23 5,001 to 10,000 Satisfactory	0	0	0
Unsatisfactory	0	0	0
24 10,001 to 50,000 Satisfactory	2	0	2
Unsatisfactory	2	0	2
25 50,001 to 100,000 Satisfactory	0	0	0
Unsatisfactory	1	0	1
26 100,001 to 500,000 Satisfactory	2	0	2
Unsatisfactory	0	0	0
27 Over 500,000 Satisfactory	0	0	0
Unsatisfactory	1	0	1

capacity. About 40 per cent of all the effluent is required to comply with standards more stringent than Royal Commission (some more stringent than 10:10), and a very high proportion of this effluent (96 per cent) is discharged to a satisfactory standard. On the other hand almost all of the effluent required to meet Royal Commission standard is unsatisfactory.

195 The largest number of discharges is to the River Lee (7) with a total of about 28 mgd of effluent (1.25 mgd unsatisfactory), but the Salmons Brook, a tributary of the Lee, receives some 40 mgd of

effluent from only two GLC discharges and the whole of this effluent is considered to be unsatisfactory. Only one other river, the Great Halingbury Brook, receives any significant amount of unsatisfactory effluent (1.2 mgd). It is evident from the returns that in dry weather there is lack of reasonable dilution in several rivers in the area especially the River Lee, which for much of its length below Luton is largely sewage effluent.

196 There are 6 discharges of effluent greater than 1 mgd and 4 of these exceed 5 mgd the largest being 35 mgd. These works together discharge about 95 per cent of all the effluent in the area and the 4 that are unsatisfactory account for almost all of the unsatisfactory effluent in the area.

197 The returns indicate that, in the period up to 1980, 4 works are likely to close down leaving 28 operating. These are expected to serve some 1.4 million people and dispose of a dry weather flow of about 95 mgd. In addition, higher effluent standards are likely to be called for at 6 works, 3 on the Pincey Brook, 2 on the River Stort and 1 on the Salmons Brook. The estimated population affected will be about 800,000 and the dry weather flow about 52.5 mgd representing around 55 per cent of the overall 1980 totals.

198 Several remedial schemes involving large expenditure are already under construction and others are imminent. □

30 Port of London Authority (including the London Excluded Area)
(see Table 33)

199 The responsibilities for pollution prevention in the area are, in practice, shared between the Port of London Authority who control direct discharges to the tidal Thames and to the tidal reaches of its tributaries and the Greater London Council who control, in consultation with the PLA, other discharges in the London Excluded Area. Pollution control is subject to special legislation but the powers are similar to those of the other river authorities in the country. The control of pollution in the tidal Thames is exercised by the PLA on the basis of the extensive study of the river carried out in the 1950's*.

200 The London Excluded Area lies between the statutory areas of the Thames and Lee Conservancy Catchment Boards and the Essex and Kent river authorities; it extends over most of Greater London, but also includes small areas beyond the boundary of Greater Luton. The area is fully developed for housing, commerce and industry.

201 The general situation in the area is that the tidal Thames receives, either directly or from the London Excluded Area, almost three-quarters of the sewage effluent discharged to all tidal rivers in England and Wales and when the effluents discharged to the Thames Conservancy and Lee Conservancy areas are added, it can be seen that about one-third of all the treated sewage effluent discharged to all the rivers in England and Wales passes down the Thames estuary.

202 There are 25 treatment works in the area and 80 per cent of these discharge the greater volume of effluent in the area (92 per cent) to tidal waters. Only 10 works produce satisfactory effluents, 2 discharging to non-tidal waters and 8 to tidal waters. The population served by and volume passed to treatment works discharging unsatisfactory effluents to tidal waters represent nearly 80 per cent of the

*Two major reports have been issued relating to pollution in the tidal Thames. The first is entitled "Effects of Polluting Discharges on the Thames Estuary" published HMSO 1964 (Water Pollution Research Technical Paper No. 11) is the report of the Thames Survey Committee and the Water Pollution Research Laboratory into their investigation to determine the effects of various factors on the distribution of dissolved oxygen and to develop methods by which this distribution could be predicted for any combination of conditions of fresh water flow, temperature and polluting load etc, that might arise in the future. The second entitled "Pollution of the Tidal Thames" published HMSO 1961 is the Ministry of Housing and Local Government report of the Departmental Committee on the Effects of Heated and other Effluents and Discharges on the Condition of the Tidal Reach of the River Thames. They considered both the existing situation and also the effects of any proposed new developments in the area.

Table 33 Details of sewage treatment works and discharges of sewage effluent

Description	Rivers Non-tidal	Tidal	All rivers
a. Sewage treatment works			
1 Total number of discharges	5	20	25
2 Number where effluent is satisfactory	2	8	10
3 Number where effluent is unsatisfactory (capacity)	3	11	14
4 Number where effluent is unsatisfactory (maintenance)	0	1	1
b. Populations			
5 Total population served	697,910	7,724,513	8,422,423
6 Total served, effluent satisfactory	386,450	1,733,591	2,120,041
7 Total served, effluent unsatisfactory	311,460	5,990,922	6,302,382
c. Dry weather flow			
8 Present dry weather flow of sewage effluent (x 1,000 gpd)	36,975	474,784	511,759
9 Percentage of industrial effluent in DWF	12	13	13
10 Dry weather flow. Discharges satisfactory (x 1,000 gpd)	17,972	100,522	118,494
11 Dry weather flow. Discharges unsatisfactory (x 1,000 gpd)	19,003	374,262	393,265
d. Number of discharges by effluent standards			
12 Royal Commission Standard			
Number	4	10	14
Number satisfactory	1	2	3
13 More stringent than RC Standard			
Number	1	3	4
Number satisfactory	1	2	3
14 Less stringent than RC Standard			
Number	0	7	7
Number satisfactory	—	4	4
e. Population by effluent standards			
15 Royal Commission Standard			
Total population	358,460	4,249,206	4,607,666
% satisfactory	13	1*	1
16 More stringent than RC Standard			
Total population	339,450	3,107,800	3,447,250
% satisfactory	100	49	54
17 Less stringent than RC Standard			
Total population	0	367,507	367,507
% satisfactory	—	55	55
f. Dry weather flow by effluent standards			
18 Royal Commission Standard			
Total DWF (x 1,000 gpd)	21,003	273,539	294,542
% satisfactory	10	0*	1
19 More stringent than RC Standard			
Total DWF (x 1,000 gpd)	15,972	186,347	202,319
% satisfactory	100	49	53
20 Less stringent than RC Standard			
Total DWF (x 1,000 gpd)	0	14,898	14,898
% satisfactory	—	60	60
g. Sewage treatment works by population ranges Present population			
21 1,000 and under Satisfactory	0	1	1
Unsatisfactory	0	1	1
22 1,001 to 5,000 Satisfactory	0	2	2
Unsatisfactory	0	0	0
23 5,001 to 10,000 Satisfactory	0	0	0
Unsatisfactory	0	1	1
24 10,001 to 50,000 Satisfactory	1	2	3
Unsatisfactory	0	3	3
25 50,001 to 100,000 Satisfactory	0	1	1
Unsatisfactory	2	3	5
26 100,001 to 500,000 Satisfactory	1	1	2
Unsatisfactory	1	1	2
27 Over 500,000 Satisfactory	0	1	1
Unsatisfactory	0	3	3

*Two discharges are satisfactory but together the populations served are only 0.52% of the total and the flow only 0.24% of the total in the "Tidal" column at lines 15 and 18 respectively.

totals to tidal waters, but the corresponding percentages for non-tidal waters are rather better, at about 50 per cent. Fourteen of the discharges are required to comply with the Royal Commission standard but only 3 of these are satisfactory representing about 1 per cent of the corresponding population served and volume of effluent that should be discharged to this standard. Of the 4 works that are required to discharge effluent to a standard more stringent than Royal Commission, 3 comply and represent just over 50 per cent of the corresponding population served and the dry weather flow.

203 Sixteen of the discharges are to the tidal Thames and they amount to 470 mgd or 92 per cent of the total flow and nearly four-fifths of this flow is unsatisfactory, almost all from 7 treatment works where the individual flows range from just over 1 mgd to as high as 214 mgd. Other rivers receiving particularly large quantities of effluent are the non-tidal River Wandle (15.5 mgd, all unsatisfactory) and its Beddington Branch (16.0 mgd, all satisfactory), Beverley Brook (3.5 mgd all unsatisfactory), Pyl Brook (2 mgd all satisfactory), and the tidal Hole Haven Creek (2.9 mgd all unsatisfactory).

204 Although, in comparison with other river authorities, the number of treatment works in this area is small there are several very large works, including the GLC treatment works at Beckton, serving a population of 3 million and the largest treatment works in the country. Only 5 of the total of 25 works serve populations less than 10,000. Sixteen treatment works discharge flows in excess of 1 mgd as follows:

Over 200 mgd	1
50–100 mgd	2
10–50 mgd	4
5–10 mgd	2
1–5 mgd	7
	16

205 Considerable changes are likely to take place over the period up to 1980, the forecast being that 3 works are likely to close down thus reducing the number of discharges from 25 to 22. The population to be served in 1980 is estimated at about 8.6 million and the dry weather flow at about 600 mgd. Higher standards are likely to be called for at 12 works, 10 of them being on the tidal Thames so that by 1980 all works should be treating sewage to at least Royal Commission standard, with 7 treating to a standard more stringent than Royal Commission. The estimated 1980 population affected is 5.5 million and the dry weather flow 409 mgd, just over two-thirds of the overall 1980 estimated flow.

206 Remedial works proposed relate mainly to the large works discharging to the tidal Thames and a few schemes are already under construction. ☐

3 Discharges of crude sewage

Introduction

1 The discharge of crude sewage to inland waters has been almost completely eliminated, largely as a result of the control exercisable by river authorities under the Rivers (Prevention of Pollution) Acts 1951 and 1961. So far as tidal stretches of rivers are concerned, however, the main powers of control are exercised under the Clean Rivers (Estuaries and Tidal Waters) Act of 1960, although a river authority may take special steps to bring an estuary under control by Order (see Appendix 2 to Volume 1). Since the 1960 Act applies only to new or altered discharges, many long-standing discharges of crude sewage continue to be made to tidal waters. This chapter deals with a survey of all those discharges of sewage (from populations of 200 or above) which are untreated or which are given only preliminary treatment such as screening, comminution or tidal storage. Discharges treated by settlement, with removal of sludge, are considered as discharges of sewage effluent and have been included in Chapter 2 of this report. Certain controlled tidal waters* are outside the scope of the survey and so crude sewage discharges to these waters are not recorded.

2 As with sewage effluent discharges, river authorities were asked to record every discharge (above the minimum) and to say whether the discharge was satisfactory or not. "Unsatisfactory" in this case would always mean that remedial works were, in the opinion of the river authority, required even although, in many cases, the river authority would not have powers to impose the standards which would make remedial works necessary. Also, as before, the river authorities were asked to give their views on the date when remedial works would be required and the same convention was followed, namely that where a discharge is at present unsatisfactory and remedial works are required, the date 1970 was recorded even although it would not normally have been possible to carry out the work at once and even although the river authority might not want to press for immediate remedial works. Some authorities might, for example, prefer to see financial resources concentrated on improving inland discharges to non-tidal waters.

3 In some cases, a sewerage authority discharges crude sewage by way of a large number of closely-spaced outfalls and sometimes the river authority elected to record these as a single discharge—usually because it would be difficult to assess how much of the flow was discharged by each. For that reason, the number of discharges recorded is less than the total number of individual outfalls. Where appropriate, further mention of this is made in the descriptions of the individual river authority areas.

Details of the survey for England and Wales

4 A national summary of the information obtained in the survey is at Table 34 and Tables 185 and 186 at the end of the Report give a river authority by river authority breakdown of some of the details.

5 There are no discharges of crude sewage in the areas of Welland and Nene RA, Avon and Dorset RA, Thames Conservancy Catchment Board and Lee Conservancy Catchment Board, nor in the Port of London Authority and London Excluded Area. Neither are there any discharges of crude sewage to canals.

6 Out of a total of 484 discharges recorded, only 14 are considered satisfactory. All of these discharge into tidal waters in the following river authority areas—1 in Kent, 1 in Sussex, 3 in Hampshire, 5 in Cornwall, 1 in Wye, 1 in Glamorgan, 1 in South West Wales and 1 in Lancashire. The number represents 3 per cent of the total number,

---
* The River Humber, The Wash, The Solent, the controlled parts of the Bristol Channel, The Menai Strait, part of Morecambe Bay, and the Solway Firth.

Table 34 Summary of information about discharges of crude sewage, England and Wales

Description	Non-tidal	Tidal	All rivers
a. Number of discharges*			
1 Number of discharges	47	437	484
2 Number where effluent is satisfactory	0	14	14
3 Number where effluent is unsatisfactory	47	423	470
b. Treatment			
4 Number of discharges where preliminary treatment is provided	0	56	56
5 Number of discharges where no treatment is provided	47	381	428
c. Control			
6 Number of discharges where control is exercised	0	30	30
7 Number of discharges where control is not exercised	47	407	454
d. Populations			
8 Total population served	38,989	4,427,591	4,466,580
9 Total served, discharge satisfactory	0	290,398	290,398
10 Total served, discharge unsatisfactory	38,989	4,137,193	4,176,182
11 Total served, preliminary treatment	0	768,644	768,644
12 Total served, no preliminary treatment	38,989	3,658,947	3,697,936
13 Total served, control exercised	0	244,971	244,971
14 Total served, no control exercised	38,989	4,182,620	4,221,609
e. Dry weather flow (x 1,000 gpd)			
15 Present dry weather flow	1,226	262,783	264,009
16 Percentage of industrial effluent in DWF	4	29	29
17 Dry weather flow, discharges satisfactory	0	16,823	16,823
18 Dry weather flow, discharges unsatisfactory	1,226	245,960	247,186
19 Dry weather flow, preliminary treatment	0	47,151	47,151
20 Dry weather flow, no preliminary treatment	1,226	215,632	216,858
21 Dry weather flow, control exercised	0	11,352	11,352
22 Dry weather flow, no control exercised	1,226	251,431	252,657
f. Forecast of 1980 situation			
23 Number of outlets still in operation	17	153	170
24 Number where no further treatment required	0	19	19
25 Estimated population discharging to outlets still in operation	40,962	3,830,393	3,871,355
26 Estimated population to outlets where no further treatment required	0	357,960	357,960
27 Estimated dry weather flow to outlets still in operation (x 1,000 gpd)	1,053	247,177	248,230
28 Estimated dry weather flow to outlets where no further treatment required (x 1,000 gpd)	0	23,901	23,901

*The actual number is greater. In some cases, a number of closely spaced outfalls are recorded as a single discharge.

and the population served and the flow passed by these outfalls represent about 6 per cent each of the totals.

7 As to be expected control is exercised over very few outfalls and the bulk of the sewage (about 80 per cent) is discharged without preliminary treatment of any kind.

8 The population served by the crude sewage outfalls is some 4.5 million as compared with about 39 million served by sewage treatment works and the dry weather flow discharged without treatment is about 264 mgd (equivalent to about 59 ghd) as compared with about 2,232 mgd of sewage effluent (see para. 12 of Chapter 2). The largest volumes of crude sewage are discharged in the areas of Northumbrian RA (about 65 mgd), Mersey and Weaver (about 65 mgd), Bristol Avon (about 20 mgd), Lancashire (about 17.5 mgd), Hampshire (about 14.5 mgd) and Glamorgan (about 12 mgd).

9 The river authorities' forecast of the future situation shows that by 1980, some 314 out of the 484 existing outfalls should have been abandoned, which indicates that during the intervening period the flow to them is likely to be diverted for disposal elsewhere. In many cases, outfalls are likely to be diverted to sea. Some form of additional treatment or pre-treatment should have been provided for the sewage that will discharge at all but 19 of those outfalls that remain. The 19 outfalls will discharge a flow of about 24 mgd, all to tidal waters. Thirteen of these outfalls already have screens or tidal storage tanks or both so that there should remain only 6 discharges completely untreated and uncontrolled, discharging, together, a flow of about 1.7 mgd.

10 There follows a description of the situation in each river authority area.

1 Northumbrian River Authority

11 A total of 56 discharges is recorded—to the Tweed, the Coquet, the Lyne, the Wansbeck, the Blyth, the Tyne, the Wear and the Tees. Several of these represent multiple discharges, notably at Berwick-upon-Tweed, Blyth, Teesside and along Tyneside—there are about 175 major outfalls to the Tyne alone—and so the actual number of outfalls is considerably more. Most of the discharges are given no preliminary treatment and all are classed as unsatisfactory.

12 The total population from which crude sewage is discharged is about 1,420,000 and the approximate flow is 65 mgd of which about 20 mgd is industrial effluent. With only two exceptions (both small discharges) all of the discharges are to tidal waters and by far the biggest discharges are to the Tyne (approximately 850,000 population and about 42 mgd, largely from the area of Tyneside Joint Sewerage Board), to the Tees (approximately 354,000 population and about 12.7 mgd, largely from Teesside CB) and to the Wear (approximately 170,000 and about 7.8 mgd from Sunderland).

13 The situation forecast for the future is that, by 1980, virtually all of these outfalls will have been intercepted and diverted either to treatment works or to sea outfalls. Only 8 discharges to the rivers are likely to remain in place of the several hundred crude discharges at present. They are likely to serve a population of about 1,410,000 and the flow will be about 85 mgd. All but one should receive partial treatment at least.

14 Major regional schemes are in preparation for the estuaries of the Tyne and the Tees and there are further proposals for Sunderland affecting the Wear estuary. Schemes are also envisaged for Berwick-upon-Tweed, Blyth and several smaller authorities.

2 Yorkshire River Authority

15 A total of 12 discharges is recorded, 6 to the Ouse, and one each to the Esk, the Hertford, the Holmes Dike, the Selby Dam, the Rusholme Dike, and the Lambwath Stream. Only one discharge receives preliminary treatment and all are classed as unsatisfactory.

16 The total population from which crude sewage is discharged is some 29,000 and of this total, 20,000 are in the Borough of Goole. The volume discharged is about 1.5 mgd, of which about 40 per cent is industrial effluent. Five of the discharges are to non-tidal waters but they account for only a very small proportion of the total flow.

17 The forecast is that two of the discharges will be discontinued and the rest treated to Royal Commission standard. It is estimated that the 1980 population discharging to these outlets (after treatment) will be about 39,000 with a corresponding flow of about 2 mgd. Remedial schemes are in preparation or under construction.

18 Discharges to the Humber are not included in the above details.

3 Trent River Authority

19 There are 10 outfalls discharging crude sewage to the tidal part of the Trent and all are from Gainsborough. The total population served is 25,000 and the flow about 1.4 mgd of which some 16 per cent is industrial effluent. No pre-treatment is provided and the discharges are all unsatisfactory. Construction of a treatment works to deal with the whole of this flow is in hand, for completion in the near future.

4 Lincolnshire River Authority

20 There are 3 discharges of crude sewage in this area, all from Boston and all to the tidal Witham Haven. The total population discharging to these outfalls is about 25,000 and the flow amounts to 1.5 mgd of which about 40 per cent is industrial effluent. The discharges receive some preliminary treatment by screening and/or storage, but all are classed as unsatisfactory.

21 The river authority indicate that the discharges will be collected and given partial treatment at a new treatment works. The 1980 population is estimated at about 30,000 and the corresponding flow is estimated at about 2.6 mgd.

22 There are known to be several discharges of crude sewage to the Humber, which are not covered by the survey.

5 Welland and Nene River Authority

23 There are no discharges of crude sewage in the area.

6 Great Ouse River Authority

24 A total of 9 discharges, all from King's Lynn, is recorded to the Ouse, the Nar, the Gaywood and the West Lynn Drain. None of the discharges is given preliminary treatment and all are classed as unsatisfactory.

25 The total population from which crude sewage is discharged is about 28,000 and the dry weather flow is about 1.7 mgd of which some 11 per cent is industrial effluent. One discharge is made to the non-tidal part of the River Gaywood and the others are to tidal waters.

26 Work is already in hand for all these discharges to be connected to sewage works for treatment over the next 2 or 3 years.

7 East Suffolk and Norfolk River Authority

27 The survey records a total of 31 discharges, 30 from Great Yarmouth to the tidal parts of the Yare and the Bure and one from Felixstowe to the Orwell estuary. The discharges from Great Yarmouth are given no pre-treatment, but tidal storage is provided at Felixstowe. All the discharges are classed as unsatisfactory.

28 The total population from which crude sewage is discharged is about 63,000 and the dry weather flow about 6.5 mgd of which nearly a half is industrial effluent. Felixstowe accounts for about 575,000 gallons per day of effluent.

29 The 1980 population is estimated at about 67,000 and dry weather flow about 10.6 mgd. There may be some grouping of the outfalls and schemes are in preparation for the introduction, over a period, of comminutors on all the outfalls that remain.

8 Essex River Authority

30 There are at present 5 discharges, one from Harwich and 4 from Tendring Rural District Council, to the tidal part of the Stour. The largest discharge—400,000 gpd from Harwich—is given preliminary treatment by screens or comminutors as, also, is one of the Tendring discharges. All the discharges are regarded as unsatisfactory.

31 The total present population discharging to the outfalls is about 16,000 and the flow about 1.8 mgd. Some 70 per cent of the flow is industrial effluent, but most of this is cooling water discharged through one of the Tendring outfalls.

32 The future proposals envisage retaining only 2 outfalls—one for the industrial flow of about 1.2 mgd from Tendring RDC and one at Harwich. The remaining flows from Tendring RDC will be diverted either to the treatment works at Lawford or to the Harwich outfall and the river authority indicate that they will expect to see partial treatment provided there. The future population to the Harwich outfall will be about 12,000 and the flow is expected to be something over half a million gallons per day.

9 Kent River Authority

33 A total of 8 discharges is recorded—one to the Thames, 2

to the Medway, 2 to Queenborough Creek, 2 to the Stour and one to the Rother. All are to tidal waters and one small discharge (to the Medway) which is provided with tidal storage is satisfactory. The main discharges are from Queenborough-in-Sheppey to the Medway and Queenborough Creek and from Sandwich to the Stour. The population discharging to the outfalls is about 26,000 and the flow is about 2.3 mgd of which some 13 per cent is industrial effluent.

34 The forecast for the future is that all but 2 of the outfalls will be abandoned by diverting either to treatment works or to the sea. The population which will be served by these 2 outfalls in the future is estimated at about 9,600 and the flow at about 1.3 mgd. However, there is a possibility that one of them—at Sandwich—may also be diverted to sea leaving only one which will serve about 1,600 population and will discharge after preliminary treatment to the Medway.

10 Sussex River Authority
35 There are at present 2 discharges to the tidal section of the River Arun, a relatively small one from Arundel and a major discharge from Littlehampton. Both outfalls are provided with tidal storage and the former was considered satisfactory at the time of the survey but water conservation considerations may require some remedial works. The total population connected is about 33,000 and the flow is about 1.8 mgd. No industrial effluent is discharged.

36 There is a proposal to divert the Littlehampton sewage to a new sea outfall.

11 Hampshire River Authority
37 The survey records 5 discharges of crude sewage in the area, all to tidal waters, one to the Hamble, 3 to Southampton Water (River Test) and one to Langstone Harbour (Hermitage Stream). The total population connected is nearly 250,000 and the flow is about 14.5 mgd of which 44 per cent is industrial effluent.

38 The principal discharges are from Portsmouth to Langstone Harbour (13.5 mgd) and from Fareham to Southampton Water (0.65 mgd). Both receive preliminary treatment and both are considered satisfactory. One other discharge of about 200,000 gpd is largely satisfactory. Two discharges totalling some 220,000 gpd are unsatisfactory.

39 The forecast for the future is that the 3 satisfactory discharges will continue as at present but it may become desirable to lengthen the outfalls in time. The other 2 outfalls will receive partial treatment locally. This, however, could all be changed by long-term proposals for main drainage in the South Hampshire area. The future population served by the outfalls is estimated at about 250,000 and the flow is likely to increase to around 17 mgd.

40 Direct discharges to The Solent are not included in the above details.

12 Isle of Wight River and Water Authority
41 There are 13 discharges of crude sewage recorded, all from Cowes, and all to the tidal section of the River Medina. None receives any preliminary treatment and all are unsatisfactory. The total population connected is 13,000 and the flow is about 0.5 mgd of which some 9 per cent is industrial effluent. A scheme is being prepared to connect all the discharges to a future outfall discharging to The Solent.

42 Existing discharges to The Solent, of which there are several, are not included in the above details.

13 Avon and Dorset River Authority
43 There are no discharges of crude sewage in the area.

14 Devon River Authority
44 The survey records 29 discharges of crude sewage but the actual number is greater as several of the individual returns cover multiple discharges. The largest discharges are to the Exe, the Teign, the Kingsbridge Stream and Estuary, the Dart, the Torridge and the Taw. All but 5 of the 28 are to tidal waters.

45 The total population from which crude sewage is discharged is about 107,000 and the flow is a little over 5 mgd of which some 5 per cent is industrial effluent. About one quarter of the sewage is given some form of pre-treatment but all the outfalls are classed as unsatisfactory. The biggest single discharges are from Exmouth to the Exe (a population of 23,500), from Barnstaple to the Taw (14,000), from Salcombe to the Kingsbridge Estuary (12,000) and from Teignmouth to the Teign (10,100).

46 A number of schemes are already under construction and others are in preparation with the aim of dealing with all these discharges, either by diverting to sea outfalls or to new sewage treatment works or, in a few cases, by providing partial treatment at the outfall. By 1980, the number of outfalls to rivers should be reduced to 13, serving a population of about 74,000 with a sewage flow of about 4.2 mgd, all at least partially treated.

15 Cornwall River Authority
47 There are 37 discharges of crude sewage recorded in this area, the largest discharges being to the Tamar, the Plym, the East Looe River, the Fowey, the Fal, the Angarrack River and the Camel. All but one of the discharges are to tidal waters. The discharge to the non-tidal river is small. Only 7 of the discharges receive any preliminary treatment and a total of 5 are considered to be satisfactory.

48 The population from which crude sewage is discharged is about 213,000 and the sewage flow is some 10.8 mgd of which about 6 per cent is industrial effluent. The effluent considered satisfactory amounts to about 2 mgd. The Tamar receives altogether about 5.5 mgd, the Plym about 2.9 mgd and the Fal some 1.2 mgd. Most of the crude sewage discharged to the Tamar and to the Plym is from Plymouth and all the sewage to the Fal is from Falmouth.

49 The 5 satisfactory discharges will continue as at present, but major works are being planned to deal with most of the others. These include schemes for diverting many of the outfalls either to other outfalls or to treatment works, and for partial treatment in places. The returns indicate that, by 1980, the number of discharges should be reduced to 19 and the flow is expected to be about 16 mgd, all of which (with very minor exceptions), should be receiving at least partial treatment.

16 Somerset River Authority
50 The survey records 27 discharges of crude sewage in this area, 9 to non-tidal and 18 to tidal waters. The largest amounts are discharged to the Parrett, the Brue and the Congresbury Yeo. Only one discharge receives pre-treatment (by screening) and all the discharges are regarded as unsatisfactory.

51 The total population connected is about 53,000 and the sewage flow is about 5.5 mgd of which nearly a half is industrial effluent. Only a very small proportion of the total flow is discharged to non-tidal waters. By far the largest quantity is discharged to the Parrett which receives about 4.2 mgd, mainly from a number of outfalls at Bridgwater. The Brue receives about 800,000 gpd from Burnham and the Congresbury Yeo about 450,000 gpd from 2 rural districts.

52. Several schemes for improving the situation are in hand including major schemes for dealing with Bridgwater and Burnham. The returns indicate that by 1980 the number of outfalls should be reduced to 5, serving a population of about 95,000 with a flow of some 8 mgd and that the whole of this flow should be receiving at least partial treatment.

53 Direct discharges to the Bristol Channel of which there are several, are not included in the above details.

17 Bristol Avon River Authority
54 There are 17 discharges of crude sewage in the area, 11 to non-tidal and 6 to tidal waters. Only one small discharge receives any preliminary treatment and all are classed as unsatisfactory.

55 The total population served by the outfalls is about 308,000 and the dry weather flow is about 19.7 mgd of which some 21 per cent is

industrial effluent. All but about 150,000 gpd of the total flow is discharged to the Avon and most of the flow is from Bristol by way of 4 outfalls.

56 Major schemes are planned for directing the Bristol sewage to the treatment works at Avonmouth and for dealing with the other outfalls. The survey indicates that the number of outfalls in the area is likely to be reduced to 3, serving a population of only just over 5,000 with a flow of about 225,000 gpd, all treated to Royal Commission standard. The diverted Bristol flow will be discharged to a watercourse controlled by Severn RA.

18 Severn River Authority
57 The survey records 8 discharges all to the tidal length of the Severn. No pre-treatment is provided and all the discharges are classed as unsatisfactory. The total population served is about 15,500 and the dry weather flow about 8.7 mgd of which about 8 mgd is industrial effluent from Bristol and about 350,000 gpd is from Lydney.

58 The forecast of the situation in 1980 indicates that only one of the existing discharges (a small one) should remain partially treated, all the others having been diverted for treatment either at existing treatment works or at new regional treatment works.

19 Wye River Authority
59 There are 4 discharges of crude sewage in the Wye area all to tidal stretches of the River Wye. One is given pre-treatment by screening and tidal storage and is considered to be satisfactory. The total population from which crude sewage is discharged is 9,150 and the flow is about 450,000 gpd of which 14 per cent is industrial effluent. The satisfactory discharge accounts for about 66,000 gpd.

60 The 3 unsatisfactory discharges, all from Chepstow, are to be collected to a single outfall where partial treatment will be provided. The 1980 population which will be served by the two remaining outfalls is estimated at about 13,500 and the flow will be about 600,000 gpd.

20 Usk River Authority
61 A total of 18 discharges is recorded, 17 to the estuary of the River Usk and one to the tidal section of the Ebbw Fawr. All 18 discharges are from Newport, one is provided with tidal storage and all are considered unsatisfactory. The population served is about 121,000 and the flow is about 6.8 mgd, 10 per cent of which is industrial effluent. All but about 0.5 mgd of the total flow is discharged to the Usk.

62 A scheme is under construction for intercepting by stages, most of these outfalls during the next 10 years and for diverting them to a new treatment works at Nash, the ultimate intention being that all should be diverted for treatment there. By 1980 it is expected that the flow will be some 11.6 mgd of which nearly 10 mgd will be receiving full treatment.

63 There are several direct discharges to the Severn Estuary, which are not included in the above details.

21 Glamorgan River Authority
64 The survey records 29 discharges of crude sewage, all to tidal waters in the Ely, the Afan and the Neath. One of the latter discharges, which is provided with tidal storage, is classed as satisfactory; the other 28 are unsatisfactory.

65 The total population from which crude sewage is discharged is about 92,000 and the flow is about 12.3 mgd of which 40 per cent is industrial effluent. Of the 29 discharges, 20 are to the Neath, mainly from the Borough of Neath and adjoining rural district and they account for about 9.5 mgd of effluent. About 0.4 mgd is discharged to the Ely from Penarth and about 2.4 mgd to the Afan from Port Talbot.

66 The forecast for the future is that only the one satisfactory discharge point, that on the Neath, will be retained. A major sea-outfall scheme for intercepting all the other crude sewage outfalls discharging to the Neath is under construction, the outfalls to the Afan will be dealt with by another sea outfall scheme and the discharges to the Ely will be diverted partly to Cardiff and partly to a sea outfall.

67 Direct discharges to the Bristol Channel, of which there are several, are not included in the above details.

22 South West Wales River Authority
68 A total of 34 discharges is recorded to 13 different rivers, the greatest quantities going into the Loughor, the Dafen, the Lliw, the Towy, the Ritec, the Western Cleddau (Milford Haven), the Teify and the Rheidol. Three are to non-tidal waters. Preliminary treatment is provided at 4 outfalls, but only one small discharge out of the 34 is considered satisfactory.

69 The population from which crude sewage is discharged is about 156,000 and the flow is about 9.2 mgd of which 1 per cent is industrial effluent. The biggest discharges are made to the Dafen (about 2.2 mgd from Llanelli), to the Lliw (1.9 mgd from Llwchwr UDC), to the Ritec (1.2 mgd from Tenby) and to Milford Haven (1.2 mgd from Milford Haven UDC and Pembroke).

70 Proposals are in hand or under construction to eliminate many of these outfalls and the river authority will be seeking at least partial treatment at those which remain. The several schemes involve new treatment works and sea outfalls. It is envisaged that, by 1980, only 11 river discharges will remain in place of the 34 existing. They will serve a population of about 115,000 with a flow of some 4.2 mgd.

71 The above figures exclude the major discharge from Swansea to the Bristol Channel and other lesser discharges to these waters.

23 Gwynedd River Authority
72 There are 20 discharges recorded in the area to the Mawddach, the Artro, the Dwyryd, the Llyfnwy, the Conway and the Llugwy. Ten are to non-tidal waters and 10 to tidal. None receives any pre-treatment and all are classed as unsatisfactory.

73 The total population from which crude sewage is discharged is about 37,000 and the flow is about 1 mgd. No industrial effluent is discharged with the sewage. The greatest quantity of sewage is discharged to the River Conway (about 640,000 gpd mostly from Conway).

74 The forecast is that by 1980 the number of discharges will have been reduced to 6 by the abandoning and grouping of discharges. The population and flow are not likely to change much, but all the flow will receive at least partial treatment.

75 The direct discharges to the Menai Straits were not included in the survey.

24 Dee and Clwyd River Authority
76 There are 5 discharges of crude sewage in the area, 4 to the Dee and 1 to the Holywell Stream; all are to tidal waters. One is provided with tidal storage and all 5 are classed as unsatisfactory.

77 The population from which crude sewage is discharged is about 34,000 and the flow is about 1.6 mgd, all domestic sewage, the Dee receiving 1.25 mgd from Flint and Hoylake and the balance being discharged to the Holywell Stream from Holywell.

78 By 1980, in place of the 5 discharges, only 2 discharges to rivers should remain, both to the Dee and dealing with a population estimated at 28,000 and a flow of about 1.4 mgd, all fully treated. A long sea outfall, which is under construction, will carry the sewage from Hoylake to sea. This outfall will also serve parts of Wallasey and Birkenhead.

25 Mersey and Weaver River Authority
79 The survey records 68 discharges in the area, all to the tidal length of the Mersey. Preliminary treatment is provided at 2 of the discharges and all of them are classed as unsatisfactory.

80 The population from which crude sewage is discharged is about 1,160,000 and the volume about 65 mgd of which some 26 per cent is industrial effluent. By far the largest quantity comes from Liverpool which discharges some 31 mgd of crude sewage. Other big discharges are from Warrington (about 8 mgd), Birkenhead (about 7 mgd), Rimrose Joint Sewerage Board (about 43 mgd), Wallasey (about 3.3 mgd), Widnes (about 3.2 mgd) and Bebington (about 3 mgd).

81 The whole question of the pollution of the Mersey Estuary is being studied by a Steering Committee to which reference has already been made in para. 174 of Chapter 2 and the Committee will make recommendations. Some major schemes are already under construction, a particularly large diversion scheme being under way at Warrington and mention has been made in para. 78 of the sea outfall to serve parts of Wallasey and Birkenhead. Schemes are under preparation by other local authorities but the future situation is rather obscure. So far as can be judged at present, 22 outfalls will remain serving about 960,000 persons and discharging about 55 mgd and the river authority have indicated that they will expect all the sewage to receive at least partial treatment.

26 Lancashire River Authority

82 A total of 30 discharges is recorded to tidal waters of 17 rivers. Some form of pre-treatment is provided at 11 of the discharges, but only one small discharge is classed as satisfactory.

83 The population served is about 233,000 and the flow is about 17.4 mgd of which some 33 per cent is industrial effluent. The largest quantities are discharged to the Lune (about 7.7 mgd), the Ribble (about 3 mgd), the Salthouse Pool (about 2.3 mgd) and the Wyre (about 1 mgd). The main discharges come from Lancaster (about 7.7 mgd), Lytham St. Annes (3 mgd) and Barrow-in-Furness (2.3 mgd).

84 The forecast is that by 1980 the number of discharges will have been reduced to 16, serving a population of about 255,000 with a flow of some 22 mgd and the river authority indicate that all of the flow should receive at least partial treatment. Several schemes for intercepting and diverting discharges of crude sewage are already under construction.

85 A few discharges to the seaward end of Morecambe Bay are not included in the above details.

27 Cumberland River Authority

86 There are 4 discharges of crude sewage recorded in the area, 3 to the Southern Esk and one to the Ehen. All 4 discharges are to tidal waters, no pre-treatment is provided and all are classed as unsatisfactory. The total population from which crude sewage is discharged is about 4,000 and the flow is about 1.5 mgd, mostly to the Ehen from the UKAEA establishment at Sellafield. Only one local authority discharges crude sewage and the amount is small.

87 The returns indicate that the local authority sewage should be collected to a single point for treatment and that the UKAEA discharge should be given partial treatment. There will be little change in population connected and flow discharged.

88 Direct discharges to the controlled waters of the Solway Firth are not included in the above details.

28 Thames Conservancy Catchment Board

29 Lee Conservancy Catchment Board

30 Port of London Authority (including the London Excluded Area)

89 There are no discharges of crude sewage in the above areas.

4 Discharges of storm sewage from unsatisfactory storm overflows

Introduction
1 The function of a storm overflow is to relieve the sewerage system, at times of rain, of flows in excess of a selected rate, the excess flow usually being discharged direct and without treatment to a watercourse. Discharges are normally intermittent, but because crude sewage is discharged, they can cause serious pollution and visual offence. Frequently the overflows are simple flood relief devices and often they are located on sewerage systems which are very old and which have inadequate capacity even for the normal sewage flow, so that they may operate for longer than they should and, in extreme cases, they may operate every day even in dry weather. Many would agree, however, that a well designed and properly set overflow discharging at a suitable point to a watercourse giving adequate dilution is an acceptable ancillary structure on a sewerage system.

2 In a recent Government report* it was estimated, on the basis of a pilot survey, that there might be between 10,000 and 12,000 storm overflows in England and Wales and it was concluded that it would be unrealistic to contemplate eliminating them in the foreseeable future.

3 Since storm overflows are subject to control by river authorities, they come within the scope of the survey and it was at first thought that the opportunity could be taken to obtain a full record of all overflows in the country. However, after discussion with a number of river authorities, it was decided that it would be impracticable to attempt to record them all. Particularly in the larger towns, where the sewerage systems may have developed over the best part of 100 years, and where many watercourses are culverted, the river authorities do not always know the locations of all the discharge points and so the survey was restricted to those overflows known to the authority and considered to be unsatisfactory.

4 What is satisfactory or unsatisfactory is often a matter of judgement since few river authorities have reached the stage where they have issued consents for all storm overflows. For consistency, therefore, the general criterion was adopted that an overflow would be recorded as unsatisfactory if the river authority thought that remedial works would have a marked beneficial effect on the receiving watercourse or, where amenity considerations were foremost, the authority thought that they could make a case at an inquiry or investigation for remedial works to be carried out.

Other intermittent and polluting discharges
5 Storm overflow discharges are not the only intermittent and polluting discharges in which river authorities have an interest. Many sewage treatment works have storm tanks and these frequently fill during storms and the effluent from them is discharged to a watercourse. Adequately designed and properly operated, however, storm tanks are an acceptable part of the sewage treatment process. No details of unsatisfactory storm tank discharges were sought, mainly because such discharges are frequently associated with inadequate capacity in the main treatment works, resulting in premature operation of the storm tanks. Such cases would be included in the survey, not as unsatisfactory storm tank discharges, but as unsatisfactory treatment works. If a treatment works were in all respects adequate except for the capacity of the storm tanks, the extent and cost of the remedial measures would usually be relatively small and so the exclusion of such cases from the survey is not a significant omission.

6 River authorities from time to time express concern about the polluting nature of some intermittent surface water discharges containing such contaminants as oil, grit, pesticides, herbicides and washings from the paved areas of industrial premises. Whilst the potential pollution from such discharges is recognised, they are not included in the survey.

Details of the survey for England and Wales
7 The survey shows that there are, in England and Wales, some 2,162 storm overflows which the river authorities consider to be unsatisfactory within the definition outlined above. Table 35 shows how they are distributed between rivers and canals in different chemical classes and Table 187 at the end of the Report gives a river authority by river authority breakdown of the details. The greatest numbers of unsatisfactory discharges are recorded in the areas of Mersey and Weaver RA (791), Yorkshire (302), Lancashire (268), Usk (176) and Severn (123). Nil returns were submitted by Welland and Nene, Great Ouse and Lee Conservancy.

Table 35 Unsatisfactory storm overflows, England and Wales

Chemical class of receiving watercourse at point of discharge	Rivers			Canals	All rivers and canals
	Non-tidal	Tidal	All rivers		
Class 1	430	26	456	6	462
Class 2	520	48	568	2	570
Class 3	236	41	277	0	277
Class 4	840	9	849	4	853
	2,026	124	2,150	12	2,162

8 River authorities were asked to indicate, where they could, what remedial works they thought would be required and, if possible, the cost. As expected the range of remedial works varied enormously. For example it was suggested that in a number of cases the simple raising of the overflow cill would be sufficient and in other cases, an effective screen to trap solids was all that would be required. Some overflows would be made satisfactory by reconstruction to modern design standards, usually at moderate cost, but there were many cases where major works of sewerage in built-up areas would be necessary to render the situation satisfactory and in some cases the indications were that schemes costing many millions of pounds would be required to eliminate the unsatisfactory overflows in a single town.

9 A description of the situation in each area follows.

1 Northumbrian River Authority
10 There are 8 unsatisfactory storm overflows recorded, 2 to the Ouse Burn and one to each of 6 other rivers, all on non-tidal stretches.

11 One unsatisfactory overflow, which is in an area where mining subsidence has caused sewer blockages, is known to operate prematurely and sometimes in dry weather. In other cases there have been complaints of discharge in dry weather and of pollution and nuisance when overflows discharge near parks and open spaces.

12 The measures necessary to overcome the problems vary considerably and frequently involve costly improvements to the sewerage system. At Newcastle-upon-Tyne, for example, two unsatisfactory overflows, and others which no doubt contribute in a smaller way to the same pollution problem, could only be eliminated by large scale re-sewering at a cost of roughly £7 million. This area has several good

---
* Ministry of Housing and Local Government. Technical Committee on Storm Overflows and the Disposal of Storm Sewage. Final Report. HMSO 1970.

examples (frequently repeated throughout the survey) of how the cost of remedial works is influenced by the general age and state of the sewerage system.

2 Yorkshire River Authority

13 There are 302 unsatisfactory storm overflows in the area, 300 discharging to 61 rivers and 2 to the Sheffield and South Yorkshire Canal. All but 3 of the river discharges are to non-tidal waters. Rivers conspicuous by the large numbers of unsatisfactory overflow discharges are the Hebble Brook (27), the Don (23), Bradford Beck (21), the Calder (20), the Worth (19), the Holme (14), the Colne (13) and the Rother (11). Areas with particularly large numbers of unsatisfactory discharges are Sheffield (33), Halifax (33), Bradford (31), Keighley (28), Elland (22), Huddersfield (17), Holmfirth (15), Rotherham (12) and Dearne Urban District (12).

14 The necessary remedial measures vary again from the simple installation of a screen on a single overflow to massive re-sewering. Examples of places where remedial works are likely to be particularly costly are Sheffield, Bradford, Keighley, Wortley Rural District and Dearne Urban District.

3 Trent River Authority

15 There are 80 unsatisfactory storm overflows in the area, 78 discharging to the non-tidal waters of 34 rivers and one each to the Birmingham Canal Navigation and the Trent and Mersey Canal. The Tame receives 9 unsatisfactory discharges, Ford Brook receives 6, and the Hockley Brook and Fowlea Brook 5 each and conspicuous areas are Walsall and Stoke-on-Trent with 11 discharges each, Birmingham and Newcastle-under-Lyme with 8 each and West Bromwich with 7.

16 No details of costs were provided but it is evident that much re-sewering will be required and the returns indicated that a number of schemes are in preparation and others under construction.

4 Lincolnshire River Authority

17 There are 12 unsatisfactory storm overflows in the area and they discharge to 10 rivers. Eleven of the discharges are to non-tidal waters. Three of the discharges are from Welton Rural District, but no other authority has more than one unsatisfactory discharge.

18 No details of costs were provided but the comments of the river authority indicated that works of sewerage and, in some cases, increased capacity at treatment works will be required.

5 Welland and Nene River Authority

6 Great Ouse River Authority

19 There are no unsatisfactory storm overflows in the above areas.

7 East Suffolk and Norfolk River Authority

20 There are 19 unsatisfactory storm overflows in the area, 12 to non-tidal and 7 to tidal waters. Norwich accounts for 12 of the discharges, all to the Wensum and Yare and Ipswich makes 3 unsatisfactory discharges to the Gipping.

21 The remedial works necessary are largely re-sewering and the returns indicated that works are in hand to deal with the main unsatisfactory overflows.

8 Essex River Authority

22 There are 29 unsatisfactory storm overflows in the area, 24 discharging to non-tidal and 5 discharging to tidal stretches in 13 rivers. The Blackwater receives 7 unsatisfactory discharges and the Roding receives 5. Braintree Rural District accounts for 5, Rochford Rural District for 4 and Epping and Ongar Rural District and Brentwood 3 each.

23 In many cases, foul sewerage to divert flows, to provide greater sewer capacity and to replace existing sewers are required and in certain cases schemes to segregate foul sewage and surface water were suggested. The few costs provided gave no indication of the likely overall costs in the area. Some works are already under construction.

9 Kent River Authority

24 There are 11 unsatisfactory storm overflows in the area, 7 discharging to non-tidal and 4 discharging to tidal stretches in 5 rivers. The Medway receives 5 discharges, Faversham Creek 3, and there is one discharge each to the Grom, Milton Creek and the East Stour. Maidstone accounts for 5 discharges and Faversham for 3.

25 The returns indicated that remedial works for all of these discharges have been prepared, some are already under construction and the rest have been approved. In most cases, the improvement of the overflow will be incidental to a main sewerage scheme.

10 Sussex River Authority

26 There are 5 unsatisfactory storm overflows in the area, 4 to non-tidal and one to tidal waters, one discharge each to the Rother, the Arun and one of its tributaries, Hurst Haven, and the Down Sewer.

27 In all cases re-sewering will be required, one scheme is already under construction and two others have been approved. No indication of overall cost is available.

11 Hampshire River Authority

28 There are 7 unsatisfactory storm overflows in the area, 6 discharging to non-tidal and one to tidal waters, one discharge each to the Wallington, the Alver, the Itchen, Pudbrook Lake, Lymore Stream, Danes Stream and its tributary.

29 The returns suggested that remedial works would in most cases involve the construction of relief sewers and the provision of screens, but the few costs given provided no indication of the overall cost in the area.

12 Isle of Wight River and Water Authority

30 There are 35 unsatisfactory storm overflows in the area 18 to non-tidal and 17 to tidal waters. The rivers receiving the largest numbers of discharges are the Medina (12), the Monkton Mead Brook (5), the Yar Western (4) and the Lukely Brook and Parkhurst Stream (3 each). Newport accounts for 14 of the discharges and Cowes and Ryde 8 each.

31 The comments of the river authority suggested that the remedial works are likely to comprise relief sewers, screening and the provision of adequate sewer capacity with tidal storage. It is evident that a number of the overflows could be improved at moderate cost but several of the remedial schemes are not priced.

13 Avon and Dorset River Authority

32 There are 14 unsatisfactory storm overflows in the area, all discharging to non-tidal waters. The River Wey receives 4 and the remainder are distributed one or two each to several rivers. Weymouth acccounts for 4 and 2 each are situated in Salisbury, Bournemouth and Bridport Rural District.

33 The comments of the river authority indicated that relief sewers are required in places, and also reconstruction of old sewers and the provision of surface water sewers. In one case, partial treatment of the overflowing sewage and a sea outfall were suggested. Unsatisfactory overflows at Salisbury, Christchurch and Bournemouth are likely to be improved as incidental benefits from sewerage schemes under construction. Certain schemes involving re-sewering and provision of surface water sewers could be costly but no estimates were available to the authority.

14 Devon River Authority

34 There are 33 unsatisfactory storm overflows in the area, 27 discharging to non-tidal and 6 to tidal waters in 11 different rivers. There are 9 discharges to the North Brook, 5 to the Exe and 3 each to the Sid, the Dart and the Malt Mill Lake. Sixteen of the unsatisfactory overflows are in Exeter, 6 in Totnes and 3 in Sidmouth.

35 In most cases new sewers will be required and in some cases additional treatment works capacity also. The returns indicated that works are already under construction and others are imminent. No estimates of cost were given but the comments suggested that some of the remedial works could be expensive, particularly in Exeter.

15 Cornwall River Authority

36 There are 83 unsatisfactory storm overflows in the area, 49 to non-tidal and 34 to tidal waters in 23 rivers. Conspicuous in the returns are the Plym (14 discharges), Tamar and two of its tributaries near Plymouth (26), Portreath Stream (6) and Penryn River (5). Fifty of the unsatisfactory overflows are in Plymouth, 14 in Camborne-Redruth, 7 in Falmouth and 5 in Truro.

37 The comments of the river authority suggested that a good deal of re-sewering will be required, particularly in Plymouth, Falmouth and Truro, and preparation is in hand for dealing with some areas. Such costs as were provided were not sufficient to indicate the overall expenditure necessary but it is likely to be high.

16 Somerset River Authority

38 The river authority reported 10 unsatisfactory storm overflows but it was their opinion that a more detailed study might have revealed others.

39 To enable these overflows to be improved or eliminated, new works of sewerage will be required in places, but no indication was given of the likely overall cost. Two schemes for improvement are, however, already in hand.

17 Bristol Avon River Authority

40 There are 36 unsatisfactory storm overflows in the area all discharging to non-tidal waters. The Bristol Avon receives 21 discharges, the Somerset Frome 5 and the Bristol Frome 3. There are 13 discharges from Bath, 5 from Frome, 5 from Bathavon Rural District and 3 each from Bristol and Chippenham.

41 In some cases, major works of re-sewering are required and several major schemes are already under construction, notably at Bath, Bristol, Frome and Trowbridge. Schemes have been prepared in other areas. The indications are that, whereas some of the schemes will be very costly, other overflows could be improved at moderate cost.

18 Severn River Authority

42 There are 123 unsatisfactory overflows discharging in the area, 121 to non-tidal waters and 2 to the Staffordshire-Worcestershire Canal. The main rivers affected are the Severn (32 discharges), the Smestow Brook (34 discharges), the Worcestershire Stour (8 discharges) and the Arrow and the Merry Hill Brook (5 discharges each). The main discharges are from Wolverhampton (41), Worcester (19), Coventry (15), Upper Stour Main Drainage Authority (7), Bridgnorth Rural District (7), Stratford-on-Avon (6) and Redditch (5).

43 The comments of the river authority indicated that many large schemes of re-sewering are required and also, in places, thorough investigation of the sewerage systems. Separation of surface water was also suggested. Several multi-million pound sewerage schemes are already under construction or proposed in the towns where there are large numbers of overflows and other schemes are in preparation. The overall costs will be very high, but, as in most similar cases, the whole of the cost will not be attributable to the elimination or improvement of the overflows.

19 Wye River Authority

44 There are 4 unsatisfactory storm overflows in the area, all in Hereford, 2 to the non-tidal length of the Wye and 2 to the Eign Brook. A scheme has been prepared which will make the necessary improvements and construction will start soon.

20 Usk River Authority

45 The survey records 176 unsatisfactory storm overflows to the non-tidal stretches of 12 rivers. The rivers receiving the largest numbers of discharges are the Sirhowy (55 discharges), Ebbw Fawr and one tributary from Cumtillery (60), Afon Lwyd (28) and the Ebbw Fach (18). The largest numbers of unsatisfactory discharges come from Pontypool (32), Western Valley Sewerage Board's trunk sewer (30), Tredegar (18), Risca (18), Magor and St. Mellons Rural District (12), Mynyddislwyn Urban District (12), Abertillery (12) and Bedwellty Urban District (10).

46 The comments of the river authority suggested that remedial works will, in many cases, be confined to re-construction of the overflows and possibly, in some cases, the adjustment of weirs. Other overflows will be improved in line with the increased carrying capacity in trunk sewers which will result from major sewerage works already under construction.

21 Glamorgan River Authority

47 There are 5 unsatisfactory storm overflows in this area, 3 in the Rhymney Valley Sewerage Board's area and discharging to the non-tidal Rhymney, one from Cardiff to the tidal Ely and one from Ogmore and Garw Rural District to the non-tidal Ogmore.

48 The overflows will be improved or eliminated, incidental to major long-term sewerage schemes already in hand or proposed.

22 South West Wales River Authority

49 There are 23 unsatisfactory storm overflows, 20 to non-tidal and 3 to tidal waters in 15 rivers. Swansea accounts for 7 of the discharges, Pembroke for 4, and the rural districts of Haverfordwest and Narberth for 3 each.

50 The remedial works required were indicated in many cases as re-construction of the overflow. In some cases, premature discharge is caused by the inability of small treatment works to receive the flows that they should, and extensions there are called for. A number of schemes are in preparation. No large scale re-sewering appears to be needed and the indications are that several of the overflows could be improved at moderate cost.

23 Gwynedd River Authority

51 The survey records 3 unsatisfactory storm overflows in Dolgellau. They discharge to the non-tidal length of the River Wnion. They will be remedied by a sewerage and sewage disposal scheme proposed by the council.

24 Dee and Clwyd River Authority

52 There are 30 unsatisfactory storm overflows in the area, 29 to non-tidal and one to tidal waters. The largest numbers of discharges are to the Dee and a small tributary at Chester (19) and the Gwenfro (6). Fourteen of these overflows are located at Chester and 5 at Wrexham.

53 Certain of the Chester overflows will be remedied by a proposed major scheme of sewerage. Other overflows to the Dee are above waterworks intakes, and the river authority suggested that an intercepting sewer is needed to eliminate them. A similar solution is suggested for Wrexham, where amenity needs are involved. The indications are that fairly extensive re-sewering is called for, and costs are likely to be high.

25 Mersey and Weaver River Authority

54 There are 791 unsatisfactory overflows recorded in the survey—more than one-third of all the unsatisfactory overflows reported in England and Wales. Of the total, 786 discharge to rivers (regarding the Manchester Ship Canal as part of the Mersey and Irwell river systems) and all but 10 of these are to non-tidal waters, 5 discharge to two canals, the St. Helens and the Manchester, Bolton and Bury Canal. More than 80 rivers altogether are directly affected and 740 unsatisfactory discharges are made to rivers in the Mersey basin.

55 Of particular note are the Medlock (119 unsatisfactory discharges), the Irk (84), the Irwell (50, including 12 to the Manchester Ship Canal), the Mersey (37, including 9 to the Manchester Ship Canal), Moston Brook (33), the Roch (32), Bordane Brook-River Glaze system (30), Corn Brook (26), Gore Brook-Chorlton Brook system (23), and the Bollin (33 including 17 on the Dams Brook, Macclesfield). Several other rivers have between 10 and 20 unsatisfactory overflows discharging to them. The overflows are spread over the whole area but Manchester stands out with more than 230 classed as unsatisfactory. Others which are prominent are Bolton with 35, Macclesfield with 34, Middleton with 29, and St. Helens, Swinton and Pendlebury, and Rochdale with 21 each.

56 The remedial measures indicated in the comments vary widely in their extent and, in some cases, simple reconstruction only is needed, but it is evident that if all the overflows are to be brought up to the standard the river authority seek, extensive re-sewering of built up areas will be required at great expense. Only a few estimates are available, and without a comprehensive examination of the problems particularly in the large towns, no reliable estimates could be made of the total cost. A number of schemes are in hand but they are small in relation to the overall problems shown up by the survey. The comments indicated, however, that a number of overflows could probably be improved at moderate cost.

26 Lancashire River Authority

57 The survey records 268 unsatisfactory storm overflows, 263 to non-tidal and 5 to tidal waters in 29 rivers, the most conspicuous in the returns being the Blakewater (40 unsatisfactory), the Darwen (36), the Calder (28), the Hyndburn (24), Walverden Water (18) and the Douglas and the Savick Brook (14 each). The greatest numbers are to be found in Blackburn (66), Burnley (44), Wigan (23), Fulwood (22), Accrington (22), Nelson (20) and Darwen (14). Thus nearly half of all the unsatisfactory overflows are concentrated in the first 3 towns listed above.

58 The comments of the river authority indicated that much re-sewering will be required coupled with the segregation of foul sewage and surface water in places. Almost everywhere the problem appears to be caused by inadequate sewer capacity. A number of schemes are already under construction and others are in preparation. Some costs were provided and they suggest that the overall expenditure is likely to be high. The indications are that something in excess of £7 million would have to be spent in the Blackburn, Burnley, Wigan, Nelson and Preston areas alone.

27 Cumberland River Authority

59 There are 20 unsatisfactory storm overflows in the area, all discharging to non-tidal waters of 9 rivers, prominent being the Eden with 6 unsatisfactory discharges and the Caldew with 3. Carlisle accounts for 8 of the discharges and Cockermouth and Appleby for 3 each.

60 Some re-sewering is indicated as being necessary, but the comments of the river authority suggested that several overflows could be improved by screening and the re-setting of the weirs, which indicates that some of the improvements could be achieved at moderate cost.

28 Thames Conservancy

61 There are 10 unsatisfactory storm overflows in the area, 9 to non-tidal waters of 4 rivers and one to the Grand Union Canal. The Misbourne receives 4 discharges and the River Wey North receives 3. Three of the discharges are from Farnham and 3 from Amersham Rural District.

62 The remedial works necessary to eliminate the unsatisfactory overflows are mainly regional or local sewerage. Several schemes have been prepared and construction has either started or is imminent in some cases. Some overflows will be improved as incidental benefits from sewerage schemes and, on the whole, costs are not likely to be high apart from the works in Amersham Rural District.

29 Lee Conservancy

63 There are no unsatisfactory storm overflows recorded in the area.

30 Port of London Authority (including the London Excluded Area)

64 The survey records 25 unsatisfactory storm overflows in the area, all discharging to tidal waters, 23 to the Thames and 2 to the Roding. All the overflows are on GLC sewers. The PLA consider that the elimination of these overflows would involve a major scheme of trunk sewering with a cost of the order of £40 million, but they also pointed out that this sum might be spent on improvements to sewage treatment with greater benefit.

5 Discharges of industrial effluent

(i) THE MAIN SURVEY

Introduction

1 Industry is the largest user of water, the quantity taken being very much more than that taken by domestic consumers, but over 90 per cent of it is used for cooling. By cooling water is meant water used for cooling in such a way that it is substantially unchanged in use; it will usually be water used in a closed cooling system. The only change in properties might be an increase in temperature and perhaps an increase in the dissolved solids content brought about by the increased concentration resulting from evaporation. Cooling water returns do not include water used for quenching purposes, which is considered to be process water.

2 Water is used in a wide variety of processes and it follows that the liquid wastes produced will also vary widely in strength and composition. Some are not unlike domestic sewage in their general composition but others contain complex organic impurities that are not found in domestic sewage. Some wastes, particularly from certain branches of engineering, may contain compounds which are toxic.

3 All of these wastes have to be disposed of and, generally speaking, this is achieved either by disposal to the open sea (about 16 per cent), or to a watercourse (about 81 per cent), or to a sewerage system for treatment along with domestic sewage at sewage treatment works (about 3 per cent).

4 The administration of river pollution control was described in Appendix 2 of Volume 1. Generally, discharges of industrial effluent to rivers are subject to the same control as discharges of sewage effluent.

5 The main survey deals with direct discharges of industrial effluent to those controlled waters which were covered by the survey. The information provided was normally limited to discharges of 5,000 gpd or more, but smaller discharges were included if the river authority considered that they were of special significance. Certain controlled tidal waters* are outside the scope of the survey and so industrial discharges to those waters are not recorded.

6 In order to obtain information about different types of industrial effluent an Industrial Effluent Classification Code was devised by the Confederation of British Industry and subsequently extended to include certain other types of discharge. The full list of code numbers is at Table 36 and these descriptions appear in an abbreviated form in the subsequent tables.

7 River authorities were asked to give details (not for publication) of the required standards and the extent to which they were achieved and then to indicate whether or not they considered each discharge satisfactory.

8 A distinction has to be drawn between the methods used for the rating of discharges of mine waters and other discharges because the former are subject to special provisions under the Rivers (Prevention of Pollution) Act 1951 (see paragraph 3 of Appendix 2 to Volume 1) which exempts them from liability for the offence of causing pollution provided the mine water is discharged in the condition in which it is raised. However mine water can be and often is heavily polluting as it may contain iron in an unoxidised state that could give rise to oxygen depletion, iron in an oxidised state that could give rise to discolouration, and toxic metals that could prevent the development of fisheries and render water unsuitable for public supply. In some

---

* The River Humber, The Wash, The Solent, the controlled parts of the Bristol Channel, the Menai Strait, part of Morecambe Bay, and the Solway Firth.

Table 36 Industrial effluent classification code

Brewing	01
Brickmaking	02
Cement making	03
Chemical and allied industries including manufacture of synthetic resin, agricultural chemicals, synthetic rubber, pharmaceuticals, pigment and colours, and tar distilling	04
Trade effluent from coal mining, including associated engineering activities such as processing, workshops, small power stations supplying power to mines, etc.	05
Distillation of ethanol	06
Electricity generation	07
Engineering, all kinds including vehicle construction	08
Food processing and manufacture	09
Gas and coke making, excluding tar distilling	10
Glass making	11
Glue and gelatine manufacture	12
General manufacturing	13
Iron and steel making	14
Laundering and dry cleaning	15
Leather tanning, fellmongering	16
Metal smelting and refining	17
Paint making	18
Paper and board making	19
Petroleum refining	20
Plastics manufacture	21
Plating and metal finishing	22
Pottery making	23
Printing ink manufacture including carbon paper, etc. and printing industry	24
Trade effluent from quarrying and mining including sand and gravel washing	25
Rubber processing	26
Soap and detergent manufacture	27
Textile manufacture, cotton and man-made fibres	28
Textile manufacture, wool	29
General farming	30
Atomic energy establishments	31
Effluent from water treatment plants	50
Drainage from disposal tips	51
Mine water discharged from derelict coal mines	52
Mine water discharged from derelict mines, other than coal mines	53
Mine water discharged from active coal mines	54
Mine water discharged from active mines, other than coal mines	55

areas, mine water gives rise to serious problems. It could therefore be misleading to include mine water figures in an overall summary and so they are recorded separately throughout. It will be noted that six code numbers refer to mining operations, 05, 25 and 52 to 55. The first two relate to process water and the last four are used to describe water raised or drained from an underground mine and discharged to a stream without being used in any process.

9 Many industrial discharges to tidal waters are not subject to control. In such cases, when a river authority considered that remedial action would be of advantage, the standard they would wish to impose was indicated as the required standard.

10 River authorities were asked to comment on the remedial work that might be required in order that unsatisfactory discharges could be made to comply with the required standards and also to suggest what sort of cost might be involved. The remedial work required varied greatly, from simple settlement to complex specialised treatment plants. In very many cases the dischargers of the industrial effluent provided river authorities with details of proposed remedial work. Details concerning costs are provided in Chapter 6.

11 River authorities generally encourage the discharge of industrial effluents to the sewer for treatment at a sewage treatment works rather than discharge directly to a watercourse. Such a procedure often reduces the amount of treatment or pre-treatment required at the premises and the local authority providing the necessary treatment at the sewage works normally requires payment for the service, usually on the basis of volume and strength of the effluent. In many cases where remedial action has been indicated as necessary, river authorities have suggested that diverting the effluent to public sewers will provide the most satisfactory solution.

12 There were many instances of individual discharges of farm effluents of less than 5,000 gpd and, as such, they should not have been included in the survey. Occasionally, however, a single stretch of a watercourse received a number of these small discharges that collectively were of significance and in these circumstances a single return was made covering the total number of farm effluents concerned. Several river authorities (Somerset especially) have remarked on problems arising from farm effluents and it is believed that others also have such difficulties.

Details of the survey for England and Wales

13 Tables 37 to 78 summarise the information collected on existing discharges of industrial effluent to controlled waters in England and Wales. As there is a considerable number of discharges of solely cooling water, these have been listed separately. There are often, in addition, substantial volumes of cooling water included in what is referred to as industrial effluent and the tables give a breakdown of industrial effluent into process water and cooling water. This enables a figure for the total volume of cooling water to be obtained.

14 A number of significant points arise from the information in Tables 37 to 46. They show that, excluding mine water, there are some 3,500 discharges of industrial effluent to the controlled waters covered by the survey and just over a half are considered to be satisfactory. There are also about 400 discharges of mine water. Table 37 shows that the largest volume (over 35 per cent) of industrial effluent excluding solely cooling water is discharged to Class 3 rivers. As with Classes 1 and 2 however, over 70 per cent is considered satisfactory compared with only 23 per cent of the discharges to Class 4 rivers.

15 The breakdown between numbers of discharges of industrial effluent consisting of process water with or without cooling water in admixture and numbers of discharges of solely cooling water is:

	Number	% Satisfactory
Industrial effluent	2,449	44%
Cooling water only	1,067	86%
Total	3,516	56%

16 Apart from mine water (398 discharges) the largest numbers of discharges of all types occur from chemical and allied industries

Table 37 Details of discharges of industrial effluent (excluding mine water) to rivers and canals in different chemical classes with percentage satisfactory.
(All volumes x 1,000 gpd)

Chemical class of water receiving discharge	*Type of discharge	Rivers						Canals		All rivers and canals	
		Non-tidal		Tidal		All rivers					
		Number	Volume	Number	Volume	Number	Volume	Number	Volume	Number	Volume
Class 1	Industrial effluent	850 (47%)	389,322 (72%)	51 (60%)	297,532 (70%)	901 (48%)	686,854 (71%)	9 (23%)	5,904 (16%)	910 (47%)	692,758 (71%)
	Solely cooling water	275 (93%)	802,572 (99%)	18 (89%)	3,004,722 (96%)	293 (93%)	3,807,294 (97%)	36 (89%)	16,633 (90%)	329 (92%)	3,823,927 (97%)
Class 2	Industrial effluent	415 (55%)	165,481 (75%)	50 (24%)	76,068 (73%)	465 (52%)	241,549 (74%)	14 (50%)	1,558 (25%)	479 (52%)	243,107 (74%)
	Solely cooling water	166 (91%)	1,221,806 (96%)	21 (100%)	1,099,063 (100%)	187 (92%)	2,320,869 (98%)	52 (90%)	24,521 (75%)	239 (92%)	2,345,390 (97%)
Class 3	Industrial effluent	310 (36%)	278,664 (58%)	108 (42%)	543,136 (86%)	418 (38%)	821,800 (77%)	33 (28%)	11,579 (49%)	451 (37%)	833,379 (76%)
	Solely cooling water	91 (80%)	954,812 (91%)	86 (88%)	4,133,967 (100%)	177 (84%)	5,088,779 (98%)	40 (82%)	127,584 (97%)	217 (83%)	5,216,363 (98%)
Class 4	Industrial effluent	457 (33%)	234,041 (37%)	123 (26%)	357,731 (14%)	580 (32%)	591,772 (23%)	29 (41%)	11,144 (28%)	609 (32%)	602,961 (23%)
	Solely cooling water	218 (79%)	1,717,577 (92%)	19 (68%)	563,825 (99%)	237 (78%)	2,281,402 (94%)	45 (67%)	20,247 (63%)	282 (76%)	2,301,649 (93%)
Total England and Wales	Industrial effluent	2,032 (45%)	1,067,508 (57%)	332 (36%)	1,274,467 (61%)	2,364 (44%)	2,341,975 (54%)	85 (40%)	30,185 (42%)	2,449 (44%)	2,372,160 (54%)
	Solely cooling water	750 (87%)	4,696,767 (93%)	144 (87%)	8,801,577 (99%)	894 (87%)	13,498,344 (97%)	173 (78%)	188,985 (35%)	1,067 (86%)	13,787,329 (96%)
	All discharges	2,782 (56%)	5,764,275 (86%)	476 (51%)	10,055,044 (94%)	3,258 (56%)	15,840,319 (91%)	258 (66%)	219,170 (36%)	3,516 (56%)	16,059,489 (90%)

* In this table "Industrial effluent" means process water with any cooling water in admixture.

(384) followed by engineering (335), food processing (279), electricity generation (259), quarrying (253) and coal mining process water (218). If discharges of solely cooling water are disregarded, the largest numbers of discharges of industrial effluent arise from quarrying (246), chemical and allied industries (222) and coal mining process water (205), whilst the largest numbers of discharges of solely cooling water come from engineering (163), chemical and allied industries (162), electricity generation (156) and food processing (115).

17 Excluding mine water, the total volume of all industrial effluent recorded in the survey is approximately 16,000 mgd of which about 90 per cent is considered satisfactory, so that about 1,600 mgd of industrial effluent is discharged to a standard which is not considered by the river authorities to be acceptable. Mine water accounts for about 207 mgd.

18 The breakdown between the volume of industrial effluent consisting of process water with or without cooling water in admixture and the volume in discharges of solely cooling water is:

	Volume	% Satisfactory
Industrial effluent	2,372 mgd	54%
Solely cooling water	13,687 mgd	96%
Total	16,059 mgd	90%

This shows that of the 1,600 mgd of effluent considered unsatisfactory, about 500 mgd is in discharges of solely cooling water and the balance is in industrial effluent consisting of process water with or without cooling water.

19 These figures also illustrate the very large quantities of water used by industry for cooling purposes, but they do not give the full amounts because there is also some 1,025 mgd of cooling water included in the volume of industrial effluent, so that a further breakdown gives the following:

	Volume
Process water	1,347 mgd
Cooling water	14,712 mgd
Total	16.059 mgd

Process water thus accounts for less than 10 per cent of the total volume of industrial effluent recorded. Table 46 shows that electricity generation accounts for nearly 13,000 mgd out of the total of about 14,712 mgd of all cooling water. These figures, of course, relate only to discharges to the controlled waters covered by the survey.

20 It is of interest to compare these figures with the figures given in Chapters 2 and 3 for sewage effluent and crude sewage discharged to the same waters (see Tables 1 and 34). Together, they amount to some 2,500 mgd or about one-sixth of the above volume.

Table 38 Number of discharges of industrial effluent and percentage satisfactory

CBI Code	Description	Rivers Non-tidal Total number	% satisfactory	Tidal Total number	% satisfactory	All rivers Total number	% satisfactory	Canals Total number	% satisfactory	All rivers and canals Total number	% satisfactory
01	Brewing	39	74	11	45	50	68	1	100	51	69
02	Brickmaking	28	79	1	100	29	79	0	—	29	79
03	Cement making	28	68	11	91	39	74	1	0	40	73
04	Chemical etc	237	56	105	38	342	50	42	67	384	52
05	Coal mining	212	73	2	50	214	73	4	75	218	73
06	Ethanol distillation	1	100	0	—	1	100	0	—	1	100
07	Electricity generation	156	93	90	98	246	95	13	31	259	91
08	Engineering	248	60	13	54	261	59	74	66	335	61
09	Food processing	211	64	51	25	262	56	17	76	279	58
10	Gas and coke	72	62	13	54	85	60	6	50	91	60
11	Glass making	9	67	0	—	9	67	13	62	22	64
12	Glue and gelatine	9	33	3	33	12	33	2	50	14	36
13	General manufacturing	68	65	5	40	73	63	10	90	83	66
14	Iron and steel	160	47	12	67	172	48	27	48	199	48
15	Laundering, dry cleaning	16	38	2	100	18	44	0	—	18	44
16	Leather tanning	19	26	7	0	26	19	2	50	28	21
17	Metal smelting	42	67	6	50	48	65	9	56	57	63
18	Paint making	10	70	1	0	11	64	1	100	12	67
19	Paper and board making	137	49	42	31	179	45	4	100	183	46
20	Petroleum refining	25	44	36	75	61	62	5	60	66	62
21	Plastics manufacture	19	84	1	100	20	85	4	75	24	83
22	Plating and metal finishing	65	55	9	22	74	51	4	100	78	54
23	Pottery making	13	38	0	—	13	38	0	—	13	38
24	Printing ink etc	6	100	0	—	6	100	1	100	7	100
25	Quarrying and mining	237	59	12	33	249	58	4	100	253	59
26	Rubber processing	46	87	0	—	46	87	3	100	49	88
27	Soap and detergent	10	70	19	5	29	28	0	—	29	28
28	Textile, cotton and man-made	141	28	10	60	151	30	10	80	161	34
29	Textile, wool	101	23	0	—	101	23	1	100	102	24
30	General farming	159	4	5	0	164	4	0	—	164	4
31	Atomic energy establishments	4	100	0	—	4	100	0	—	4	100
50	Water treatment	200	73	8	75	208	73	0	—	208	73
51	Disposal tip drainage	54	34	1	0	55	33	0	—	55	33
	Total excl. mine waters	2,782	56	476	51	3,258	56	258	66	3,516	56
	Total mine waters	380		8		388		10		398	
	Total England and Wales	3,162		484		3,646		268		3,914	

Note: A breakdown of the numbers in this table is given in Tables 39 and 40.

Table 39 Number of discharges of industrial effluent (excluding discharges of solely cooling water) and percentage satisfactory[1]

CBI Code	Description	Rivers Non-tidal Total number	Rivers Non-tidal % satisfactory	Rivers Tidal Total number	Rivers Tidal % satisfactory	All rivers Total number	All rivers % satisfactory	Canals Total number	Canals % satisfactory	All rivers and canals Total number	All rivers and canals % satisfactory
01	Brewing	19	47	7	14	26	38	0	—	26	38
02	Brickmaking	23	74	0	—	23	74	0	—	23	74
03	Cement making	22	59	4	75	26	62	1	0	27	59
04	Chemical etc	115	35	91	34	206	35	16	50	222	36
05	Coal mining	201	72	1	100	202	72	3	66	205	72
06	Ethanol distillation	0	—	0	—	0	—	0	—	0	—
07	Electricity generation	73	90	26	92	99	90	4	50	103	90
08	Engineering	137	38	8	25	145	35	27	30	172	38
09	Food processing	122	48	38	13	160	39	4	25	164	39
10	Gas and coke	54	53	7	43	61	52	0	—	61	52
11	Glass making	5	45	0	—	5	40	4	25	9	33
12	Glue and gelatine	4	0	2	0	6	0	0	—	6	0
13	General manufacturing	39	46	5	40	44	45	0	—	44	45
14	Iron and steel	109	42	9	46	118	43	13	38	131	143
15	Laundering, dry cleaning	12	17	2	100	14	29	0	—	14	29
16	Leather tanning	15	7	7	0	22	45	1	0	23	4
17	Metal smelting	26	58	4	50	30	57	3	66	33	58
18	Paint making	5	60	1	0	6	50	0	—	6	50
19	Paper and board making	105	33	34	15	139	29	2	100	141	30
20	Petroleum refining	22	36	29	75	51	59	2	0	53	56
21	Plastics manufacture	4	50	0	—	4	50	0	—	4	50
22	Plating and metal finishing	44	43	9	22	53	40	0	—	53	40
23	Pottery making	11	27	0	—	11	27	0	—	11	27
24	Printing ink etc	2	100	0	—	2	100	0	—	2	100
25	Quarrying and mining	230	58	12	33	242	57	4	100	246	58
26	Rubber processing	13	62	0	—	13	62	0	—	13	62
27	Soap and detergent	5	40	17	6	22	14	0	—	22	14
28	Textile, cotton and man-made	118	14	6	33	124	15	1	0	125	15
29	Textile, wool	87	10	0	—	87	10	0	—	87	10
30	General farming	158	3	5	0	163	3	0	—	163	3
31	Atomic energy establishments	4	100	0	—	4	100	0	—	4	100
50	Water treatment	194	72	7	71	201	72	0	—	201	72
51	Disposal tip drainage	54	33	1	0	55	33	0	—	55	33
	Total excl. mine waters	2,032	45	332	36	2,364	44	85	40	2,449	44
	Total mine waters	380		8		388		10		398	
	Total England and Wales	2,412		340		2,752		95		2,847	

Note: The numbers in this table are included in the totals in Table 38.

Table 40 Number of discharges of solely cooling water and percentage satisfactory

CBI Code	Description	Rivers Non-tidal Total number	% satisfactory	Tidal Total number	% satisfactory	All rivers Total number	% satisfactory	Canals Total number	% satisfactory	All rivers and canals Total number	% satisfactory
01	Brewing	20	100	4	100	24	100	1	100	25	100
02	Brickmaking	5	100	1	100	6	100	0	—	6	100
03	Cement making	6	100	7	100	13	100	0	—	13	100
04	Chemical etc	122	75	14	64	136	71	26	77	162	75
05	Coal mining	11	100	1	0	12	92	1	100	13	92
06	Ethanol distillation	1	100	0	—	1	100	0	—	1	100
07	Electricity generation	83	96	64	100	147	98	9	22	156	94
08	Engineering	111	89	5	100	116	90	47	89	163	90
09	Food processing	89	65	13	62	102	84	13	92	115	84
10	Gas and coke	18	89	6	67	24	83	6	50	30	77
11	Glass making	4	100	0	—	4	100	9	78	13	85
12	Glue and gelatine	5	60	1	100	6	67	2	50	8	63
13	General manufacturing	29	90	0	—	29	90	10	90	39	90
14	Iron and steel	51	57	3	100	54	59	14	57	68	59
15	Laundering, dry cleaning	4	100	0	—	4	100	0	—	4	100
16	Leather tanning	4	100	0	—	4	100	1	100	5	100
17	Metal smelting	16	81	2	50	18	78	6	50	24	71
18	Paint making	5	80	0	—	5	80	1	100	6	83
19	Paper and board making	32	100	8	100	40	100	2	100	42	100
20	Petroleum refining	3	100	7	71	10	80	3	100	13	85
21	Plastics manufacture	15	93	1	100	16	94	4	75	20	90
22	Plating and metal finishing	21	81	0	—	21	81	4	100	25	84
23	Pottery making	2	100	0	—	2	100	0	—	2	100
24	Printing ink etc	4	100	0	—	4	100	1	100	5	100
25	Quarrying and mining	7	100	0	—	7	100	0	—	7	100
26	Rubber processing	33	97	0	—	33	97	3	100	36	97
27	Soap and detergent	5	100	2	0	7	71	0	—	7	71
28	Textile, cotton and man-made	23	100	4	100	27	100	9	76	36	97
29	Textile, wool	14	100	0	—	14	100	1	100	15	100
30	General farming	1	100	0	—	1	100	0	—	1	100
31	Atomic energy establishments	0	—	0	—	0	—	0	—	0	—
50	Water treatment	6	100	1	100	7	100	0	—	7	100
51	Disposal tip drainage	0	—	0	—	0	—	0	—	0	—
	Total England and Wales	750	87	144	87	894	87	173	78	1,067	86

Note: The numbers in this table are included in the totals in Table 38.

Table 41 Volume of all industrial effluent and percentage satisfactory

(All volumes x 1,000 gpd)

CBI Code	Description	Rivers Non-tidal Total volume	Rivers Non-tidal % satisfactory	Rivers Tidal Total volume	Rivers Tidal % satisfactory	All rivers Total volume	All rivers % satisfactory	Canals Total volume	Canals % satisfactory	All rivers and canals Total volume	All rivers and canals % satisfactory
01	Brewing	4,364	93	2,159	14	6,523	67	25	100	6,548	67
02	Brickmaking	2,464	90	40	100	2,504	91	0	—	2,504	91
03	Cement making	8,531	95	1,081	99	9,612	95	3	0	9,615	95
04	Chemical etc	270,549	43	370,899	26	641,448	33	23,526	61	664,974	34
05	Coal mining	35,494	72	57	12	35,551	72	292	68	35,843	72
06	Ethanol distillation	20	100	0	—	20	100	0	—	20	100
07	Electricity generation	4,656,615	97	8,545,447	99·9	13,202,062	99	114,557	6	13,316,619	98
08	Engineering	29,280	55	79,681	99	108,961	87	24,328	90	133,289	88
09	Food processing	93,934	71	66,890	49	160,824	62	6,653	90	167,477	63
10	Gas and coke	32,024	59	33,237	96	65,261	78	1,428	27	66,689	76
11	Glass making	3,439	67	0	—	3,439	67	1,975	74	5,414	69
12	Glue and gelatine	7,362	37	618	78	7,980	40	1,042	4	9,022	36
13	General manufacturing	17,389	76	795	3	18,184	73	2,669	99·7	20,853	76
14	Iron and steel	101,886	32	110,319	56	212,205	44	31,413	52	243,618	45
15	Laundering, dry cleaning	1,077	47	120	96	1,197	78	0	—	1,197	78
16	Leather tanning	1,739	23	458	0	2,197	18	30	67	2,227	19
17	Metal smelting	35,782	87	636	39	36,418	86	2,947	40	39,365	82
18	Paint making	1,865	90	10	0	1,875	90	13	100	1,888	90
19	Paper and board making	200,560	39	201,095	45	401,655	43	363	100	402,018	43
20	Petroleum refining	15,675	62	594,665	81	610,340	80	1,134	85	611,474	80
21	Plastics manufacture	2,479	87	25	100	2,504	87	634	61	3,138	82
22	Plating and metal finishing	6,862	50	877	7	7,739	45	177	100	7,916	46
23	Pottery making	245	40	0	—	245	40	0	—	245	40
24	Printing ink etc	568	100	0	0	568	100	9	100	577	100
25	Quarrying and mining	116,504	57	17,407	0·2	133,911	50	767	100	134,678	50
26	Rubber processing	21,940	88	0	—	21,940	88	2,335	100	24,275	89
27	Soap and detergent	5,879	47	7,243	0·4	13,122	21	0	—	13,122	21
28	Textile, cotton and man-made	32,207	25	39,825	0·5	72,032	11	2,750	98	74,782	14
29	Textile, wool	6,037	36	0	—	6,037	36	100	100	6,137	37
30	General farming	3,583	62	45	0	3,628	45	0	—	3,628	61
31	Atomic energy establishments	2,016	0	0	—	2,016	0	0	—	2,016	0
50	Water treatment	41,261	87	2,365	97	43,626	87	0	—	43,626	87
51	Disposal tip drainage	4,645	34	50	0	4,695	33	0	—	4,695	33
	Total excl. mine waters	5,764,275	86	10,076,044	93	15,840,319	91	219,170	36	16,059,489	90
	Total for mine waters	197,047		5,665		202,712		4,083		206,795	
	Total England and Wales	5,961,322		10,081,709		16,043,031		223,253		16,266,284	

Note: A breakdown of this table is given in Tables 42 and 43.

Table 42 Volume of industrial effluent, excluding discharges of solely cooling water and percentage satisfactory

(All volumes x 1,000 gpd)

CBI Code	Description	Rivers Non-tidal Total volume	% satisfactory	Tidal Total volume	% satisfactory	All rivers Total volume	% satisfactory	Canals Total volume	% satisfactory	All rivers and canals Total volume	% satisfactory
01	Brewing	1,955	85	1,859	5	3,814	44	0	—	3,814	44
02	Brickmaking	2,243	98	0	—	2,243	98	0	—	2,243	98
03	Cement making	6,130	93	560	99	6,690	93	3	0	6,693	93
04	Chemical etc	92,742	22	310,824	13	403,566	15	7,472	51	411,038	15
05	Coal mining	32,263	69	7	100	32,270	69	287	68	32,557	69
06	Ethanol distillation	0	—	0	—	0	—	0	—	0	—
07	Electricity generation	381,767	99	248,096	100	629,863	99	1,515	74	631,378	99
08	Engineering	17,142	31	943	39	18,085	32	6,369	75	24,454	43
09	Food processing	51,600	67	24,182	6	75,782	47	665	2	76,447	49
10	Gas and coke	25,778	51	8,216	96	33,994	62	0	—	33,994	62
11	Glass making	1,464	22	0	—	1,464	22	260	92	1,724	33
12	Glue and gelatine	3,637	0	138	0	3,775	0	0	—	3,775	0
13	General manufacturing	13,512	75	795	31	14,307	69	0	—	14,307	69
14	Iron and steel	53,562	32	90,901	47	144,463	41	10,368	77	154,831	39
15	Laundering, dry cleaning	300	33	120	100	420	38	0	—	420	38
16	Leather tanning	1,352	0	458	0	1,810	0	10	0	1,820	0
17	Metal smelting	11,629	61	370	57	11,999	61	2,013	31	14,012	57
18	Paint making	1,550	89	10	0	1,560	88	0	—	1,560	88
19	Paper and board making	148,093	17	82,779	10	230,872	13	285	100	231,157	13
20	Petroleum refining	8,241	27	449,828	99	458,069	99	170	0	458,239	99
21	Plastics manufacture	329	33	0	—	329	33	0	—	329	33
22	Plating and metal finishing	5,676	43	877	7	6,553	39	0	—	6,553	39
23	Pottery making	241	39	0	—	241	39	0	—	241	39
24	Printing ink etc	377	100	0	—	377	100	0	—	377	100
25	Quarrying and mining	116,010	57	17,407	0	133,417	49	767	100	134,184	50
26	Rubber processing	4,671	47	0	—	4,671	47	0	—	4,671	47
27	Soap and detergent	3,809	18	5,093	1	8,902	8	0	—	8,902	8
28	Textile, cotton and man-made	28,860	16	28,675	4	57,535	17	1	0	57,536	17
29	Textile, wool	4,634	16	0	—	4,634	16	0	—	4,634	16
30	General farming	3,553	62	45	0	3,598	61	0	—	3,598	61
31	Atomic energy establishments	2,016	100	0	—	2,016	100	0	—	2,016	100
50	Water treatment	37,727	85	2,234	97	39,961	86	0	—	39,961	86
51	Disposal tip drainage	4,645	34	50	0	4,695	33	0	—	4,695	33
	Total excl. mine waters	1,067,508	57	1,274,467	54	2,341,975	54	30,185	42	2,372,160	54
	Total mine waters	197,047		5,665		202,712		4,083		206,795	
	Total England and Wales	1,264,555		1,280,132		2,544,687		34,268		2,578,955	

Note: These volumes consist of process water together with any cooling water that may be present. Discharges of solely cooling water are not included; they are set down in Table 43. The volumes in this table are included in the volumes in Table 41. A breakdown of the volumes in this table into "process water" and "cooling water" is given in Tables 44 and 45.

Table 43 Volume of solely cooling water and percentage satisfactory

(All volumes x 1,000 gpd)

CBI Code	Description	Rivers Non-tidal Total volume	% satisfactory	Tidal Total volume	% satisfactory	All rivers Total volume	% satisfactory	Canals Total volume	% satisfactory	All rivers and canals Total volume	% satisfactory
01	Brewing	2,409	100	300	100	2,709	100	25	100	2,734	100
02	Brickmaking	221	100	40	100	261	100	0	—	261	100
03	Cement making	2,401	100	521	100	2,922	100	0	—	2,922	100
04	Chemical etc	177,807	54	60,075	97	237,882	65	16,054	65	253,936	65
05	Coal mining	3,231	100	50	0	3,281	98	5	100	3,286	98
06	Ethanol distillation	20	100	0	—	20	100	0	—	20	100
07	Electricity generation	4,274,848	97	8,297,351	100	12,572,199	99	113,042	5	12,685,241	98
08	Engineering	12,138	88	78,738	100	90,876	98	17,959	98	108,835	98
09	Food processing	42,334	77	42,708	73	850,042	75	5,988	100	91,030	77
10	Gas and coke	6,246	89	25,021	96	31,267	94	1,428	27	32,695	91
11	Glass making	1,975	100	0	—	1,975	100	1,715	71	3,690	86
12	Glue and gelatine	3,725	74	480	100	4,205	77	1,042	4	5,247	62
13	General manufacturing	3,877	86	0	—	3,877	86	2,669	100	6,546	92
14	Iron and steel	48,324	31	19,418	100	67,742	51	21,045	74	88,787	56
15	Laundering, dry cleaning	777	100	0	—	777	100	0	—	777	100
16	Leather tanning	387	100	0	—	387	100	20	100	407	100
17	Metal smelting	24,153	99	266	15	24,419	98	934	58	25,353	96
18	Paint making	315	97	0	—	315	97	13	100	328	97
19	Paper and board making	52,467	100	118,316	100	170,783	100	78	100	170,861	100
20	Petroleum refining	7,434	100	144,837	26	152,271	29	964	100	153,235	30
21	Plastics manufacture	2,150	95	25	100	2,175	95	634	61	2,809	88
22	Plating and metal finishing	1,186	81	0	—	1,186	81	177	100	1,363	83
23	Pottery making	4	100	0	—	4	100	0	—	4	100
24	Printing ink etc	191	100	0	—	191	100	9	100	200	100
25	Quarrying and mining	494	100	0	—	494	100	0	—	494	100
26	Rubber processing	17,269	99	0	—	17,269	99	2,335	100	19,604	99
27	Soap and detergent	2,070	100	2,150	0	4,220	49	0	—	4,220	49
28	Textile, cotton and man-made	3,347	100	11,150	100	14,497	100	2,749	98	17,246	100
29	Textile, wool	1,403	100	0	—	1,403	100	100	100	1,503	100
30	General farming	30	100	0	—	30	100	0	—	30	100
31	Atomic energy establishments	0	—	0	—	0	—	0	—	0	—
50	Water treatment	3,534	100	131	100	3,665	100	0	—	3,665	100
51	Disposal tip drainage	0	—	0	—	0	—	0	—	0	—
	Total England and Wales	4,696,767	93	8,801,577	99	13,498,344	97	188,985	35	13,687,329	96

Note: The volumes in this table are included in Table 41.

Table 44 Volume of process water discharged as part of industrial effluent recorded in Table 42

(All volumes x 1,000 gpd)

CBI Code	Description	Rivers Non-tidal	Tidal	All rivers	Canals	All rivers and canals
01	Brewing	1,913	983	2,896	0	2,896
02	Brickmaking	2,243	0	2,243	0	2,243
03	Cement making	6,039	560	6,599	3	6,602
04	Chemical etc	53,610	104,261	157,871	6,205	164,076
05	Coal mining	31,223	7	31,230	287	31,517
06	Ethanol distillation	0	0	0	0	0
07	Electricity generation	285,147	73,036	358,183	1,471	359,654
08	Engineering	14,740	909	15,649	6,302	21,951
09	Food processing	42,466	20,855	63,321	665	63,986
10	Gas and coke	18,751	1,118	19,869	0	19,869
11	Glass making	1,440	0	1,440	260	1,700
12	Glue and gelatine	3,523	138	3,661	0	3,661
13	General manufacturing	11,979	695	12,674	0	12,674
14	Iron and steel	41,724	81,926	123,650	8,946	132,596
15	Laundering, dry cleaning	300	120	420	0	420
16	Leather tanning	1,349	458	1,807	10	1,817
17	Metal smelting	11,243	368	11,611	1,321	12,932
18	Paint making	1,523	10	1,533	0	1,533
19	Paper and board making	92,617	71,742	164,359	285	164,644
20	Petroleum refining	6,479	81,429	87,908	86	87,994
21	Plastics manufacture	183	0	183	0	183
22	Plating and metal finishing	4,903	226	5,129	0	5,129
23	Pottery making	241	0	241	0	241
24	Printing ink etc	213	0	213	0	213
25	Quarrying and mining	114,627	17,407	132,034	767	132,801
26	Rubber processing	3,381	0	3,381	0	3,381
27	Soap and detergent	879	4,491	5,370	0	5,370
28	Textile, cotton and man-made	27,332	26,575	53,907	1	53,908
29	Textile, wool	4,605	0	4,605	0	4,605
30	General farming	2,889	45	2,934	0	2,934
31	Atomic energy establishments	1,668	0	1,668	0	1,668
50	Water treatment	37,562	2,234	39,796	0	39,796
51	Disposal tip drainage	4,645	50	4,695	0	4,695
	Total excluding mine waters	831,437	489,643	1,321,080	26,609	1,347,689
	Total mine waters	197,047	5,665	202,712	4,083	206,795
	Total England and Wales	1,028,484	495,308	1,523,792	30,692	1,554,484

Table 45 Volume of cooling water discharged as part of industrial effluent recorded in Table 42

(All volumes x 1,000 gpd)

CBI Code	Description	Rivers Non-tidal	Rivers Tidal	Rivers All rivers	Canals	All rivers and canals
01	Brewing	42	876	918	0	918
02	Brickmaking	0	0	0	0	0
03	Cement making	91	0	91	0	91
04	Chemical etc	39,132	206,563	245,695	1,267	246,962
05	Coal mining	1,040	0	1,040	0	1,040
06	Ethanol distillation	0	0	0	0	0
07	Electricity generation	96,620	175,060	271,680	44	271,724
08	Engineering	2,402	34	2,436	67	1,503
09	Food processing	9,134	3,327	12,461	0	12,461
10	Gas and coke	7,027	7,098	14,125	0	14,125
11	Glass making	24	0	24	0	24
12	Glue and gelatine	114	0	114	0	114
13	General manufacturing	1,533	100	1,633	0	1,633
14	Iron and steel	11,838	8,975	20,813	1,422	22,235
15	Laundering, dry cleaning	0	0	0	0	0
16	Leather tanning	3	0	3	0	3
17	Metal smelting	386	2	388	692	1,080
18	Paint making	27	0	27	0	27
19	Paper and board making	55,476	11,037	66,513	0	66,513
20	Petroleum refining	1,762	368,399	370,161	84	370,245
21	Plastics manufacture	146	0	146	0	146
22	Plating and metal finishing	773	651	1,424	0	1,424
23	Pottery making	0	0	0	0	0
24	Printing ink etc	164	0	164	0	164
25	Quarrying and mining	1,383	0	1,383	0	1,383
26	Rubber processing	1,290	0	1,290	0	1,290
27	Soap and detergent	2,930	602	3,532	0	3,532
28	Textile, cotton and man-made	1,528	2,100	3,628	0	3,628
29	Textile, wool	29	0	29	0	29
30	General farming	664	0	664	0	664
31	Atomic energy establishments	348	0	348	0	348
50	Water treatment	165	0	165	0	165
51	Disposal tip drainage	0	0	0	0	0
	Total England and Wales	236,071	784,824	1,020,895	3,576	1,024,471

Table 46 Total volume of all cooling water consisting of solely cooling water (Table 43) and cooling water, if any, discharged as part of industrial effluent (Table 45)

(All volumes x 1,000 gpd)

CBI Code	Description	Rivers			Canals	All rivers and canals
		Non-tidal	Tidal	All rivers		
01	Brewing	2,451	1,176	3,627	25	3,652
02	Brickmaking	221	40	261	0	261
03	Cement making	2,492	521	3,013	0	3,013
04	Chemical etc	216,939	266,638	483,577	17,321	500,898
05	Coal mining	4,271	50	4,321	5	4,326
06	Ethanol distillation	20	0	20	0	20
07	Electricity generation	4,371,468	8,472,411	12,843,879	113,086	12,956,965
08	Engineering	14,540	78,772	93,312	18,026	111,338
09	Food processing	51,468	46,035	97,503	5,988	103,491
10	Gas and coke	13,273	32,119	45,392	1,428	46,820
11	Glass making	1,999	0	1,999	1,715	3,714
12	Glue and gelatine	3,839	480	4,319	1,042	5,361
13	General manufacturing	5,410	100	5,510	2,669	8,179
14	Iron and steel	60,162	28,393	88,555	22,467	111,022
15	Laundering, dry cleaning	777	0	777	0	777
16	Leather tanning	390	0	390	20	410
17	Metal smelting	24,539	268	24,807	1,626	26,433
18	Paint making	342	0	342	13	355
19	Paper and board making	107,943	129,353	237,296	78	237,374
20	Petroleum refining	9,196	513,236	522,432	1,048	523,480
21	Plastics manufacture	2,296	25	2,321	634	2,955
22	Plating and metal finishing	1,959	651	2,610	177	2,787
23	Pottery making	4	0	4	0	4
24	Printing ink etc	355	0	355	9	364
25	Quarrying and mining	1,877	0	1,877	0	1,877
26	Rubber processing	18,559	0	18,559	2,335	20,894
27	Soap and detergent	5,000	2,752	7,752	0	7,752
28	Textile, cotton and man-made	4,875	13,250	18,125	2,749	20,874
29	Textile, wool	1,432	0	1,432	100	1,532
30	General farming	694	0	694	0	694
31	Atomic energy establishments	348	0	348	0	348
50	Water treatment	3,699	131	3,830	0	3,830
51	Disposal tip drainage	0	0	0	0	0
	Total England and Wales	4,932,838	9,586,401	14,519,239	192,561	14,711,800

Table 47 Discharges of industrial effluent in each river authority area in order of volume (× 1,000 gpd)

Process water (industrial effluent from which any cooling water content is deducted)				Total cooling water (solely cooling water discharges to which any cooling water content of industrial effluent is added)				Total industrial effluent discharged	
Non-tidal rivers River Authority		Tidal rivers River Authority		Canals River Authority		Non-tidal rivers River Authority		Tidal rivers River Authority		Canals River Authority		All rivers and canals River Authority	
Yorkshire	125,130	Northumbrian	86,284	Trent	15,967	Trent	1,317,847	Port of London Authority (including the London Excluded Area)	3,288,894	Mersey and Weaver**	124,804	Port of London Authority (including the London Excluded Area)	3,386,124
Trent	112,336	Port of London Authority (including the London Excluded Area)	77,106	Mersey and Weaver**	4,625	Yorkshire	1,288,425	Hampshire	1,769,774	Trent	40,632	Mersey and Weaver	1,872,644
Mersey and Weaver	111,268	Mersey and Weaver	62,577	Severn	4,586	Mersey and Weaver	784,767	Severn	1,005,980	Severn	10,160	Hampshire	1,791,524
Cornwall*	63,466	Kent	51,679	Yorkshire	696	Severn	485,030	Mersey and Weaver	784,603	Yorkshire	5,104	Trent	1,714,666
Thames Conservancy	55,668	Dee and Clwyd	45,351	Thames Conservancy	695	Great Ouse	294,435	Northumbrian	699,878	Lancashire	3,937	Severn	1,524,697
Glamorgan	33,779	Cornwall*	30,541	South West Wales	30	Thames Conservancy	253,957	Kent	514,517	Thames Conservancy	3,497	Yorkshire	1,469,575
Bristol Avon	21,259	Hampshire	18,083	Port of London Authority (including the London Excluded Area)	10	Lee Conservancy	122,754	East Suffolk and Norfolk	340,575	South West Wales	3,068	Northumbrian	797,772
Lancashire	19,215	Glamorgan	16,113	Dee and Clwyd	0	Glamorgan	100,224	Essex	334,008	Welland and Nene	1,314	Kent	574,332
Devon	15,875	Trent	14,374	Lancashire	0	Cumberland	81,657	Trent	213,510	Dee and Clwyd	42	Essex	351,087
Severn	15,120	Lancashire	12,425	Welland and Nene	0	Welland and Nene	64,085	Usk	195,999		3	East Suffolk and Norfolk	348,852
Usk	14,561	South West Wales*	10,488			Bristol Avon	45,505	Devon	160,000			Thames Conservancy	313,388
Essex	13,869	Severn	8,877			Lancashire	40,751	Lancashire	126,880			Great Ouse	307,814
Northumbrian	9,047	Yorkshire	7,731			Port of London Authority (including the London Excluded Area)		Yorkshire	45,135			Usk	214,350
Great Ouse	7,479	Great Ouse	6,252			Essex	8,693	Isle of Wight	30,000			Lancashire	202,768
Lincolnshire	5,424	Lincolnshire	5,810			Devon	6,555	Cornwall	23,180			Cornwall	188,766
Welland and Nene	4,780	Welland and Nene	4,745			Kent	5,760	Glamorgan	19,060			Devon	182,949
Somerset	4,682	East Suffolk and Norfolk	4,156			South West Wales	5,582	South West Wales	19,050			Glamorgan	169,176
Dee and Clwyd	4,636	Bristol Avon	2,005			Northumbrian	4,060	Bristol Avon	11,220			South West Wales	136,503
Hampshire	3,287	Welland and Nene	1,908			Lincolnshire	3,879	Dee and Clwyd	3,343			Lee Conservancy	122,804
East Suffolk and Norfolk	2,862	Usk	1,625			Somerset	3,573	Welland and Nene	517			Cumberland	84,659
Cumberland	2,282	Essex	1,477			Usk	3,000	Avon and Dorset	160			Bristol Avon	79,888
Kent	2,554	Devon	1,314			Cornwall	2,857	Great Ouse	90			Welland and Nene	71,332
Port of London Authority (including the London Excluded Area)		Cumberland	720			Wye	2,500	Lincolnshire	28			Dee and Clwyd	55,200
Wye	1,261	Avon and Dorset	640			Dee and Clwyd	1,867	Cumberland	0			Avon and Dorset	42,987
Avon and Dorset*	1,236	Sussex	46			East Suffolk and Norfolk	1,259	Gwynedd	0			Isle of Wight	30,197
Isle of Wight	230	Lincolnshire	27			Avon and Dorset	957	Lee Conservancy	0			Somerset	12,427
Sussex	176	Isle of Wight	20			Hampshire	380	Somerset	0			Lincolnshire	9,052
Lee Conservancy	148	Gwynedd	0			Isle of Wight	1	Sussex	0			Wye	3,736
Sussex	50	Lee Conservancy	0			Gwynedd	0	Thames Conservancy	0			Sussex	194
Gwynedd	26	Thames Conservancy	0			Sussex	0	Wye	0			Gwynedd	26
		Wye	0										
Turbine water	172,000	Turbine water	25,000										
Total	831,437	Total	489,643	Total	26,609	Total	4,932,838	Total	9,586,401	Total	192,561	Total	16,059,489

* This total excludes turbine water

** Attention is drawn to Paragraph 136 of Chapter 5 regarding discharges to the Manchester Ship Canal.

21 Table 47 lists the river authorities in order of (a) the amount of process water discharged and (b) the amount of cooling water discharged, the latter being a summation of volumes in discharges of solely cooling water and the cooling water included with process water in other discharges (see Table 46).

22 So far as process waters discharged to non-tidal rivers are concerned, the greatest amounts are discharged in the areas of the Yorkshire River Authority, the Trent River Authority and the Mersey and Weaver River Authority, all in excess of 100 mgd. The greatest amount of process water discharged to tidal waters occurs in the area of the Northumbrian River Authority, about 86 mgd, closely followed by the next highest amount of 77 mgd which occurs in the Port of London Authority (including the London Excluded Area). The Mersey and Weaver River Authority follows with 63 mgd and the Kent River Authority with 52 mgd. Of the 7 river authorities in whose areas discharges of process water take place to canals, the largest amount is in the Trent River Authority, of nearly 16 mgd, followed by the Mersey and Weaver River Authority and the Severn River Authority, both nearly 5 mgd.

23 So far as cooling water discharged to non-tidal waters is concerned, the greatest amounts are discharged in the area of Trent River Authority, Yorkshire River Authority, Mersey and Weaver River Authority and Severn River Authority in that order, the amounts ranging from about 1,320 mgd down to about 485 mgd. By far the greatest amount of cooling water discharged to tidal waters occurs in the Port of London Authority (including the London Excluded Area) with nearly 3,300 mgd. Hampshire River Authority follows with about 1,770 mgd and Severn River Authority with about 1,000 mgd. Of the 10 river authorities in whose areas discharges of cooling water takes place to canals the largest amount is in the Mersey and Weaver River Authority Area, of nearly 125 mgd, followed by the Trent River Authority with nearly 41 mgd and the Port of London Authority (including the London Excluded Area) with about 10 mgd.

24 Taking all the effluents together the greatest amounts discharged are as follows:

Authority	Approximate Volume discharged
Port of London Authority (including the London Excluded Area)	3,386 mgd
Mersey and Weaver River Authority	1,873 mgd
Hampshire River Authority	1,791 mgd
Trent River Authority	1,715 mgd
Severn River Authority	1,525 mgd
Yorkshire River Authority	1,470 mgd

25 Table 48 gives information about discharges of mine water and these occur in 14 river authority areas with Northumbrian River Authority having the greatest amount of about 74 mgd, followed by Yorkshire River Authority with about 48 mgd. These two stand out far ahead of all others, the next in order being Cornwall River Authority with about 18 mgd.

Table 48 Discharges of mine water to rivers and canals. All volumes x 1,000 gpd

River authority	Rivers			Canals	All rivers and canals
	Non-tidal	Tidal	All rivers		
Northumbrian	69,811	3,937	73,748		73,748
Yorkshire	45,052		45,052	2,683	47,735
Cornwall	18,173		18,173		18,173
South West Wales	13,122	1,728	14,850		14,850
Trent	14,281		14,281		14,281
Usk	9,397		9,397		9,397
Glamorgan	8,971		8,971		8,971
Severn	6,299		6,299		6,299
Mersey and Weaver	4,225		4,225	580	4,805
Lancashire	2,410		2,410	820	3,230
Kent	2,230		2,230		2,230
Cumberland	2,200		2,200		2,200
Bristol Avon	600		600		600
Dee and Clwyd	276		276		276
Total, England and Wales	197,047	5,665	202,712	4,083	206,795

26 A description of the situation in each river authority area follows. For each authority, a table provides broad information about the discharges, with separate information for process water and cooling water. More detailed information about the different types of industrial effluent discharged in each river authority area is given in Tables 188 to 217 at the end of the Report.

1 Northumbrian River Authority

Table 49 Numbers and volumes of discharges of industrial effluent (excluding mine water discharges, codes 52 to 55 inclusive)

Discharges	Rivers		
	Non-tidal	Tidal	All rivers
1 Total industrial effluent excluding solely cooling water			
Number	42	40	82
% satisfactory	43	30	37
2 Solely cooling water			
Number	5	7	12
% satisfactory	100	100	100
3 All discharges (1+2)			
Number	47	47	94
% satisfactory	49	48	48
4 Total industrial effluent excluding solely cooling water			
Volume x 1,000 gpd	7,731	281,101	288,832
% satisfactory	54	9	10
5 Process water (part of total effluent, 4–6)			
Volume x 1,000 gpd	7,731	86,284	94,015
6 Cooling water (part of total effluent)			
Volume x 1,000 gpd	0	194,817	194,817
7 Solely cooling water			
Volume x 1,000 gpd	3,879	505,061	508,940
% satisfactory	100	100	100
8 Total cooling water (6+7)			
Volume x 1,000 gpd	3,879	699,878	703,757
9 Process water and cooling water (5+8)			
Total volume x 1,000 gpd	11,610	786,162	797,772

27 Excluding mine waters a total of 94 discharges, 45 of which are classed as being satisfactory, discharge about 798 mgd of industrial effluent. Of this quantity about 704 mgd, or 88 per cent, consists of cooling water. Some 538 mgd of the total volume discharged is classed as satisfactory.

28 There are 19 industrial classes covered in the discharges, the largest discharge coming from electricity generation with 505 mgd. Other significant discharges are from the chemical and allied industries (216 mgd), iron and steel making (46 mgd), textile manufacture, cotton and man-made fibres, (17 mgd), petroleum refining (5 mgd), gas and coke making (2 mgd), engineering (1 mgd) and water treatment (1 mgd); these, together with the discharge from electricity generation, constitute over 99 per cent of the total volume discharged. About 33 per cent of all the effluent from the chemical and allied industries and 23 per cent of all the effluent from the textile (cotton and man-made fibres) industries is discharged in this river authority area.

29 The large quantity of cooling water discharged tends to hide the fact that 94 mgd of process water (industrial effluent with no cooling water present) is discharged to controlled waters and, ignoring turbine process water*, this is the fourth highest river authority total. Of this total the industries mentioned above, excluding electricity generation, account for 91 mgd.

30 There is a flow of about 74 mgd of mine water discharged, the largest total for any river authority. Of this total about 69 mgd comes from active coal mines and pumping stations maintained to protect active coal mines.

---

*In drawing up the table of process water in order of volume (Table 47) turbine water is listed separately because although it is process water, it is discharged substantially unchanged. If included as process water it could give a misleading impression of the extent of the problem.

31 In the case of about 1 mgd of industrial effluent requiring remedial action, the river authority have suggested that this might take the form of diverting to public sewers, for treatment at a sewage treatment works.

32 For further details of industrial discharges in this area, see Table 188. ☐

2 Yorkshire River Authority
(see Table 50)

33 Excluding mine waters, a total of 821 outlets discharge about 1,470 mgd of industrial effluent; 409 outlets and about 1,318 mgd are regarded as being satisfactory. Cooling water makes up 91 per cent of the total volume discharged, mostly to non-tidal rivers. About 132 mgd of process water (industrial effluent without any cooling water content) is discharged to controlled waters, the third largest total among the river authorities, if turbine process water is ignored (see footnote on page 55).

34 The discharges cover a wide range of industry and are spread over 29 of the 33 industrial classes. The largest discharge is that of 1,321 mgd from electricity generation, accounting for about 90 per cent of all the industrial effluent discharged in the area. Of the other discharges the principal ones come from the chemical and allied industries (30 mgd), iron and steel making (23 mgd), paper and board making (9 mgd), gas and coke production (7 mgd), quarrying and mining, excluding coal mining (7 mgd), potable water treatment (5 mgd) and coal mining (5 mgd).

35 About 48 mgd of mine water is discharged to non-tidal rivers and canals, including 33 mgd from active and 11 mgd from derelict coal mines.

36 In the case of about 9 mgd of industrial effluent requiring remedial action, the river authority have suggested that this might take the form of diverting to public sewers, for treatment at a sewage treatment works.

37 There are known to be several large discharges of industrial effluent to the River Humber, which are not covered by the survey.

38 For further details of industrial discharges in this area, see Table 189. ☐

Table 50 Numbers and volumes of discharges of industrial effluent (excluding mine water discharges, codes 52 to 55 inclusive)

Discharges	Rivers			Canals	All rivers and canals
	Non-tidal	Tidal	All rivers		
1 Total industrial effluent excluding solely cooling water					
Number	537	13	550	18	568
% satisfactory	37	38	37	17	36
2 Solely cooling water					
Number	217	3	220	33	253
% satisfactory	81	67	80	91	81
3 All discharges (1+2)					
Number	754	16	770	51	821
% satisfactory	50	44	49	65	50
4 Total industrial effluent excluding solely cooling water					
Volume x 1,000 gpd	134,716	8,637	143,353	852	144,205
% satisfactory	63	23	60	20	60
5 Process water (part of total effluent, 4-6)					
Volume x 1,000 gpd	125,130	6,252	131,382	696	132,078
6 Cooling water (part of total effluent)					
Volume x 1,000 gpd	9,586	2,385	11,971	156	12,127
7 Solely cooling water					
Volume x 1,000 gpd	1,278,839	42,750	1,321,589	3,781	1,325,370
% satisfactory	92	98	93	52	93
8 Total cooling water (6+7)					
Volume x 1,000 gpd	1,288,425	45,135	1,333,560	3,937	1,337,497
9 Process water and cooling water (5+8)					
Total volume x 1,000 gpd	1,413,555	51,387	1,464,942	4,633	1,469,575

3 Trent River Authority
(see Table 51)

39 A total of 513 discharges, 363 of which are regarded as being satisfactory, discharge about 1,715 mgd of industrial effluent (excluding mine water), mostly to non-tidal waters. Of this total about 1,572 mgd (92 per cent) consists of cooling water. Some 1,630 mgd of the total discharge is classed as satisfactory. The large quantity of cooling water discharged (the largest discharge of a river authority to non-tidal waters) tends to hide the fact that there is about 143 mgd of process water (industrial effluent without any cooling water content) discharged to controlled waters, which is the second highest total for river authorities, if turbine process water is ignored (see footnote on page 55).

40 There are 26 industrial classes represented among the discharges; the largest volume comes from electricity generation and totals about 1,476 mgd, or 86 per cent of the total discharge. Other major discharges come from the chemical and allied industries (85 mgd), iron and steel making (40 mgd), metal smelting and refining (27 mgd), engineering (16 mgd), coal mining (16 mgd), paper and board making (10 mgd), quarrying and mining excluding coal mining (9 mgd), food processing (9 mgd), rubber processing (7 mgd), water treatment (4 mgd), gas and coke production (4 mgd). About 69 per cent of all the effluent from metal smelting and 27 per cent of all the effluent from rubber processing is discharged in this river authority area.

41 There is about 14 mgd of mine water discharged, 8 mgd from derelict and 6 mgd from active coal mines, all to non-tidal rivers.

42 A total of about 57 mgd of industrial effluent is discharged to canals, mainly from iron and steel making, engineering, chemical and allied industries and electricity generation.

43 In the case of about 2 mgd of industrial effluents requiring remedial action, the river authority has suggested that this might take the form of diverting to public sewers for treatment at a sewage treatment works.

44 For further details of industrial discharges in this area, see Table 190. ☐

Table 51 Numbers and volumes of discharges of industrial effluent (excluding mine water discharges, codes 52 to 55 inclusive)

Discharges	Rivers			Canals	All rivers and canals
	Non-tidal	Tidal	All rivers		
1 Total industrial effluent excluding solely cooling water					
Number	282	9	291	43	334
% satisfactory	71	33	69	44	66
2 Solely cooling water					
Number	113	4	117	62	179
% satisfactory	87	100	87	68	80
3 All discharges (1+2)					
Number	395	13	408	105	513
% satisfactory	75	54	74	57	71
4 Total industrial effluent excluding solely cooling water					
Volume x 1,000 gpd	200,373	14,374	214,747	17,541	232,288
% satisfactory	86	90	86	47	83
5 Process water (part of total effluent, 4-6)					
Volume x 1,000 gpd	112,336	14,374	126,710	15,967	142,677
6 Cooling water (part of total effluent)					
Volume x 1,000 gpd	88,037	0	88,037	1,574	89,611
7 Solely cooling water					
Volume x 1,000 gpd	1,229,810	213,510	1,443,320	39,058	1,482,378
% satisfactory	98	100	98	74	97
8 Total cooling water (6+7)					
Volume x 1,000 gpd	1,317,847	213,510	1,531,357	40,632	1,571,989
9 Process water and cooling water (5+8)					
Total volume x 1,000 gpd	1,430,183	227,884	1,658,067	56,599	1,714,666

4 Lincolnshire River Authority

Table 52 Numbers and volumes of discharges of industrial effluent (excluding mine water discharges, codes 52 to 55 inclusive)

Discharges	Rivers		
	Non-tidal	Tidal	All rivers
1 Total industrial effluent excluding solely cooling water			
Number	20	1	21
% satisfactory	50	0	48
2 Solely cooling water			
Number	8	0	8
% satisfactory	63	—	63
3 All discharges (1+2)			
Number	28	1	29
% satisfactory	54	0	52
4 Total industrial effluent excluding solely cooling water			
Volume x 1,000 gpd	5,781	55	5,836
% satisfactory	11	0	11
5 Process water (part of total effluent 4-6)			
Volume x 1,000 gpd	5,424	27	5,451
6 Cooling water (part of total effluent)			
Volume x 1,000 gpd	357	28	385
7 Solely cooling water			
Volume x 1,000 gpd	3,216	0	3,216
% satisfactory	25	—	25
8 Total cooling water (6+7)			
Volume x 1,000 gpd	3,573	28	3,601
9 Process water and cooling water (5+8)			
Total volume x 1,000 gpd	8,997	55	9,052

45 A total of 29 outlets, 15 of which are classed as satisfactory discharge 9 mgd of industrial effluent, about 1.5 mgd being regarded as satisfactory. Rather more than one-third of the total discharge consists of cooling water.

46 Thirteen different kinds of industry are represented in the survey, iron and steel making and petroleum refining accounting for some 75 per cent of the total volume discharged. Most of the discharges are made to non-tidal rivers.

47 The authority point out that problems associated with farm effluents are significant but details are not recorded in the survey because, individually, the volumes discharged are below the minimum considered for the survey, and it was not possible to make a collective entry. A good deal of expenditure could be involved.

48 In the case of about 100,000 gpd of industrial effluent requiring remedial action, the authority have suggested that this might take the form of diverting to public sewers for treatment at a sewage treatment works.

49 For further details of industrial discharges in this area, see Table 191. ☐

5 Welland and Nene River Authority

(see Table 53 over page)

50 A total of 31 outlets discharge 71 mgd of industrial effluent of which 97 per cent is to non-tidal waters. Altogether 19 outlets and 65 mgd are considered to be satisfactory. Cooling water accounts for about 90 per cent of the total volume.

51 By far the largest quantity is used in electricity generation, equal to slightly more than 75 per cent of the total volume. The iron and steel industry discharges 6 mgd (80 per cent cooling water) and the food processing industry discharges 4 mgd. The only other discharge of significance is from the chemical and allied industries with 2 mgd, mostly cooling water.

52 One small discharge of cooling water is made to a canal.

53 In the case of about 3.5 mgd of industrial effluents requiring remedial action, the river authority have suggested that this might take the form of diverting to public sewers for treatment at a sewage treatment works.

54 For further details of industrial discharges in this area, see Table 192. ☐

(5 Welland and Nene River Authority continued)

Table 53 Numbers and volumes of discharges of industrial effluent (excluding mine water discharges, codes 52 to 55 inclusive)

Discharges	Rivers			Canals	All rivers and canals
	Non-tidal	Tidal	All rivers		
1 Total industrial effluent excluding solely cooling water					
Number	17	4	21	0	21
% satisfactory	53	25	48	—	48
2 Solely cooling water					
Number	9	0	9	1	10
% satisfactory	89	—	89	100	90
3 All discharges (1+2)					
Number	26	4	30	1	31
% satisfactory	65	25	60	100	61
4 Total industrial effluent excluding solely cooling water					
Volume x 1,000 gpd	9,774	2,425	12,199	0	12,199
% satisfactory	78	0	63	—	63
5 Process water (part of total effluent 4-6)					
Volume x 1,000 gpd	4,780	1,908	6,688	0	6,688
6 Cooling water (part of total effluent)					
Volume x 1,000 gpd	4,994	517	5,511	0	5,511
7 Solely cooling water					
Volume x 1,000 gpd	59,091	0	59,091	42	59,133
% satisfactory	99.7	—	99.7	100	99.7
8 Total cooling water (6+7)					
Volume x 1,000 gpd	64,085	517	64,602	42	64,644
9 Process water and cooling water (5+8)					
Total volume x 1,000 gpd	68,865	2,425	71,290	42	71,332

6 Great Ouse River Authority

Table 54 Numbers and volumes of discharges of industrial effluent (excluding mine water discharges, codes 52 to 55 inclusive)

Discharges	Rivers		
	Non-tidal	Tidal	All rivers
1 Total industrial effluent excluding solely cooling water			
Number	23	5	28
% satisfactory	78	40	71
2 Solely cooling water			
Number	11	0	11
% satisfactory	100	—	100
3 All discharges (1+2)			
Number	34	5	39
% satisfactory	85	40	79
4 Total industrial effluent excluding solely cooling water			
Volume x 1,000 gpd	11,048	5,900	16,948
% satisfactory	95	10	67
5 Process water (part of total effluent 4-6)			
Volume x 1,000 gpd	7,479	5,810	13,289
6 Cooling water (part of total effluent)			
Volume x 1,000 gpd	3,569	90	3,659
7 Solely cooling water			
Volume x 1,000 gpd	290,866	0	290,866
% satisfactory	100	—	100
8 Total cooling water (6+7)			
Volume x 1,000 gpd	294,435	90	294,525
9 Process water and cooling water (5+8)			
Total volume x 1,000 gpd	301,914	5,900	307,814

55 A total of 39 outlets, 31 of which are regarded as satisfactory, discharge about 308 mgd of industrial effluent. Of this volume about 302 mgd is regarded as being satisfactory. Cooling water accounts for 96 per cent of the total volume.

56 The largest discharge comes from electricity generation, which accounts for 290 mgd, equal to about 94 per cent of the total. Of the remaining 12 industries covered, food processing and manufacturing discharges about 9 mgd and quarrying and mining, paper and board making and chemical and allied industries discharge volumes of between 1 mgd and 5 mgd each. About 98 per cent of the total discharge takes place to non-tidal rivers and there are no discharges recorded to canals.

57 In the case of 15,000 gpd of industrial effluents requiring remedial action the river authority have suggested that this might take the form of diverting to public sewers for treatment at a sewage treatment works.

58 For further details of industrial discharges in this area, see Table 193. □

7 East Suffolk and Norfolk River Authority

Table 55 Numbers and volumes of discharges of industrial effluent (excluding mine water discharges, codes 52 to 55 inclusive)

Discharges	Rivers		
	Non-tidal	Tidal	All rivers
1 Total industrial effluent excluding solely cooling water			
Number	23	17	40
% satisfactory	61	35	50
2 Solely cooling water			
Number	10	9	19
% satisfactory	80	100	89
3 All discharges (1+2)			
Number	33	26	59
% satisfactory	67	58	63
4 Total industrial effluent excluding solely cooling water			
Volume x 1,000 gpd	3,462	4,156	7,618
% satisfactory	41	19	29
5 Process water (part of total effluent 4-6)			
Volume x 1,000 gpd	2,862	4,156	7,018
6 Cooling water (part of total effluent)			
Volume x 1,000 gpd	600	0	600
7 Solely cooling water			
Volume x 1,000 gpd	659	340,575	341,234
% satisfactory	85	100	100
8 Total cooling water (6+7)			
Volume x 1,000 gpd	1,259	340,575	341,834
9 Process water and cooling water (5+8)			
Total volume x 1,000 gpd	4,121	344,731	348,852

59 A total of 59 discharges, 37 of which are classed as being satisfactory, discharge about 349 mgd. Of this total, about 342 mgd (98 per cent) consists of cooling water. Some 98 per cent of the volume discharged is classed as satisfactory.

60 The largest discharge comes from electricity generation accounting for about 335 mgd, or 96 per cent of the total discharge, all taking place to tidal rivers. The only other industrial discharge of significance is 10 mgd from food processing and manufacturing.

61 In the case of 335,000 gallons of industrial effluents requiring remedial action the authority have suggested that this might take the form of diverting to public sewers for treatment at a sewage treatment works.

62 For further details of industrial discharges in this area, see Table 194. ☐

8 Essex River Authority

Table 56 Numbers and volumes of discharges of industrial effluent (excluding mine water discharges, codes 52 to 55 inclusive)

Discharges	Rivers		
	Non-tidal	Tidal	All rivers
1 Total industrial effluent excluding solely cooling water			
Number	30	6	36
% satisfactory	40	0	39
2 Solely cooling water			
Number	8	1	9
% satisfactory	75	100	78
3 All discharges (1+2)			
Number	38	7	45
% satisfactory	53	14	47
4 Total industrial effluent excluding solely cooling water			
Volume x 1,000 gpd	9,761	1,485	11,246
% satisfactory	53	0	46
5 Process water (part of total effluent 4-6)			
Volume x 1,000 gpd	9,047	1,477	10,524
6 Cooling water (part of total effluent)			
Volume x 1,000 gpd	714	8	722
7 Solely cooling water			
Volume x 1,000 gpd	5,841	334,000	339,841
% satisfactory	99	100	99.9
8 Total cooling water (6+7)			
Volume x 1,000 gpd	6,555	334,008	340,563
9 Process water and cooling water (5+8)			
Total volume x 1,000 gpd	15,602	335,485	351,087

63 A total of 45 outlets, 21 of which are classed as satisfactory, discharge about 351 mgd of industrial effluent of which about 345 mgd is regarded as being satisfactory. Some 97 per cent of the total consists of cooling water.

64 Discharges from electricity generation amount to about 336 mgd (96 per cent of the total volume discharged) and take place mostly to tidal rivers. Of the remaining 13 industries covered, those with significant discharges are chemical and allied industries (6 mgd), quarrying and mining excluding coal mining (3 mgd) and potable water treatment (1 mgd).

65 In the case of about 1 mgd of industrial effluent requiring remedial action, the river authority have suggested that this might take the form of diverting to public sewers for treatment at a sewage treatment works.

66 For further details of industrial discharges in this area, see Table 195. ☐

9 Kent River Authority
(see Table 57)

67 Excluding mine waters a total of 70 outlets discharge about 574 mgd of industrial effluent, mainly to tidal rivers, with 55 outlets and about 458 mgd being regarded as satisfactory. Cooling water accounts for 90 per cent of the total discharge.

68 The discharges from 3 industries collectively add up to about 99 per cent of the total, electricity generation with 313 mgd, paper and board making with 107 mgd and petroleum refining with 150 mgd. About 26 per cent of all the effluent from paper and board making and some 25 per cent of all the effluent from petroleum refining is discharged in this river authority area.

Table 57 Numbers and volumes of discharges of industrial effluent (excluding mine water discharges, codes 52 to 55 inclusive)

Discharges	Rivers		
	Non-tidal	Tidal	All rivers
1 Total industrial effluent excluding solely cooling water			
Number	17	20	37
% satisfactory	71	65	62
2 Solely cooling water			
Number	21	12	33
% satisfactory	100	92	97
3 All discharges (1+2)			
Number	38	32	70
% satisfactory	87	69	79
4 Total industrial effluent excluding solely cooling water			
Volume x 1,000 gpd	2,612	69,679	72,291
% satisfactory	85	85	85
5 Process water (part of total effluent 4-6)			
Volume x 1,000 gpd	2,554	51,679	54,233
6 Cooling water (part of total effluent)			
Volume x 1,000 gpd	58	18,000	18,058
7 Solely cooling water			
Volume x 1,000 gpd	5,524	496,517	502,041
% satisfactory	100	79	79
8 Total cooling water (6+7)			
Volume x 1,000 gpd	5,582	514,517	520,099
9 Process water and cooling water (5+8)			
Total volume x 1,000 gpd	8,136	566,196	574,332

69 There is a discharge of about 2 mgd of mine water, mainly from active coal mines, to non-tidal rivers.

70 In the case of about 1.4 mgd of industrial effluent requiring remedial action, the river authority have suggested that this might take the form of diverting to public sewers for treatment at a sewage treatment works.

71 For further details of industrial discharges in this area, see Table 196. ☐

10 Sussex River Authority
(see Table 58 over page)

72 The quantity of industrial effluent discharged to watercourses is very small, amounting to only 194,000 gpd of which 100,000 gpd is classed as satisfactory. Out of 11 discharges, 4 are classed as satisfactory. Potable water treatment and plating and metal finishing account for most of the volume discharged.

73 In the case of about 76,000 gpd of industrial effluent requiring remedial action, the river authority have suggested that this might take the form of diverting to public sewers for treatment at a sewage treatment works.

74 For further details of industrial discharges in this area, see Table 197. ☐

(10 Sussex River Authority continued)

Table 58 Numbers and volumes of discharges of industrial effluent (excluding mine water discharges, codes 52 to 55 inclusive)

Discharges	Rivers		
	Non-tidal	Tidal	All rivers
1 Total industrial effluent excluding solely cooling water			
Number	10	1	11
% satisfactory	40	0	36
2 Solely cooling water			
Number	0	0	0
% satisfactory	—	—	—
3 All discharges (1+2)			
Number	10	1	11
% satisfactory	40	0	36
4 Total industrial effluent excluding solely cooling water			
Volume x 1,000 gpd	148	46	194
% satisfactory	70	0	53
5 Process water (part of total effluent 4-6)			
Volume x 1,000 gpd	148	46	194
6 Cooling water (part of total effluent)			
Volume x 1,000 gpd	0	0	0
7 Solely cooling water			
Volume x 1,000 gpd	0	0	0
% satisfactory	—	—	—
8 Total cooling water (6+7)			
Volume x 1,000 gpd	0	0	0
9 Process water and cooling water (5+8)			
Total volume x 1,000 gpd	148	46	194

11 Hampshire River Authority

Table 59 Numbers and volumes of discharges of industrial effluent (excluding mine water discharges, codes 52 to 55 inclusive)

Discharges	Rivers		
	Non-tidal	Tidal	All rivers
1 Total industrial effluent excluding solely cooling water			
Number	10	8	18
% satisfactory	40	75	55
2 Solely cooling water			
Number	5	3	8
% satisfactory	80	100	88
3 All discharges (1+2)			
Number	15	11	26
% satisfactory	53	82	65
4 Total industrial effluent excluding solely cooling water			
Volume x 1,000 gpd	3,401	165,855	169,256
% satisfactory	90	99.9	99.9
5 Process water (part of total effluent 4-6)			
Volume x 1,000 gpd	3,287	18,083	21,370
6 Cooling water (part of total effluent)			
Volume x 1,000 gpd	114	147,772	147,886
7 Solely cooling water			
Volume x 1,000 gpd	266	1,622,002	1,622,268
% satisfactory	62	100	100
8 Total cooling water (6+7)			
Volume x 1,000 gpd	380	1,769,774	1,770,154
9 Process water and cooling water (5+8)			
Total volume x 1,000 gpd	3,667	1,787,857	1,791,524

75 There are 26 discharges of industrial effluent recorded of which 17 are satisfactory. The total volume discharged is 1,792 mgd of which over 99 per cent is satisfactory. Cooling water accounts for 91 per cent of the total which is the second highest quantity discharged in a river authority area.

76 The largest volume comes from electricity generation, amounting to 1,622 mgd and originates from 2 discharges only with outfalls to tidal rivers. The remaining 170 mgd is very largely made up of effluent from petroleum refining with about 151 mgd (138 mgd of which is cooling water), chemical and allied industries with about 15 mgd (10 mgd of which is cooling water) and paper and board making with 3 mgd (largely process water). About 25 per cent of all the effluent from petroleum refining is discharged in this river authority area.

77 In the case of about 17,000 gpd of industrial effluent requiring remedial action, the river authority have suggested that this might take the form of diverting to public sewers for treatment at a sewage treatment works.

78 For further details of industrial discharges in this area, see Table 198.

12 Isle of Wight River and Water Authority

Table 60 Numbers and volumes of discharges of industrial effluent (excluding mine water discharges, codes 52 to 55 inclusive)

Discharges	Rivers		
	Non-tidal	Tidal	All rivers
1 Total industrial effluent excluding solely cooling water			
Number	6	1	7
% satisfactory	33	100	43
2 Solely cooling water			
Number	0	1	1
% satisfactory	—	100	100
3 All discharges (1+2)			
Number	6	2	8
% satisfactory	33	100	50
4 Total industrial effluent excluding solely cooling water			
Volume x 1,000 gpd	177	20	197
% satisfactory	83	100	85
5 Process water (part of total effluent 4-6)			
Volume x 1,000 gpd	176	20	196
6 Cooling water (part of total effluent)			
Volume x 1,000 gpd	1	0	1
7 Solely cooling water			
Volume x 1,000 gpd	0	30,000	30,000
% satisfactory	—	100	100
8 Total cooling water (6+7)			
Volume x 1,000 gpd	1	30,000	30,001
9 Process water and cooling water (5+8)			
Total volume x 1,000 gpd	177	30,020	30,197

79 Apart from electricity generation, which accounts for over 99 per cent of the total discharge of about 30 mgd, the quantity of industrial effluent discharged is very small. Most of the effluent is discharged to tidal waters and with one or two minor exceptions, is satisfactory.

80 In the case of 15,000 gpd of industrial effluent requiring remedial action, the river authority have suggested that this might take the form of diverting to public sewers for treatment at a sewage treatment works.

81 For further details of industrial discharges in this area, see Table 199.

13 Avon and Dorset River Authority
(see Table 61)

82 Of 11 discharges recorded, 5 are satisfactory and of some 43 mgd of industrial effluent from them, 98 per cent is satisfactory. Most of the effluent is discharged to non-tidal rivers.

83 Electricity generation effluent accounts for 95 per cent of the total and is of special interest because it is process water used in

Table 61 Numbers and volumes of discharges of industrial effluent (excluding mine water discharges, codes 52 to 55 inclusive)

Discharges	Rivers		
	Non-tidal	Tidal	All rivers
1 Total industrial effluent excluding solely cooling water			
Number	9	1	10
% satisfactory	44	0	40
2 Solely cooling water			
Number	1	0	1
% satisfactory	100	—	100
3 All discharges (1+2)			
Number	10	1	11
% satisfactory	50	0	45
4 Total industrial effluent excluding solely cooling water			
Volume x 1,000 gpd	41,687	800	42,487
% satisfactory	99.8	0	98
5 Process water (part of total effluent 4-6)			
Volume x 1,000 gpd	41,230	640	41,870
6 Cooling water (part of total effluent)			
Volume x 1,000 gpd	457	160	617
7 Solely cooling water			
Volume x 1,000 gpd	500	0	500
% satisfactory	100	—	100
8 Total cooling water (6+7)			
Volume x 1,000 gpd	957	160	1,117
9 Process water and cooling water (5+8)			
Total volume x 1,000 gpd	42,187	800	42,987

driving turbine generators. Water treatment and food processing and manufacturing together very largely make up the remaining discharge of 2 mgd.

84 In the case of 11,000 gpd of industrial effluent requiring remedial action, the river authority have suggested that this might take the form of diverting to public sewers for treatment at a sewage treatment works.

85 For further details of industrial discharges in this area, see Table 200. □

14 Devon River Authority

Table 62 Numbers and volumes of discharges of industrial effluent (excluding mine water discharges, codes 52 to 55 inclusive)

Discharges	Rivers		
	Non-tidal	Tidal	All rivers
1 Total industrial effluent excluding solely cooling water			
Number	54	5	59
% satisfactory	30	20	29
2 Solely cooling water			
Number	20	1	21
% satisfactory	90	100	90
3 All discharges (1+2)			
Number	74	6	80
% satisfactory	46	33	45
4 Total industrial effluent excluding solely cooling water			
Volume x 1,000 gpd	16,469	1,314	17,783
% satisfactory	61	15	58
5 Process water (part of total effluent 4-6)			
Volume x 1,000 gpd	15,875	1,314	17,189
6 Cooling water (part of total effluent)			
Volume x 1,000 gpd	594	0	594
7 Solely cooling water			
Volume x 1,000 gpd	5,166	160,000	165,166
% satisfactory	82	100	99
8 Total cooling water (6+7)			
Volume x 1,000 gpd	5,760	160,000	165,760
9 Process water and cooling water (5+8)			
Total volume x 1,000 gpd	21,635	161,314	182,949

86 A total of 80 outlets discharge about 183 mgd of industrial effluent and 36 outlets and about 175 mgd are considered to be satisfactory. Cooling water accounts for 91 per cent of the total.

87 Electricity generation accounts for by far the largest volume of some 168 mgd, equal to 91 per cent of the total and mostly discharged to tidal waters. Of the remaining 15 mgd, food processing and manufacturing contributes 5 mgd, paper and board making, 5 mgd and quarrying and mining, 3 mgd. Most of the discharges, other than that from electricity generation, are to non-tidal rivers.

88 In the case of 359,000 gpd of industrial effluent requiring remedial action, the river authority have suggested that this might take the form of diverting to public sewers for treatment at a sewage treatment works.

89 For further details of industrial discharges in this area, see Table 201. □

15 Cornwall River Authority

Table 63 Numbers and volumes of discharges of industrial effluent (excluding mine water discharges, codes 52 to 55 inclusive)

Discharges	Rivers		
	Non-tidal	Tidal	All rivers
1 Total industrial effluent excluding solely cooling water			
Number	93	8	101
% satisfactory	38	38	37
2 Solely cooling water			
Number	2	3	5
% satisfactory	100	100	100
3 All discharges (1+2)			
Number	95	11	106
% satisfactory	40	55	42
4 Total industrial effluent excluding solely cooling water			
Volume x 1,000 gpd	113,601	51,921	165,522
% satisfactory	63	42	57
5 Process water (part of total effluent 4-6)			
Volume x 1,000 gpd	111,466	51,541	163,007
6 Cooling water (part of total effluent)			
Volume x 1,000 gpd	2,135	380	2,515
7 Solely cooling water			
Volume x 1,000 gpd	444	22,800	23,244
% satisfactory	100	100	100
8 Total cooling water (6+7)			
Volume x 1,000 gpd	2,579	23,180	25,759
9 Process water and cooling water (5+8)			
Total volume x 1,000 gpd	114,045	74,721	188,766

90 Of 106 discharges and about 189 mgd of industrial effluent discharged, excluding mine waters, 44 discharges and about 118 mgd are regarded as satisfactory. About 26 mgd (14 per cent) of the total discharge consists of cooling water. The volume of process water (163 mgd) is one of the largest volumes discharged in a river authority area, but this includes 69 mgd of turbine water and it will be noted that in Table 47 this is entered separately.

91 Electricity generation is responsible for the largest volume with nearly 92 mgd, 53 per cent discharging to non-tidal waters, followed by discharges of over 65 mgd from quarrying and mining (excluding coal mining) and 27 mgd from the chemical and allied industries. Nearly a half of all the industrial effluent from quarrying and mining in England and Wales is discharged in this river authority area.

92 Some 18 mgd of mine water is discharged mainly to non-tidal rivers, with about 14 mgd originating from active and 4 mgd from derelict mines, other than coal mines. A scheme to deal with part or

all of the industrial effluent from quarrying and mining is under consideration.

93 For further details of industrial discharges in this area, see Table 202. □

16 Somerset River Authority

Table 64 Numbers and volumes of discharges of industrial effluent (excluding mine water discharges, codes 52 to 55 inclusive)

Discharges	Rivers		
	Non-tidal	Tidal	All rivers
1 Total industrial effluent excluding solely cooling water			
Number	127	7	134
% satisfactory	2	14	2
2 Solely cooling water			
Number	0	0	0
% satisfactory	—	—	—
3 All discharges (1+2)			
Number	127	7	134
% satisfactory	2	14	2
4 Total industrial effluent excluding solely cooling water			
Volume x 1,000 gpd	7,682	4,745	12,427
% satisfactory	1	53	21
5 Process water (part of total effluent 4-6)			
Volume x 1,000 gpd	4,682	4,745	9,427
6 Cooling water (part of total effluent)			
Volume x 1,000 gpd	3,000	0	3,000
7 Solely cooling water			
Volume x 1,000 gpd	0	0	0
% satisfactory	—	—	—
8 Total cooling water (6+7)			
Volume x 1,000 gpd	3,000	0	3,000
9 Process water and cooling water (5+8)			
Total volume x 1,000 gpd	7,682	4,745	12,427

94 A total of 134 discharges from 7 categories of industry discharge about 12 mgd of industrial effluent; only 3 mgd from 3 discharge points are regarded as being satisfactory. Of the total volume discharged, 24 per cent consists of cooling water.

95 The largest volume comes from paper and board making, with about 5 mgd, and the only other discharges of consequence are from chemical and allied industries (2.5 mgd), textile manufacturing, cotton and man-made fibres, (2.5 mgd) and general farming (1 mgd).

96 In the case of 105,000 gpd of industrial effluent requiring remedial action, the river authority have suggested that this might take the form of diverting to public sewers for treatment at a sewage treatment works.

97 For further details of industrial discharges in this area, see Table 203. □

17 Bristol Avon River Authority
(see Table 65)

98 Excluding mine waters there are 43 discharges of industrial effluent recorded from 11 categories of industry and 22 are classed as satisfactory. The total volume discharged amounts to almost 80 mgd of which about 55 mgd is satisfactory. About 71 per cent of the total volume discharged consists of cooling water.

99 The largest volume discharged, about 43 mgd, comes from paper and board making. Other significant discharges are from food processing (about 18 mgd), rubber processing (9 mgd), chemical and allied industries (about 3 mgd) and brewing (2 mgd). About 27 per

Table 65 Numbers and volumes of discharges of industrial effluent (excluding mine water discharges, codes 52 to 55 inclusive)

Discharges	Rivers		
	Non-tidal	Tidal	All rivers
1 Total industrial effluent excluding solely cooling water			
Number	19	6	25
% satisfactory	26	33	28
2 Solely cooling water			
Number	16	2	18
% satisfactory	88	50	83
3 All discharges (1+2)			
Number	35	8	43
% satisfactory	54	38	51
4 Total industrial effluent excluding solely cooling water			
Volume x 1,000 gpd	51,182	2,925	54,107
% satisfactory	59	37	57
5 Process water (part of total effluent 4-6)			
Volume x 1,000 gpd	21,259	2,005	23,264
6 Cooling water (part of total effluent)			
Volume x 1,000 gpd	29,923	920	30,843
7 Solely cooling water			
Volume x 1,000 gpd	15,481	10,300	25,781
% satisfactory	93	97	95
8 Total cooling water (6+7)			
Volume x 1,000 gpd	45,404	11,220	56,624
9 Process water and cooling water (5+8)			
Total volume x 1,000 gpd	66,663	13,225	79,888

cent of the effluent from the brewing industry in England and Wales, containing as is normal about 50 per cent of cooling water, is discharged in this river authority area.

100 Some 600,000 gpd of mine water, mainly from active coal mines, is discharged to non-tidal waters.

101 In the case of 54 mgd of industrial effluent requiring remedial action, the river authority have suggested that this might take the form of diverting to public sewers for treatment at a sewage treatment works.

102 For further details of industrial effluent in this area, see Table 204. □

18 Severn River Authority
(see Table 66)

103 There are 163 discharges from 22 categories of industry, excluding mine waters, amounting to almost 1,525 mgd with 109 discharges and 1,512 mgd regarded as being satisfactory. Cooling water accounts for 1,496 mgd or 98 per cent of the total discharge.

104 Electricity generation accounts for about 1,492 mgd, equal to 98 per cent of the total volume discharged. Of the remaining 33 mgd the principal discharges are from food processing and manufacture (9 mgd), the chemical and allied industries (8 mgd), iron and steel (5 mgd), engineering (4 mgd), water treatment (3 mgd) and quarrying and mining, excluding coal mining (2.5 mgd).

105 There is over 6 mgd of mine water discharged to non-tidal waters, of which 5 mgd is from derelict coal mines.

106 In the case of 586,000 gpd of industrial effluent requiring remedial action, the river authority have suggested that this might take the form of diverting to public sewers for treatment at a sewage treatment works.

107 For further details of industrial discharges in this area, see Table 205. □

Table 66 Numbers and volumes of discharges of industrial effluent (excluding mine water discharges, codes 52 to 55 inclusive)

Discharges	Rivers			Canals	All rivers and canals
	Non-tidal	Tidal	All rivers		
1 Total industrial effluent excluding solely cooling water					
Number	85	8	93	8	101
% satisfactory	52	88	55	38	53
2 Solely cooling water					
Number	48	2	50	12	62
% satisfactory	90	100	90	92	92
3 All discharges (1+2)					
Number	133	10	143	20	163
% satisfactory	68	90	67	70	67
4 Total industrial effluent excluding solely cooling water					
Volume x 1,000 gpd	29,534	10,857	40.391	5,419	45,810
% satisfactory	81	99.9	86	2	76
5 Process water (part of total effluent 4-6)					
Volume x 1,000 gpd	15,120	8,877	23,997	4,586	28,583
6 Cooling water (part of total effluent)					
Volume x 1,000 gpd	14,414	1,980	16,394	833	17,227
7 Solely cooling water					
Volume x 1,000 gpd	470,616	1,004,000	1,474,616	4,271	1,478,887
% satisfactory	99.9	100	99.9	100	99.9
8 Total cooling water (6+7)					
Volume x 1,000 gpd	485,030	1,005,980	1,491,010	5,104	1,496,114
9 Process water and cooling water (5+8)					
Total volume x 1,000 gpd	500,150	1,014,857	1,515,007	9.690	1,524,697

19 Wye River Authority

Table 67 Numbers and volumes of discharges of industrial effluent (excluding mine water discharges, codes 52 to 55 inclusive)

Discharges	Rivers		
	Non-tidal	Tidal	All rivers
1 Total industrial effluent excluding solely cooling water			
Number	6	0	6
% satisfactory	50	—	50
2 Solely cooling water			
Number	2	0	2
% satisfactory	100	—	100
3 All discharges (1+2)			
Number	8	0	8
% satisfactory	63	—	63
4 Total industrial effluent excluding solely cooling water			
Volume x 1,000 gpd	1,236	0	1,236
% satisfactory	90	—	90
5 Process water (part of total effluent 4-6)			
Volume x 1,000 gpd	1,236	0	1,236
6 Cooling water (part of total effluent)			
Volume x 1,000 gpd	0	0	0
7 Solely cooling water			
Volume x 1,000 gpd	2,500	0	2,500
% satisfactory	100	—	100
8 Total cooling water (6+7)			
Volume x 1,000 gpd	2,500	0	2,500
9 Process water and cooling water (5+8)			
Total volume x 1,000 gpd	3,736	0	3,736

108 Of 3.7 mgd discharged from 8 outlets, 3.6 mgd and 5 outlets are regarded as being satisfactory.

109 The largest volume comes from the food processing and manufacturing industry and this, together with discharges from brewing and water treatment, account for 99 per cent of the total.

110 For further details of industrial discharges in this area, see Table 206. ☐

20 Usk River Authority

Table 68 Numbers and volumes of discharges of industrial effluent (excluding mine water discharges, codes 52 to 55 inclusive)

Discharges	Rivers		
	Non-tidal	Tidal	All rivers
1 Total industrial effluent excluding solely cooling water			
Number	36	5	41
% satisfactory	30	40	29
2 Solely cooling water			
Number	5	2	7
% satisfactory	80	100	86
3 All discharges (1+2)			
Number	41	7	48
% satisfactory	36	57	39
4 Total industrial effluent excluding solely cooling water			
Volume x 1,000 gpd	14,324	1,625	15,949
% satisfactory	8	17	9
5 Process water (part of total effluent 4-6)			
Volume x 1,000 gpd	13,869	1,625	15,494
6 Cooling water (part of total effluent)			
Volume x 1,000 gpd	455	0	455
7 Solely cooling water			
Volume x 1,000 gpd	2,402	195,999	198,401
% satisfactory	99	100	99.9
8 Total cooling water (6+7)			
Volume x 1,000 gpd	2,857	195,999	198,856
9 Process water and cooling water (5+8)			
Total volume x 1,000 gpd	16,726	197,624	214,350

111 Excluding mine waters there are 48 discharges of industrial effluent of which 19 are satisfactory and about 200 mgd out of 214 mgd discharged is considered to be satisfactory. Cooling water accounts for 93 per cent of the total volume.

112 Of the 10 industrial classes recorded, the largest discharge of 197 mgd comes from electricity generation. The only other discharges of significance are those from the iron and steel industry, totalling about 12 mgd.

113 There is about 9 mgd of mine water discharged, mainly from active coal mines, to non-tidal waters.

114 In the case of 630,000 gpd of industrial effluent requiring remedial action, the river authority have suggested that this might take the form of diverting to public sewers for treatment at a sewage treatment works.

115 There are known to be several large discharges of industrial effluent to the Severn Estuary, which are not covered by the survey.

116 For further details of industrial discharges in this area, see Table 207. ☐

21 Glamorgan River Authority
(see Table 69)

117 Excluding mine waters there are 72 outlets discharging a total of 169 mgd of industrial effluent, with 46 outlets and about 150 mgd of effluent regarded as being satisfactory. The discharges contain some 119 mgd of cooling water.

118 Of the 13 industries covered, the largest volume of nearly 100 mgd comes from electricity generation. Of the remainder, significant discharges come from the iron and steel industry (26 mgd), paper and board making (13 mgd), coal mining (12 mgd), gas and coke making (8 mgd), glue and gelatine manufacture (3.5 mgd) and chemical and allied industries (3 mgd).

119 There is nearly 9 mgd of mine water discharged, including almost 7 mgd from active coal mines, all to non-tidal rivers.

120 In the case of 12 mgd of industrial effluent requiring remedial action, the river authority have suggested that this might take the form of diverting to public sewers for treatment at a sewage treatment works and, in addition, it is proposed to divert 220,000 gpd of effluent to a new sea outfall.

121 There are known to be several large discharges of industrial effluent to the Bristol Channel, which are not covered by the survey.

122 For further details of industrial discharges in this area, see Table 208. ☐

Table 69 Numbers and volumes of discharges of industrial effluent (excluding mine water discharges, codes 52 to 55 inclusive)

Discharges	Rivers		
	Non-tidal	Tidal	All rivers
1 Total industrial effluent excluding solely cooling water			
Number	59	6	65
% satisfactory	64	33	62
2 Solely cooling water			
Number	5	2	7
% satisfactory	100	50	86
3 All discharges (1+2)			
Number	64	8	72
% satisfactory	67	38	64
4 Total industrial effluent excluding solely cooling water			
Volume x 1,000 gpd	33,779	16,113	49,892
% satisfactory	76	30	61
5 Process water (part of total effluent 4-6)			
Volume x 1,000 gpd	33,779	16,113	49,892
6 Cooling water (part of total effluent)			
Volume x 1,000 gpd	0	0	0
7 Solely cooling water			
Volume x 1,000 gpd	100,224	19,060	119,284
% satisfactory	100	98	99.9
8 Total cooling water (6+7)			
Volume x 1,000 gpd	100,224	19,060	119,284
9 Process water and cooling water (5+8)			
Total volume x 1,000 gpd	134,003	35,173	169,176

22 South West Wales River Authority

Table 70 Numbers and volumes of discharges of industrial effluent (excluding mine water discharges, codes 52 to 55 inclusive)

Discharges	Rivers			Canals	All rivers and canals
	Non-tidal	Tidal	All rivers		
1 Total industrial effluent excluding solely cooling water					
Number	43	13	56	1	57
% satisfactory	58	69	60	0	60
2 Solely cooling water					
Number	8	1	9	3	12
% satisfactory	100	100	100	100	100
3 All discharges (1+2)					
Number	51	14	65	4	69
% satisfactory	65	71	66	75	67
4 Total industrial effluent excluding solely cooling water					
Volume x 1,000 gpd	97,561	14,638	112,199	30	112,229
% satisfactory	95	94	95	0	95
5 Process water (part of total effluent 4-6)					
Volume x 1,000 gpd	97,561	14,488	112,049	30	112,079
6 Cooling water (part of total effluent)					
Volume x 1,000 gpd	0	150	150	0	150
7 Solely cooling water					
Volume x 1,000 gpd	4,060	18,900	22,960	1,314	24,274
% satisfactory	100	100	100	100	100
8 Total cooling water (6+7)					
Volume x 1,000 gpd	4,060	19,050	23,110	1,314	24,424
9 Process water and cooling water (5+8)					
Total volume x 1,000 gpd	101,621	33,538	135,159	1,344	136,503

123 There are 69 discharges of industrial effluent, amounting to about 137 mgd, with 46 discharges and about 131 mgd regarded as satisfactory, excluding mine waters.

124 Of the 16 industries covered, the largest volume of nearly 106 mgd comes from electricity generation and it is of interest that this is process water used in driving turbine generators. (It should be noted

that in Table 47, this quantity of water is entered separately). Of the remaining 15 industries represented, significant discharges come from metal smelting (10 mgd), petroleum refining (7 mgd), iron and steel making (7 mgd) and food processing (4 mgd).

125 About 15 mgd of mine water is discharged, 13 mgd from derelict mines other than coal mines and about 2 mgd from active coal mines. (In respect of certain of these discharges the river authority have applied for an Order under section 2 of the 1951 Act).

126 In the case of about 250,000 gpd of industrial effluent requiring remedial action, the river authority have suggested that this might take the form of diverting to public sewers for treatment at a sewage treatment works.

127 For further details of industrial discharges in this area, see Table 209.

23 Gwynedd River Authority
(see Table 71)

128 There is only one industrial discharge and it is proposed to divert this to a sea outfall. For further details see Table 210.

Table 71 Numbers and volumes of discharges of industrial effluent (excluding mine water discharges, codes 52 to 55 inclusive)

Discharges	Rivers		
	Non-tidal	Tidal	All rivers
1 Total industrial effluent excluding solely cooling water			
Number	1	0	1
% satisfactory	0	—	0
2 Solely cooling water			
Number	0	0	0
% satisfactory	—	—	—
3 All discharges (1+2)			
Number	1	0	1
% satisfactory	0	—	0
4 Total industrial effluent excluding solely cooling water			
Volume x 1,000 gpd	26	0	26
% satisfactory	0	—	0
5 Process water (part of total effluent 4-6)			
Volume x 1,000 gpd	26	0	26
6 Cooling water (part of total effluent)			
Volume x 1,000 gpd	0	0	0
7 Solely cooling water			
Volume x 1,000 gpd	0	0	0
% satisfactory	—	—	—
8 Total cooling water (6+7)			
Volume x 1,000 gpd	0	0	0
9 Process water and cooling water (5+8)			
Total volume x 1,000 gpd	26	0	26

24 Dee and Clwyd River Authority

Table 72 Numbers and volumes of discharges of industrial effluent (excluding mine water discharges, codes 52 to 55 inclusive)

Discharges	Rivers			Canals	All rivers and canals
	Non-tidal	Tidal	All rivers		
1 Total industrial effluent excluding solely cooling water					
Number	48	7	55	0	55
% satisfactory	54	43	53	—	53
2 Solely cooling water					
Number	7	1	8	1	9
% satisfactory	100	0	88	100	88
3 All discharges (1+2)					
Number	55	8	63	1	64
% satisfactory	60	38	57	100	58
4 Total industrial effluent excluding solely cooling water					
Volume x 1,000 gpd	5,066	48,644	53,710	0	53,710
% satisfactory	61	81	79	—	79
5 Process water (part of total effluent 4-6)					
Volume x 1,000 gpd	4,636	45,351	49,987	0	49,987
6 Cooling water (part of total effluent)					
Volume x 1,000 gpd	430	3,293	3,723	0	3,723
7 Solely cooling water					
Volume x 1,000 gpd	1,437	50	1,487	3	1,490
% satisfactory	100	0	97	100	97
8 Total cooling water (6+7)					
Volume x 1,000 gpd	1,867	3,343	5,210	3	5,213
9 Process water and cooling water (5+8)					
Total volume x 1,000 gpd	6,503	48,694	55,197	3	55,200

129 There are 64 outlets discharging over 55 mgd of industrial effluent, excluding mine waters, with 37 outlets and nearly 44 mgd regarded as being satisfactory. Cooling water accounts for only about 5 mgd or less than 10 per cent of the total industrial flow, very much less than with most other river authorities.

130 Although discharges are recorded from 15 industries, about 90 per cent of the total volume comes from only three of these—iron and steel making (38 mgd), textile manufacture, cotton and man-made fibres, (9 mgd) and chemical and allied industries (2 mgd).

131 Some 276,000 gpd of mine water, evenly divided between active coal mines and active mines other than coal mines, is discharged to non-tidal waters.

132 In the case of 40,000 gpd requiring remedial action, the river authority have suggested that this might take the form of diverting to public sewers for treatment at a sewage treatment works. Furthermore, a scheme for providing an additional potable water supply, which involves constructing a Dee barrage, is now under consideration and if such a scheme were to go ahead, about 46 mgd of industrial effluent would need to be diverted to a new outfall in the estuary below the barrage.

133 For further details of industrial discharges in this area, see Table 211.

25 Mersey and Weaver River Authority
(see Table 73)

134 Excluding mine waters there are 473 outlets discharging about 1,873 mgd of industrial effluent and 223 discharges and 1,598 mgd are regarded as being satisfactory; 1,694 mgd of cooling water accounts for 90 per cent of the total and of this, 1,503 mgd consists of discharges of solely cooling water. These large quantities tend to mask what is of more significance, namely that nearly 178 mgd of process water (industrial effluent without any cooling water content) is discharged in the area. More process water is discharged to controlled waters in the Mersey and Weaver RA area than in any other.

135 The largest volume is from electricity generation, with 1,310 mgd (70 per cent of the total), but of importance are the discharges from the chemical and allied industries (198 mgd), petroleum refining (176 mgd), paper and board making (61 mgd), iron and steel making (37 mgd), textile manufacture, cotton and man-made fibres, (23 mgd), food processing (16 mgd), general manufacturing (12 mgd), soap and detergent manufacture (10 mgd), engineering (9 mgd), rubber processing (4 mgd) and glue and gelatine manufacture (3 mgd). In all there are 30 different classes of industrial discharges. The river authority area receives discharges from a number of industries which are a significant proportion of the total volumes for England and Wales. Amongst these are the chemical and allied industries (30 per cent), glass making (25 per cent), glue and gelatine manufacture (31 per cent), general manufacturing (58 per cent), leather tanning and fellmongering (29 per cent), petroleum refining (29 per cent), soap and detergent manufacture (74 per cent) and textile manufacture, cotton and man-made fibres, (31 per cent).

136 Over 129 mgd of industrial effluent (including nearly 124 mgd of cooling water) is discharged to canals, which is the largest volume in any river authority area and over double the next largest total. The industries mainly responsible are electricity generation (98 mgd), the chemical and allied industries (about 9 mgd), engineering (7 mgd) and iron and steel making (over 6 mgd). It should be noted that, for the purpose of the survey, the Manchester Ship Canal is considered to be a part of the river systems of the Mersey and the Irwell and details of the many discharges of industrial effluent to it are recorded under the details for rivers.

137 Some 5 mgd of mine water is discharged, 3.5 mgd from active and 1.5 mgd from derelict coal mines, mainly to non-tidal waters.

138 In the case of about 25 mgd of industrial effluent requiring remedial action, the river authority have suggested that this might take the form of diverting to public sewers for treatment at a sewage treatment works.

139 For further details of industrial discharges in this area, see Table 212. ☐

Table 73 Numbers and volumes of discharges of industrial effluent (excluding mine water discharges, codes 52 to 55 inclusive)

Discharges	Rivers			Canals	All rivers and canals
	Non-tidal	Tidal	All rivers		
1 Total industrial effluent excluding solely cooling water					
Number	247	86	333	11	344
% satisfactory	38	24	34	36	34
2 Solely cooling water					
Number	85	15	100	29	129
% satisfactory	82	67	80	90	82
3 All discharges (1+2)					
Number	332	101	433	40	473
% satisfactory	49	31	45	75	47
4 Total industrial effluent excluding solely cooling water					
Volume x 1,000 gpd	155,847	208,196	364,043	5,638	369,681
% satisfactory	23	78	55	15	54
5 Process water (part of total effluent 4-6)					
Volume x 1,000 gpd	111,268	62,577	173,845	4,625	178,470
6 Cooling water (part of total effluent)					
Volume x 1,000 gpd	44,579	145,619	190,198	1,013	1,991.211
7 Solely cooling water					
Volume x 1,000 gpd	740,188	638,984	1,379,172	123,791	1,502,963
% satisfactory	87	99	92	99	93
8 Total cooling water (6+7)					
Volume x 1,000 gpd	784,767	784,603	1,569,370	124,804	1,694,174
9 Process water and cooling water (5+8)					
Total volume x 1,000 gpd	896,035	847,180	1,743,215	129,429	1,872,644

26 Lancashire River Authority
(see Table 74)

140 A total of 130 outlets, 68 of which are classed as satisfactory, discharge nearly 203 mgd of industrial effluent excluding mine waters. Of this total 174 mgd is regarded as being satisfactory. About 84 per cent of the total volume consists of cooling water.

141 There are 18 different categories of industry represented in the survey, the largest quantity coming from electricity generation with 146 mgd (72 per cent of the total). Other significant discharges are those from paper and board making (18 mgd), textile manufacture, cotton and man-made fibres, (15 mgd) and the chemical and allied industries (12 mgd). Nearly 20 per cent of all textile manufacture, cotton and man-made fibre, effluents and 26 per cent of all disposal tip effluents for England and Wales are discharged in this river authority area.

142 About 3 mgd of mine water, including 1.8 mgd from derelict coal mines is discharged to non-tidal waters.

143 In the case of about 13 mgd of industrial effluent requiring remedial action, the river authority have suggested that this might take the form of diverting to public sewers for treatment at a sewage treatment works.

144 For further details of industrial discharges in this area, see Table 213. ☐

Table 74 Numbers and volumes of discharges of industrial effluent (excluding mine water discharges, codes 52 to 55 inclusive)

Discharges	Rivers			Canals	All rivers and canals
	Non-tidal	Tidal	All rivers		
1 Total industrial effluent excluding solely cooling water					
Number	72	13	85	0	85
% satisfactory	32	31	32	—	32
2 Solely cooling water					
Number	25	9	34	11	45
% satisfactory	96	100	97	73	91
3 All discharges (1+2)					
Number	97	22	119	11	130
% satisfactory	48	59	50	73	52
4 Total industrial effluent excluding solely cooling water					
Volume x 1,000 gpd	20,173	16,030	36,203	0	36,203
% satisfactory	12	28	22	—	22
5 Process water (part of total effluent 4-6)					
Volume x 1,000 gpd	19,215	12,425	31,640	0	31,640
6 Cooling water (part of total effluent)					
Volume x 1,000 gpd	958	3,605	4,563	0	4,563
7 Solely cooling water					
Volume x 1,000 gpd	39,793	123,275	163,068	3,497	166,565
% satisfactory	99	100	99.7	63	99.7
8 Total cooling water (6+7)					
Volume x 1,000 gpd	40,751	126,880	167,631	3,497	171,128
9 Process water and cooling water (5+8)					
Total volume x 1,000 gpd	59,966	139,305	199,271	3,497	202,768

27 Cumberland River Authority

Table 75 Numbers and volumes of discharges of industrial effluent (excluding mine water discharges, codes 52 to 55 inclusive)

Discharges	Rivers		
	Non-tidal	Tidal	All rivers
1 Total industrial effluent excluding solely cooling water			
Number	23	1	24
% satisfactory	35	0	33
2 Solely cooling water			
Number	6	0	6
% satisfactory	100	—	100
3 All discharges (1+2)			
Number	29	1	30
% satisfactory	48	0	47
4 Total industrial effluent excluding solely cooling water			
Volume x 1,000 gpd	2,282	720	3,002
% satisfactory	50	0	38
5 Process water (part of total effluent 4-6)			
Volume x 1,000 gpd	2,282	720	3,002
6 Cooling water (part of total effluent)			
Volume x 1,000 gpd	0	0	0
7 Solely cooling water			
Volume x 1,000 gpd	81,657	0	81,657
% satisfactory	100	—	100
8 Total cooling water (6+7)			
Volume x 1,000 gpd	81,657	0	81,657
9 Process water and cooling water (5+8)			
Total volume x 1,000 gpd	83,939	720	84,659

145 Of the 30 discharges and nearly 85 mgd of effluent, 14 discharges and almost 83 mgd are classed as satisfactory, excluding mine waters.

146 Electricity generation accounts for the largest volume with about 78 mgd, equal to 92 per cent of the total. The only other discharges of significance are those from food processing and manufacture (4 mgd) and chemical and allied industries (2 mgd).

147 There is over 2 mgd of mine water discharged from active mines other than coal mines to non-tidal waters in the area.

148 In the case of about 751,000 gpd of industrial effluent requiring remedial action, the river authority have suggested that this might take the form of diverting to public sewers for treatment at a sewage treatment works.

149 For further details of industrial discharges in this area, see Table 214. □

28 Thames Conservancy Catchment Board
(see Table 76 over page)

150 A total of 177 outlets, 163 of which are classed as satisfactory, discharge over 313 mgd of industrial effluent and of this total 257 mgd consists of cooling water. Nearly 282 mgd of the total volume discharged is regarded as being satisfactory.

151 Of the 22 different categories of industry represented, electricity generation accounts for over 186 mgd followed by paper and board making (55 mgd), quarrying and mining, excluding coal mining (23 mgd), water treatment (18 mgd), food processing and manufacture (7 mgd), cement making (6 mgd), engineering (5 mgd), general manufacturing (3 mgd) and gas and coke making (3 mgd). The river authority area receives discharges of effluents from a number of industries which make up a significant proportion of the total volume for England and Wales. Amongst these industries are cement making, (64 per cent), printing ink and printing industry (87 per cent), all the atomic energy establishments, and water treatment (41 per cent).

152 About 4 mgd of industrial effluent is discharged to canals.

153 In the case of 189,000 gpd of industrial effluent requiring remedial action, the river authority have suggested that this might take the form of diverting to public sewers for treatment at a sewage treatment works.

154 For further details of industrial discharges in this area, see Table 215. □

(28 Thames Conservancy Catchment Board—continued)

Table 76 Numbers and volumes of discharges of industrial effluent (excluding mine water discharges, codes 52 to 55 inclusive)

Discharges	Rivers			Canals	All rivers and canals
	Non-tidal	Tidal	All rivers		
1 Total industrial effluent excluding solely cooling water					
Number	90	0	90	3	93
% satisfactory	84	—	84	100	85
2 Solely cooling water					
Number	75	0	75	9	84
% satisfactory	95	—	95	100	95
3 All discharges (1+2)					
Number	165	0	165	12	177
% satisfactory	92	—	92	100	92
4 Total industrial effluent excluding solely cooling water					
Volume x 1,000 gpd	86,310	0	86,310	695	87,005
% satisfactory	64	—	64	100	64
5 Process water (part of total effluent 4-6)					
Volume x 1,000 gpd	55,668	0	55,668	695	56,363
6 Cooling water (part of total effluent)					
Volume x 1,000 gpd	30,642	0	30,642	0	30,642
7 Solely cooling water					
Volume x 1,000 gpd	223,315	0	223,315	3,068	226,383
% satisfactory	99.8	—	99.8	100	99.8
8 Total cooling water (6+7)					
Volume x 1,000 gpd	253,957	0	253,957	3,068	257,025
9 Process water and cooling water (5+8)					
Total volume x 1,000 gpd	309,625	0	309,625	3,763	313,388

29 Lee Conservancy Catchment Board

Table 77 Numbers and volumes of discharges of industrial effluent (excluding mine water discharges, codes 52 to 55 inclusive)

Discharges	Rivers		
	Non-tidal	Tidal	All rivers
1 Total industrial effluent excluding solely cooling water			
Number	1	0	1
% satisfactory	100	—	100
2 Solely cooling water			
Number	21	0	21
% satisfactory	100	—	100
3 All discharges (1+2)			
Number	22	0	22
% satisfactory	100	—	100
4 Total industrial effluent excluding solely cooling water			
Volume x 1,000 gpd	500	0	500
% satisfactory	100	—	100
5 Process water (part of total effluent 4-6)			
Volume x 1,000 gpd	50	0	50
6 Cooling water (part of total effluent)			
Volume x 1,000 gpd	450	0	450
7 Solely cooling water			
Volume x 1,000 gpd	122,304	0	122,304
% satisfactory	100	—	100
8 Total cooling water (6+7)			
Volume x 1,000 gpd	122,754	0	122,754
9 Process water and cooling water (5+8)			
Total volume x 1,000 gpd	122,804	0	122,804

155 A total of 22 outlets discharge nearly 123 mgd of industrial effluent. All the outlets are regarded as being satisfactory and consist almost entirely of cooling water with only 50,000 gpd being process water.

156 The largest volume discharged comes from electricity generation with 112 mgd, and the only other significant discharges are from paper and board making (6 mgd) and chemical and allied industries (3 mgd).

157 It is of interest to note that this is the only authority which has no industrial effluent requiring remedial action.

158 For further details of industrial effluents in this area, see Table 216.

30 Port of London Authority (including the London Excluded Area)
(see Table 78)

159 A total of 139 outlets from 17 different industrial classes discharge about 3,386 mgd of industrial effluent with 106 discharges and about 3,320 mgd classed as satisfactory. This discharge of 3,386 mgd is the largest volume discharged in any river authority area, cooling water accounting for 98 per cent of the total.

160 Electricity generation accounts for about 3,007 mgd (89 per cent of the total volume discharged) and other important discharges are from petroleum refining (119 mgd), engineering (79 mgd), paper and board making (61 mgd), chemical and allied industries (42 mgd), gas and coke making (32 mgd), food processing and manufacture (28 mgd) and quarrying and mining (14 mgd).

161 The area receives discharges from several industries which make up significant proportions of the total volumes for England and Wales, namely electricity generation (27 per cent), engineering (60 per cent) and gas and coke making (48 per cent).

162 In the case of about 2 mgd of industrial effluent requiring remedial action, the authority have suggested that this might take the form of diverting to public sewers for treatment at a sewage treatment works.

163 For further details of industrial discharges in this area, see Table 217.

Table 78 Numbers and volumes of discharges of industrial effluent (excluding mine water discharges, codes 52 to 55 inclusive)

Discharges	Rivers			Canals	All rivers and canals
	Non-tidal	Tidal	All rivers		
1 Total industrial effluent excluding solely cooling water					
Number	2	41	43	1	44
% satisfactory	100	46	49	100	49
2 Solely cooling water					
Number	17	66	83	12	95
% satisfactory	94	88	89	92	89
3 All discharges (1+2)					
Number	19	107	126	13	139
% satisfactory	95	72	75	92	76
4 Total industrial effluent excluding solely cooling water					
Volume x 1,000 gpd	1,265	342,206	343,471	10	343,481
% satisfactory	100	86	86	100	86
5 Process water (part of total effluent 4-6)					
Volume x 1,000 gpd	1,261	77,106	78,367	10	78,377
6 Cooling water (part of total effluent)					
Volume x 1,000 gpd	4	265,100	265,104	0	265,104
7 Solely cooling water					
Volume x 1,000 gpd	8,689	3,023,794	3,032,483	10,160	3,042,643
% satisfactory	99	99.9	99.9	53	99.8
8 Total cooling water (6+7)					
Volume x 1,000 gpd	8,693	3,288,894	3,297,587	10,160	3,307,747
9 Process water and cooling water (5+8)					
Total volume x 1,000 gpd	9,954	3,366,000	3,375,954	10,170	3,386,124

(ii) THE CONFEDERATION OF BRITISH INDUSTRY SURVEY

Details of the survey for England and Wales

164 As there was no comprehensive information on the costs and methods of disposal of industrial effluent, the CBI agreed to undertake a supplementary survey of this subject which would cover, as far as possible, all discharges whether to rivers, canals, the sea, local authority sewers or other means of disposal.

165 A questionnaire (a copy of which is reproduced at Appendix 2) was designed by the CBI who organised both its distribution and collection on behalf of their members. Some returns were made through trade associations and others were received from single companies or groups of companies. Non-members of the CBI, including the nationalised industries, were separately approached by the Department, to whom the necessary information was readily made available.

166 There is no means of knowing to what extent the coverage of the returns was complete and a statistical exercise is to be carried out based on those forms which were returned to see whether or not some assessment of completeness of return can be made. It is considered that details of most of the large discharges will have been made available and also those involving significant costs, as industry is not generally disinclined to make these figures known.

167 The CBI survey and the main survey both considered discharges to controlled waters and it is of interest to compare these returns. This is done at Table 79.

Table 79 Comparison of the CBI survey and main survey returns

Volume discharged x 1,000 gpd to:

	Non-tidal rivers	Tidal rivers	All rivers	Canals	All rivers and canals
CBI survey	4,804,516	8,080,387	12,884,903	716,505	13,601,408
Main survey*	5,764,275	10,076,044	15,840,319	219,170	16,059,489

* Excluding mine water, not included in the CBI survey.

168 It will be seen that the figures for discharges to rivers for the CBI survey give totals which, as might be expected, are rather less than in the main survey where the coverage might have been more complete. There were, however, significant discrepancies in complete returns (as in the case of electricity generation) and a few totals for an industry that were higher in the CBI survey than in the main survey. The reasons are being investigated. It will also be seen that the figure for discharges to canals is greater in the CBI survey than in the main survey. This difference is also being investigated but it is thought that it is partly due to the fact that in the main survey the Manchester Ship Canal (in the Mersey & Weaver River Authority area) was regarded as a river.

169 Shortly before publication, figures were received relating to the public sector of the iron and steel industry previously omitted from their return. The additional figures for costs were provided only in the form of a summary and could not be broken down into the detail required in all the tables; they are therefore included only in Tables 80 and 84. The additional figures for quantity were not allocated to individual river authorities and are therefore included only in Tables 81 to 83, and 85.

170 The discharge of industrial process water (industrial effluent without any cooling water that may be present) to controlled waters either directly or indirectly through public sewage treatment works was fairly evenly divided between non-tidal stretches, 35 per cent, tidal stretches, 32 per cent, and public sewers, 29 per cent, with only 4 per cent of the discharge made to canals. If all cooling water present is included then a much larger proportion of 58 per cent of the discharge was to tidal stretches, with 34 per cent to non-tidal stretches, 5 per cent to canals, and 3 per cent only to public sewers. It is of interest to note that of the total industrial effluent disposed of by all means, 16 per cent was by pipeline to uncontrolled waters.

171 The costs of disposal vary of course with the type of waste and not only with the method of disposal; the most expensive methods of disposal have to be used for the most difficult wastes.

172 The total amount which industrialists forecast would be spent on treatment plant in the two years following the survey is about

£40.5m. The capital expenditure on treatment works for effluents discharged to controlled waters was forecast to be about £18m. This is very much less than the main survey estimate of about £38.7m (Table 91) for the cost of remedial work required to bring effluent discharges up to the standards which the river authorities regard as satisfactory. It should be noted, however, that the forecast of expenditure suggested by the main survey, to provide for remedial work on all unsatisfactory discharges, is not strictly comparable with the forecast for 1969 and 1970 from the CBI survey, as the latter can be regarded as part only of a programme of expenditure. Although the survey showed that only 5 per cent of the industrial discharges were made to public sewers, the forecast of expenditure of £19m for 1969/70 is higher than the corresponding forecast for expenditure on direct discharges to rivers. This is in line with the trend to divert discharges of process waters from rivers to public sewers.

173 Tables 80–90 summarise the information collected on existing discharges of industrial effluent to both controlled and uncontrolled waters, to public sewers, and disposed of by incineration. Table 80 sets down the principal figures relating to quantity and cost. Tables 81–85 break the figures down into greater detail by industries while Tables 86–90 give the same information but on a river authority basis.

Table 80   CBI survey—Summary of discharges of industrial effluent and costs of disposal

	Volume x 1,000 gpd		Capital expenditure 1960/68 £1,000s	Forecast capital expenditure 1969/70 £1,000s	Operating costs 1968 £1,000s
	Process	Total			
Discharges to controlled waters					
Non-tidal rivers	404,691	4,804,516	23,458	10,839	5,999
Tidal rivers	381,375	8,080,357	9,029	4,723	1,503
Canals	44,754	716,505	985	1,443	315
Sludge disposal	—	—	—	—	1,487
Total	830,820	13,601,408	33,472	17,005	9,304
Discharges to public sewers					
Effluent	337,085	413,658	29,722	19,091	8,481
Sludge disposal	—	—	—	—	1,116
Total	337,085	413,658	29,722	19,091	9,597
Discharges by pipe line to uncontrolled waters		2,739,862	1,880	2,553	467
Disposal by contractor (including sea disposal)		4,259	—	—	1,299
Disposal by incineration		228	923	571	46
Total for the public sector of the iron and steel industry*		—**	5,200	1,300	730
Total for England and Wales		16,759,415	71,197	40,520	21,443

* No further breakdown for costs was provided.
** The volume is included in the overall summary.

The CBI Survey
Table 81 Information (by industries) about discharges to non-tidal rivers and canals

Industry		Industrial effluent discharged to non-tidal rivers			Capital expenditure on treatment plant 1960–68 £ thousands 4	Forecast of capital expenditure on treatment plant 1969 and 1970 £ thousands 5	Operating costs of treatment plant in 1968 £ thousands 6	Industrial effluent discharged to canals			Capital expenditure on treatment plant 1960–68 £ thousands 10	Forecast of capital expenditure on treatment plant 1969 and 1970 £ thousands 11	Operating cost of treatment plant in 1968 £ thousands 12	CBI Code
CBI Code	Description	Average daily volume in 1968 (×1,000 gpd)						Average daily volume in 1968 (×1,000 gpd)						
		Total effluent 1 (2+3)	Process water 2	Cooling water 3				Total effluent 7 (8+9)	Process water 8	Cooling water 9				
01 Brewing		3,704	809	2,895	149	41	11.2	277	31	246	0	0	0	01
02 Brickmaking		262	258	4	14	5	0.6	0	0	0	0	0	0	02
03 Cement making		236	10	226	10	0	1.5	0	0	0	0	0	0	03
04 Chemical and allied industries		159,493	88,869	70,624	2,268	1,680	866.7	33,985	4,049	29,936	170	1,082	82.9	04
05 Coal mining		33,125	31,288	1,837	9,470	3,857	4,074.2	76	76	0	30	16	17.3	05
06 Ethanol distillation		0	0	0	0	0	0	0	0	0	0	0	0	06
07 Electricity generation		4,166,113	80,273	4,085,840	4,583	415	29.8	390,348	2,426	387,922	25	14	60.4	07
08 Engineering		36,670	5,061	31,609	793	381	135.2	4,313	544	3,769	231	145	28.7	08
09 Food processing		26,851	13,755	13,096	1,171	779	140.8	16,276	1,176	15,100	73	25	3.9	09
10 Gas and coke		12,457	6,032	6,425	747	511	114.5	1,390	558	832	77	0	4.6	10
11 Glass making		1,674	943	731	78	58	7.6	12,080	444	11,636	0	0	0	11
12 Glue and gelatine		682	37	645	0	0	0	1,414	0	1,414	0	0	0	12
13 General manufacturing		4,596	82	4,514	32	11	4.7	70	0	70	0	0	0	13
14 Iron and steel		88,519	54,076	34,443	1,511	1,067	145.7	42,131	23,061	19,070	230	56	54.0	14
15 Laundering and dry cleaning		NR	NR	NR	NR	NR	NR	NR	NR	NR	NR	NR	NR	15
16 Leather tanning		687	642	45	19	62	10.4	618	9	609	0	0	0	16
17 Metal smelting		4,388	4,080	308	90	205	9.2	585	25	560	0	3	0	17
18 Paint making		5	5	0	0	0	0	0	0	0	0	0	0	18
19 Paper and board making		148,570	84,931	63,639	1,745	1,141	261.0	43,884	11,446	32,438	0	0	0	19
20 Petroleum refining		508	456	52	109	64	31.6	168,530	830	167,700	146	100	62.0	20
21 Plastics manufacture		121	0	121	0	0	0	108	0	108	0	0	0.1	21
22 Plating and metal finishing		577	498	79	136	20	9.0	2	0	2	0	2	0	22
23 Pottery making		2	2	0	0	0	0.2	80	79	1	3	0	0.6	23
24 Printing ink etc		0	0	0	0	0	0	0	0	0	0	0	0	24
25 Quarrying and mining		872	152	720	38	25	50.0	0	0	0	0	0	0	25
26 Rubber processing		12,005	3,189	8,816	134	102	14.3	108	0	108	0	0	0	26
27 Soap and detergent		288	0	288	0	0	0	130	0	130	0	0	0	27
28 Textile, cotton and man-made		91,991	20,326	71,665	361	415	80.6	100	0	100	0	0	0	28
29 Textile, wool		10,120	8,917	1,203	0	0	0	0	0	0	0	0	0	29
30 General farming		0	0	0	0	0	0	0	0	0	0	0	0	30
Total England and Wales		4,804,516	404,691	4,399,825	23,458	10,839	5,998.8	716,505	44,754	671,751	985	1,443	314.5	

Note: NR, no return received

The CBI Survey
Table 82 Information (by industries) about discharges to controlled tidal rivers and uncontrolled waters

Industry			Industrial effluent discharged to controlled tidal waters						Industrial effluent disposed of by pipeline to uncontrolled waters				
			Average daily volume in 1968 (×1,000 gpd)			Capital expenditure on treatment plant 1960–68 £ thousands	Forecast of capital expenditure on treatment plant 1969 and 1970 £ thousands	Operating cost of treatment plant in 1968 £ thousands	Average daily volume of effluent in 1968 ×1,000 gpd	Capital expenditure on pretreatment plant 1960–68 £ thousands	Forecast of capital expenditure on pretreatment plant 1969 and 1970 £ thousands	Operating costs in 1968 £ thousands	
CBI Code	Description		Total effluent 13 (14+15)	Process water 14	Cooling water 15	16	17	18	19	20	21	22	CBI Code
01	Brewing		42	17	25	0	0	0	133	0	4	0.1	01
02	Brickmaking		0	0	0	0	0	0	5	10	0	0.5	02
03	Cement making		0	0	0	0	0	0	0	0	0	0	03
04	Chemical and allied industries		340,945	98,773	242,172	4,106	1,596	902.3	515,668	424	1,415	137.5	04
05	Coal mining		1	1	0	0	0	22.9	758	113	115	94.4	05
06	Ethanol distillation		6	0	6	0	0	0	0	0	0	0	06
07	Electricity generation		6,757,880	32,538	6,725,342	322	87	165.1	2,138,015	463	3	27.8	07
08	Engineering		612	128	484	51	22	7.2	1,035	14	5	3.0	08
09	Food processing		46,258	20,185	26,073	745	186	62.5	15,634	108	322	38.6	09
10	Gas and coke		52,583	12,622	39,961	473	56	43.6	5,920	0	0	4.0	10
11	Glass making		0	0	0	0	0	0	0	0	0	0	11
12	Glue and gelatine		1,139	1,083	56	0	50	0.1	108	4	7	3.7	12
13	General manufacturing		237	90	147	11	46	3.5	1,472	5	2	0.5	13
14	Iron and steel		434,220	147,029	287,191	2,918	680	32.0	24,637	114	0	117.5	14
15	Laundering and dry cleaning		NR	NR	NR	NR	NR	NR	NR	NR	NR	NR	15
16	Leather tanning		170	130	40	0	0	0.3	816	0	2	1.0	16
17	Metal smelting		875	518	357	12	167	5.1	13	0	0	0	17
18	Paint making		0	0	0	0	0	0	0	0	0	0	18
19	Paper and board making		132,872	56,624	76,248	1,292	1,550	129.3	28,122	472	578	26.3	19
20	Petroleum refining		307,342	9,742	297,600	1,658	412	126.0	0	0	0	0	20
21	Plastics manufacture		972	262	710	0	22	1.1	0	0	0	0	21
22	Plating and metal finishing		92	52	40	0	8	0.2	0	0	0	0	22
23	Pottery making		0	0	0	0	0	0	30	0	0	0	23
24	Printing ink etc		0	0	0	0	0	0	0	0	0	0	24
25	Quarrying and mining		0	0	0	0	0	0	2	0	0	0.1	25
26	Rubber processing		0	0	0	0	0	0	0	0	0	0	26
27	Soap and detergent		2,510	10	2,500	15	15	0	4,600	20	100	1.4	27
28	Textile, cotton and man-made		1,619	1,559	60	26	5	1.5	2,894	133	0	11.0	28
29	Textile, wool		0	0	0	0	0	0	0	0	0	0	29
30	General farming		12	12	0	0	1	9.0	0	0	0	0	30
	Total England and Wales		8,080,387	381,375	7,699,012	11,629	4,903	1,503.6	2,739,862	1,880	2,553	467.4	

Note: NR, no return received

The CBI Survey
Table 83 Information (by industries) about discharges to sewers and about industrial material disposed of by contractors or by incineration

Industry		Industrial effluent discharged to sewers								Industrial materials disposed of by contractors in 1968					Industrial effluent disposed of by incineration			
CBI Code	Description	Average daily volume in 1968 (×1,000 gpd)			Capital expenditure on pre-treatment plant 1960-68	Forecast of capital expenditure on pre-treatment plant 1969 and 1970	Operating costs 1968			Removal and disposal of untreated effluents (including sea disposal)		Cost of contractors disposal services (sludge from treatment processes)		Average daily volume of effluent in 1968	Capital expenditure on pre-treatment plant 1960-68	Forecast of capital expenditure on pre-treatment plant 1969 and 1970	Operating costs in 1968	CBI Code
		Total effluent	Process water	Cooling water			Of plant excluding contractors services	In payments to local authorities		Average daily volume	Cost of contractors services	Prior to disposal by public sewer	Prior to disposal to controlled waters					
		23 (24+25)	24	25	£ thousands 26	£ thousands 27	£ thousands 28	£ thousands 29		×1,000 gpd 30	£ thousands 31	£ thousands 32	£ thousands 33	×1,000 gpd 34	£ thousands 35	£ thousands 36	£ thousands 37	
01	Brewing	24,042	21,181	2,861	189	415	9.6	275.4		228	1.9	5.8	1.1	1	0	0	0.2	01
02	Brickmaking	240	240	0	0	0	0	0		0	0	0	0	0	0	0	0	02
03	Cement making	118	0	118	0	0	0	0		0	0	0	0	0	0	0	0	03
04	Chemical and allied industries	35,999	29,270	6,729	9,585	5,820	1,184.7	2,152.8		3,483	1,008.5	296.8	603.0	24	339	504	39.1	04
05	Coal mining	104	104	0	5	0	0.2	1.2		313	95.2	0	346.3	0	0	0	0	05
06	Ethanol distillation	636	524	112	44	501	10.5	44.7		0	0	0.1	0	0	0	0	0	06
07	Electricity generation	7,019	2,097	4,922	10	21	67.5	11.0		1	0.6	1.3	263.2	0	0	0	0	07
08	Engineering	66,729	51,273	15,456	3,711	2,604	461.8	316.4		21	96.3	144.7	24.6	0	1	0	0.5	08
09	Food processing	61,625	53,767	7,858	5,495	1,775	271.7	884.9		84	41.3	101.8	17.1	0	0	0	0.6	09
10	Gas and coke	8,649	6,738	1,911	779	8	338.8	215.0		8	5.7	5.8	29.7	90	0	42	1.5	10
11	Glass making	1,486	660	826	69	146	39.7	16.6		1	1.0	2.0	3.6	0	0	0	0	11
12	Glue and gelatine	1,132	1,128	4	79	116	1.7	25.8		0	0	1.2	1.2	0	0	20	0	12
13	General manufacturing	6,179	4,672	1,507	158	127	52.5	58.7		1	2.8	12.1	3.2	22	3	0	0.2	13
14	Iron and steel	20,178	9,066	11,112	570	148	53.9	27.1		6	6.5	7.5	41.7	0	0	0	1.2	14
15	Laundering and dry cleaning	NR	NR	NR	NR	NR	NR	NR		NR	NR	NR	NR	NR	NR	NR	NR	15
16	Leather tanning	6,261	6,124	137	208	221	69.2	188.1		5	2.2	65.0	5.9	25	1	5	0.5	16
17	Metal smelting	1,161	484	677	90	92	30.7	3.3		1	0.9	6.6	0.5	1	0	0	0	17
18	Paint making	94	80	14	0	4	0.2	10.7		0	0	0.4	0.1	0	0	0	0	18
19	Paper and board making	35,997	33,917	2,080	4,508	5,469	126.5	232.8		98	6.0	382.2	57.9	1	577	0	2.0	19
20	Petroleum refining	0	0	0	18	5	0.5	1.0		5	18.0	0	77.2	0	0	0	0	20
21	Plastics manufacture	638	128	510	42	10	7.2	5.8		0	0	0.6	0	0	0	0	0	21
22	Plating and metal finishing	16,693	15,429	1,264	521	387	128.5	189.8		1	5.1	14.2	0.8	25	0	0	0	22
23	Pottery making	157	156	1	8	2	1.0	2.7		0	0	1.2	1.0	0	0	0	0	23
24	Printing ink etc	448	418	30	0	0	0.2	0.3		0	0	0.1	0	0	0	0	0	24
25	Quarrying and mining	0	0	0	0	0	0	0		0	0	0	0	0	0	0	0	25
26	Rubber processing	12,919	1,632	11,287	45	39	15.2	79.0		0	2.2	1.4	2.6	0	1	0	0	26
27	Soap and detergent	2,477	483	1,994	25	36	10.7	11.2		2	3.5	2.0	0.6	40	1	0	0.6	27
28	Textile, cotton and man-made	89,370	84,394	4,976	1,074	459	87.8	382.9		1	0.8	34.9	6.1	0	0	0	0	28
29	Textile, wool	13,307	13,120	187	2,481	686	162.0	211.4		0	0.1	28.4	0	0	0	0	0	29
30	General farming	0	0	0	0	0	0	0		2	0	0	0	0	0	0	0	30
	Total England and Wales	413,658	337,085	76,573	29,722	19,091	3,132.3	5,348.6		4,259	1,298.6	1,116.1	1,487.4	228	923	571	46.4	

Note : NR, no return received

The CBI Survey
Table 84 Summary (by industries) of financial information

Industry		Capital expenditure on treatment plant installed 1960–1968 £ thousands				Forecast of capital expenditure on treatment plant for 1969 and 1970 £ thousands			
		For discharges to:			Total expenditure all treatment plant	For discharges to:			Total expenditure all treatment plant
CBI Code	Description	Controlled waters	Controlled waters and sewers	Uncontrolled waters and disposal by incineration		Controlled waters	Controlled waters and sewers	Uncontrolled waters and disposal by incineration	
		38 (4+16+10)	39 (26+38)	40 (20+35)	41 (39+40)	42 (5+11+17)	43 (42+27)	44 (21+36)	45 (43+44)
01	Brewing	149	338	0	338	41	456	4	460
02	Brickmaking	14	14	10	24	5	5	0	5
03	Cement making	10	10	0	10	0	0	0	0
04	Chemical and allied industries	6,544	16,129	763	16,892	4,358	10,178	1,919	12,097
05	Coal mining	9,500	9,505	113	9,618	3,873	3,873	115	3,988
06	Ethanol distillation	0	44	0	44	0	501	0	501
07	Electricity generation	4,930	4,940	463	5,403	516	537	3	540
08	Engineering	1,075	4,786	15	4,801	548	3,152	5	3,157
09	Food processing	1,989	7,484	108	7,592	990	2,765	322	3,087
10	Gas and coke	1,297	2,076	0	2,076	567	575	42	617
11	Glass making	78	147	0	147	58	204	0	204
12	Glue and gelatine	0	79	4	83	50	166	27	193
13	General manufacturing	43	201	8	209	57	184	2	186
14	Iron and steel	2,059	2,637	114	7,951***	1,623	1,771	0	3,071**
15	Laundering and dry cleaning	NR	NR	NR	NR	NR	NR	NR	NR
16	Leather tanning	19	227	1	228	62	283	7	290
17	Metal smelting	102	192	0	192	375	467	0	467
18	Paint making	0	0	0	0	0	0	4	4
19	Paper and board making	3,037	7,545	1,049	8,594	2,691	8,160	578	8,738
20	Petroleum refining	1,913	1,931	0	1,931	576	581	0	581
21	Plastics manufacture	0	42	0	42	22	32	0	32
22	Plating and metal finishing	136	657	0	657	30	417	0	417
23	Pottery making	3	11	0	11	0	2	0	2
24	Printing ink etc	0	0	0	0	0	0	0	0
25	Quarrying and mining	38	38	0	38	25	25	0	25
26	Rubber processing	134	179	1	180	102	141	0	141
27	Soap and detergent	15	40	21	61	15	51	100	151
28	Textile, cotton and man-made	387	1,461	133	1,594	420	879	0	879
29	Textile, wool	0	2,481	0	2,481	0	686	0	686
30	General farming	0	0	0	0	1	1	0	1
	Total England and Wales	33,472	63,194	2,803	71,197	17,005	36,096	3,124	40,520

Note: NR, No return received.
* The total shown includes the sum of £730,000 for which no breakdown was available.
** The total shown includes the sum of £1,300,000 for which insufficient breakdown was available.
*** The total shown includes the sum of £5,200,000 for which insufficient breakdown was available.

Treatment plant operating costs for 1968
£ thousands

Controlled waters excluding contractors services	Controlled waters including contractors services for sludge disposal	Controlled waters and sewers excluding contractors services	Controlled waters and sewers, contractors costs for sludge disposal	Controlled waters and sewers including contractors services, excluding local authority charges	Controlled waters and sewers including contractors services and local authority charges	Uncontrolled waters and disposal by incineration	Total for all methods of disposal excluding local authority charges	Total for all methods of disposal including local authority charges	CBI Code
46 (6+12+18)	47 (33+46)	48 (28+46)	49 (32+33)	50 (48+49)	51 (50+29)	52 (37+34+22)	53 (50+52)	54 (51+52)	
11.2	12.3	20.8	6.9	27.7	303.1	2.2	29.9	305.3	01
0.6	0.6	0.6	0	0.6	0.6	0.5	1.1	1.1	02
1.5	1.5	1.5	0	1.5	1.5	0	1.5	1.5	03
1,851.9	2,454.9	3,036.6	899.8	3,936.4	6,089.2	1,185.1	5,121.5	7,274.3	04
4,114.4	4,460.7	4,114.6	346.3	4,460.9	4,462.1	189.6	4,650.5	4,651.7	05
0	0	10.5	0.1	10.6	55.3	0	10.6	55.3	06
255.3	518.5	322.8	264.5	587.3	598.3	28.4	615.7	626.7	07
171.1	195.7	632.9	169.3	802.2	1,118.6	99.8	902.0	1,218.4	08
207.2	224.3	478.9	118.9	597.8	1,482.7	80.5	678.3	1,563.2	09
162.7	192.4	501.5	35.5	537.0	752.0	11.2	548.2	763.2	10
7.6	11.2	47.3	5.6	52.9	69.5	1.0	53.9	70.5	11
0.1	1.3	1.8	2.4	4.2	30.0	3.7	7.9	33.7	12
8.2	11.4	60.7	15.3	76.0	134.7	3.5	79.5	138.2	13
231.7	273.4	285.6	49.2	334.8	361.9	125.2	460.0	1,217.1*	14
NR	NR	NR	NR	NR	NR	NR	NR	NR	15
10.7	16.6	79.9	70.9	150.8	338.9	3.7	154.5	342.6	16
14.3	14.8	45.0	7.1	52.1	55.4	0.9	53.0	56.3	17
0	0.1	0.2	0.5	0.7	11.4	0	0.7	11.4	18
390.3	448.2	516.8	440.1	956.9	1,189.7	34.3	991.2	1,224.0	19
219.6	296.8	220.1	77.2	297.3	298.3	18.0	315.3	316.3	20
1.2	1.2	8.4	0.6	9.0	14.8	0	9.0	14.8	21
9.2	10.0	137.7	15.0	152.7	342.5	5.1	157.8	347.6	22
0.8	1.8	1.8	2.2	4.0	6.7	0	4.0	6.7	23
0	0	0.2	0.1	0.3	0.6	0	0.3	0.6	24
50.0	50.0	50.0	0	50.0	50.0	0.1	50.1	50.1	25
14.3	16.9	29.5	4.0	33.5	112.5	2.2	35.7	114.7	26
0	0.6	10.7	2.6	13.3	24.5	5.5	18.8	30.0	27
82.1	88.2	169.9	41.0	210.9	593.8	11.8	222.7	605.6	28
0	0	162.0	28.4	190.4	401.8	0.1	190.5	401.9	29
9.0	9.0	0.9	0	0.9	0.9	0	0.9	0.9	30
7,816.9	9,304.3	10,949.2	2,603.5	13,552.7	18,901.3	1,812.4	15,365.1	21,443.7	

The CBI Survey
Table 85   Summary (by industries) of daily volumes of effluent discharged in 1968 (x 1,000 gpd)

CBI Code	Description	To controlled waters			To controlled waters plus sewers			Total effluent to other than controlled waters or sewers	Total effluent
		Total effluent	Cooling water	Process water	Total effluent	Cooling water	Process water		
		55 (1+7+13)	56 (3+9+15)	57 (2+8+14)	58 (23+55)	59 (25+56)	60 (24+57)	61 (19+30+34)	62 (58+61)
01	Brewing	4,023	3,166	857	28,065	6,027	22,038	362	28,427
02	Brickmaking	262	4	258	502	4	498	5	507
03	Cement making	236	226	10	354	344	10	0	354
04	Chemical and allied industries	534,423	342,732	191,691	570,422	349,461	220,961	519,175	1,089,597
05	Coal mining	33,202	1,837	31,365	33,306	1,837	31,469	1,071	34,377
06	Ethanol distillation	6	6	0	642	118	524	0	642
07	Electricity generation	11,314,341	11,199,104	115,237	11,321,360	11,204,026	117,334	2,138,016	13,459,376
08	Engineering	41,595	35,862	5,733	108,324	51,318	57,006	1,056	109,380
09	Food processing	89,385	54,269	35,116	151,010	62,127	88,883	15,718	166,728
10	Gas and coke	66,430	47,218	19,212	75,079	49,129	25,950	6,018	81,097
11	Glass making	13,754	12,367	1,387	15,240	13,193	2,047	1	15,241
12	Glue and gelatine	3,235	2,115	1,120	4,367	2,119	2,248	108	4,475
13	General manufacturing	4,903	4,731	172	11,082	6,238	4,844	1,495	12,577
14	Iron and steel	564,870	340,704	224,166	585,048	351,816	233,232	24,643	609,691
15	Laundering and dry cleaning	NR	NR	NR	NR	NR	NR	NR	NR
16	Leather tanning	1,475	694	781	7,736	831	6,905	846	8,582
17	Metal smelting	5,848	1,225	4,623	7,009	1,902	5,107	14	7,023
18	Paint making	5	0	5	99	14	85	0	99
19	Paper and board making	325,326	172,325	153,001	361,323	174,405	186,918	28,221	389,544
20	Petroleum refining	476,380	465,352	11,028	476,380	465,352	11,028	5	476,385
21	Plastics manufacture	1,201	939	262	1,839	1,449	390	0	1,839
22	Plating and metal finishing	671	121	550	17,364	1,385	15,979	26	17,390
23	Pottery making	82	1	81	239	2	237	30	269
24	Printing ink etc	0	0	0	448	30	418	0	448
25	Quarrying and mining	872	720	152	872	720	152	2	874
26	Rubber processing	12,113	8,924	3,189	25,032	20,211	4,821	0	25,032
27	Soap and detergent	2,928	2,918	10	5,405	4,912	493	4,642	10,047
28	Textile, cotton and man-made	93,710	71,825	21,885	183,080	76,801	106,279	2,895	185,975
29	Textile, wool	10,120	1,203	8,917	23,427	1,390	22,037	0	23,427
30	General farming	12	0	12	12	0	12	0	12
	Total England and Wales	13,601,408	12,770,588	830,820	14,015,066	12,847,161	1,167.905	2,744,349	16,759.415

Note:—NR, no return received

The CBI Survey
Table 86 Information (by River Authorities) about discharges to non-tidal rivers and canals

River authority	Industrial effluent discharged to non-tidal rivers						Industrial effluent discharged to canals					
	Average daily volume in 1968 (x 1,000 gpd)			Capital expenditure on treatment plant 1960–68	Forecast of capital expenditure on treatment plant 1969 and 1970	Operating costs of treatment plant in 1968	Average daily volume in 1968 (x 1,000 gpd)			Capital expenditure on treatment plant 1960–68	Forecast of capital expenditure on treatment plant 1969 and 1970	Operating costs of treatment plant in 1968
	Total effluent	Process water	Cooling water	£ thousands	£ thousands	£ thousands	Total effluent	Process water	Cooling water	£ thousands	£ thousands	£ thousands
	1 (2+3)	2	3	4	5	6	7 (8+9)	8	9	10	11	12
1 Northumbrian	6,069	4,377	1,692	131	286	131.6	0	0	0	0	0	0
2 Yorkshire	1,255,477	60,125	1,195,352	7,839	2,460	1,494.1	12,899	135	12,764	0	0	0
3 Trent	1,624,493	66,330	1,558,163	6,480	3,225	2,025.2	16,653	2,180	14,473	224	158	86.4
4 Lincolnshire	23,043	6,063	16,980	213	55	47.5	0	0	0	0	0	0.4
5 Welland and Nene	65,252	1,500	63,752	51	44	19.8	0	0	0	0	0	0
6 Great Ouse	136,419	4,645	131,774	279	543	72.4	59	59	0	30	0	2.7
7 East Suffolk and Norfolk	1,289	961	328	135	13	6.8	0	0	0	0	0	0
8 Essex	8,494	1,253	7,241	131	307	29.2	0	0	0	0	0	0
9 Kent	8,362	2,353	6,009	250	45	12.6	0	0	0	0	0	0
10 Sussex	0	0	0	0	0	0	0	0	0	0	0	0
11 Hampshire	4,262	4,106	156	49	60	32.4	0	0	0	0	0	0
12 Isle of Wight	0	0	0	0	0	0	3,600	0	3,600	0	0	0
13 Avon and Dorset	59	30	29	2	17	1.0	0	0	0	0	0	0
14 Devon	9,497	3,941	5,556	146	46	40.0	1	0	1	0	0	0
15 Cornwall	84	84	0	0	0	0	0	0	0	0	0	0
16 Somerset	11,594	7,122	4,472	65	58	5.0	0	0	0	0	0	0
17 Bristol Avon	50,253	16,964	33,289	233	85	14.5	0	0	0	0	0	0
18 Severn	174,185	4,562	169,623	798	379	67.6	485	205	280	4	51	0
19 Wye	0	0	0	0	0	0	0	0	0	0	0	0
20 Usk	5,015	607	4,408	307	98	107.2	0	0	0	0	0	0
21 Glamorgan	56,653	8,887	47,766	2,373	532	459.3	70	0	70	0	0	0
22 South West Wales	4,704	4,531	173	618	268	1,863.0	500	20	480	25	10	0
23 Gwynedd	843,288	9	843,279	75	4	3.5	0	0	0	0	0	0
24 Dee and Clwyd	2,236	1,995	241	27	102	177.3	0	0	0	0	0	0
25 Mersey and Weaver	197,369	59,039	138,330	2,419	1,652	867.6	542,064	17,801	524,263	655	1,223	219.2
26 Lancashire	93,821	50,963	42,858	497	400	110.4	2,840	26	2,814	0	0	1.0
27 Cumberland	36,791	2,057	34,734	83	7	4.1	0	0	0	0	0	0
28 Thames Conservancy	113,307	51,740	61,567	237	143	92.1	2,880	1,093	1,787	2	1	1.1
29 Lee Conservancy	1,108	75	1,033	0	10	0.3	91,771	235	91,536	45	0	3.7
30 Port of London Authority (including London Excluded Area)	9,392	3,372	6,020	20	0	0	4,683	0	4,683	0	0	0
Total England and Wales (excluding public sector of the Iron and Steel Industry)	4,742,516	367,691	4,374,825	23,458	10,839	5,998.8	678,505	21,754	656,751	985	1,443	314.5

The CBI Survey
Table 87 Information (by River Authorities) about discharges to controlled tidal waters and uncontrolled tidal waters

	Industrial effluent discharged to controlled tidal waters						Industrial effluent disposed of by pipeline to uncontrolled waters				
River authority	Average daily volume in 1968 (x1,000 gpd)			Capital expenditure on treatment plant 1960–68	Forecast of capital expenditure on treatment plant 1969 and 1970	Operating cost of treatment plant in 1968	Average daily volume of effluent in 1968	Capital expenditure on pretreatment plant 1960–68	Forecast of capital expenditure on pretreatment plant 1969 and 1970	Operating costs in 1968	
	Total effluent	Process water	Cooling water	£ thousands	£ thousands	£ thousands	x1,000 gpd	£ thousands	£ thousands	£ thousands	
	13 (14+15)	14	15	16	17	18	19	20	21	22	
1 Northumbrian	574,235	71,257	502,978	1,410	329	242.2	920,257	360	146	101.6	1
2 Yorkshire	135,590	6,565	129,025	137	221	67.0	4,540	37	200	1.5	2
3 Trent	110,430	7,020	103,410	139	554	9.9	0	0	0	0	3
4 Lincolnshire	19,425	13,620	5,805	925	244	64.8	3,206	191	12	22.2	4
5 Welland and Nene	1,820	1,813	7	94	0	10.8	0	0	0	0	5
6 Great Ouse	6,488	6,249	239	168	152	64.2	603	120	144	38.0	6
7 East Suffolk and Norfolk	177,718	9,973	167,745	311	41	24.3	163,421	4	0	0	7
8 Essex	322,873	524	322,349	46	27	3.2	17	1	0	0	8
9 Kent	237,046	26,662	210,384	1,025	875	96.5	18	99	0	25.9	9
10 Sussex	500	139	361	4	0	3.3	267,017	6	0	0	10
11 Hampshire	167,431	9,957	157,474	237	131	139.2	465,478	4	7	3.7	11
12 Isle of Wight	12	12	0	0	8	0.2	0	0	0	0	12
13 Avon and Dorset	711	211	500	2	25	3.0	171,270	1	3	1.8	13
14 Devon	21,168	144	21,024	21	0	1.7	109	8	1	1.4	14
15 Cornwall	33,975	3,925	30,050	54	0	1.3	5,640	4	0	0	15
16 Somerset	593,057	2,269	590,788	15	40	35.0	0	0	0	0	16
17 Bristol	7,491	31	7,460	79	0	0.5	26	0	0	0.3	17
18 Severn	773,279	12,937	760,342	812	215	74.4	0	0	0	0	18
19 Wye	0	0	0	0	6	0	0	0	0	0	19
20 Usk	412,622	8,139	404,483	168	0	35.1	0	0	0	0	20
21 Glamorgan	448	435	13	186	0	3.7	331,870	3	966	1.0	21
22 South West Wales	270,304	8,501	261,803	156	10	82.1	32	0	0	0	22
23 Gwynedd	680	100	580	11	1	2.5	328,085	126	0	0.1	23
24 Dee and Clwyd	2,135	2,096	39	22	51	2.1	54	114	0	117.5	24
25 Mersey and Weaver	411,138	24,131	387,007	1,390	625	212.5	20,655	279	623	33.2	25
26 Lancashire	101,862	7,143	94,719	263	6	59.9	4,050	183	52	83.1	26
27 Cumberland	710	710	0	37	22	10.0	15,602	175	65	5.0	27
28 Thames Conservancy	2,206	4	2,202	20	0	0	1	0	0	0.1	28
29 Lee Conservancy	201	81	120	21	0	0	0	0	0	0	29
30 Port of London Authority (including London Excluded Area)	3,374,832	48,727	3,326,105	1,276	1,140	254.2	22,911	165	334	31.0	30
Total England and Wales, (excluding public sector of Iron and Steel Industry)	7,760,387	273,375	7,487,012	9,029	4,723	1,503.6	2,724,862	1,880	2,553	467.4	

The CBI Survey
Table 88 Information (by River Authorities) about discharges to sewers and about industrial material disposed of by contractors, or by incineration

		Industrial effluent discharged to sewers								Industrial materials disposed of by contractors in 1968						Industrial effluents disposed of by incineration			
		Average daily volume in 1968 (x 1,000 gpd)			Capital expenditure on pre-treatment plant 1960–68	Forecast of capital expenditure on pre-treatment plant 1969 and 1970	Operating costs 1968			Removal and disposal of untreated effluents (including sea disposal)		Cost of contractors disposal services from treatment processes		Cost of contractors disposal services (sludge prior to treatment processes)		Average daily volume of effluent in 1968	Capital expenditure on pre-treatment plant 1960–68	Forecast of capital expenditure on pre-treatment plant 1969 and 1970	Operating costs in 1968
		Total effluent	Process water	Cooling water			Of plant excluding contractors services	In payments to local authorities		Average daily volume	Cost of contractors services	Prior to disposal to public sewer	Prior to disposal to controlled waters						
	River authority	23 (24+25)	24	25	£ thousands 26	£ thousands 27	£ thousands 28	£ thousands 29		x1,000 gpd 30	£ thousands 31	£ thousands 32	£ thousands 33			x1,000 gpd 34	£ thousands 35	£ thousands 36	£ thousands 37	
1	Northumbrian	8,598	3,755	4,843	199	285	53.3	3.1		447	111.3	129.2	26.4			92	0	32	2.2	1
2	Yorkshire	28,768	21,065	7,703	1,396	944	313.9	481.6		32	5.6	45.4	512.4			18	182	165	17.0	2
3	Trent	75,668	64,135	11,533	6,568	3,089	537.3	1,027.2		3,325	63.6	86.2	68.3			46	1	9	0.6	3
4	Lincolnshire	1,852	1,722	130	81	43	11.1	6.0		0	0.3	1.8	2.6			0	0	0	0	4
5	Welland and Nene	2,667	2,570	97	193	127	63.0	44.4		1	0.9	18.7	4.3			0	0	0	0.5	5
6	Great Ouse	2,390	1,992	398	301	226	112.9	21.6		0	1.8	2.8	3.8			1	0	0	0	6
7	East Suffolk and Norfolk	2,128	1,962	166	87	197	16.0	49.8		5	7.9	6.1	19.3			0	0	0	0	7
8	Essex	6,670	4,090	2,580	150	277	33.5	76.5		7	13.3	9.2	4.4			1	20	0	3.0	8
9	Kent	6,207	5,934	273	305	127	46.3	82.4		50	6.7	37.4	0.3			0	587	12	1.6	9
10	Sussex	628	550	78	1	5	1.7	2.2		0	0	0.5	1.0			0	0	0	0	10
11	Hampshire	1,360	747	613	118	96	15.3	25.8		57	82.8	5.1	16.3			0	1	0	0.2	11
12	Isle of Wight	12	12	0	0	0	1.0	0		0	0	0	0			0	0	0	0	12
13	Avon and Dorset	283	275	8	28	17	1.4	53.0		0	0.1	0.8	2.5			0	0	0	0	13
14	Devon	977	951	26	226	20	50.9	68.7		1	2.3	365.3	7.6			1	0	0	2.0	14
15	Cornwall	215	210	5	1	15	9.0	0		0	0	0.3	0			0	0	0	0	15
16	Somerset	3,971	2,961	1,010	46	17	1.9	108.2		0	0	29.0	0.1			0	0	0	0	16
17	Bristol Avon	22,080	16,850	5,230	126	16	42.8	127.5		0	0.2	1.2	2.1			25	1	5	0.6	17
18	Severn	6,099	5,277	822	925	204	103.1	59.6		7	16.2	19.6	17.1			0	0	0	0	18
19	Wye	14,138	14,138	0	4	6	6.8	13.2		0	0	1.1	0			0	0	0	0	19
20	Usk	442	347	95	202	82	7.3	4.0		6	3.1	3.2	35.3			19	0	0	0	20
21	Glamorgan	7,833	2,933	4,900	80	1,257	78.4	3.3		13	726.8	73.0	64.3			24	23	268	3.2	21
22	South West Wales	299	207	92	26	21	5.8	0		0	0	1.0	8.7			0	0	0	0	22
23	Gwynedd	388	29	359	12	4	0.5	0		0	0.1	0.1	0			0	0	0	0	23
24	Dee and Clwyd	531	263	268	35	14	18.9	5.1		6	0.2	2.1	8.2			0	0	0	0	24
25	Mersey and Weaver	77,209	64,247	12,962	1,231	1,274	357.9	677.8		49	124.7	161.3	620.1			0	2	20	0.3	25
26	Lancashire	26,647	15,249	11,398	618	375	101.5	160.0		8	40.1	22.2	16.6			2	106	20	15.1	26
27	Cumberland	2,149	1,560	589	8	1	1.4	0		0	0.5	0.7	0.3			0	0	0	0	27
28	Thames Conservancy	27,619	23,096	4,523	1,049	296	238.3	261.8		114	12.4	64.9	16.2			0	0	0	0	28
29	Lee Conservancy	8,540	6,955	1,585	206	298	143.5	96.5		108	22.4	9.1	1.4			0	0	0	0	29
30	Port of London Authority (including London Excluded Area)	69,290	68,003	1,287	15,500	9,758	757.6	1,899.3		23	55.4	18.8	27.8			0	0	40	0	30
	Total England and Wales (excluding public sector of Iron and Steel Industry)	405,658	332,085	73,573	29,722	19,091	3,132.3	5,348.6		4,259	1,298.6	1,116.1	1,487.4			228	923	571	46.4	

The CBI Survey
Table 89 Summary (by River Authorities) of financial information

River authority	Capital expenditure on treatment plant installed 1960–1968 (£ thousands)				Forecast of capital expenditure on treatment plant for 1969 and 1970 (£ thousands)			
	For discharges to:			Total expenditure all treatment plant	For discharges to:			Total expenditure all treatment plant
	Controlled waters	Controlled waters and sewers	Uncontrolled waters and disposal by incineration		Controlled waters	Controlled waters and sewers	Uncontrolled waters and disposal by incineration	
	38 (4+16+10)	39 (26+38)	40 (20+35)	41 (39+40)	42 (5+11+17)	43 (27+42)	44 (21+36)	45 (43+44)
1 Northumbrian	1,541	1,740	360	2,100	615	900	178	1,078
2 Yorkshire	7,976	9,372	219	9,591	2,681	3,625	365	3,990
3 Trent	6,843	13,411	1	13,412	3,937	7,026	9	7,035
4 Lincolnshire	1,138	1,219	191	1,410	299	342	12	354
5 Welland and Nene	145	338	0	338	44	171	0	171
6 Great Ouse	477	778	120	898	695	921	144	1,065
7 East Suffolk and Norfolk	446	533	4	537	54	251	0	251
8 Essex	177	327	21	348	334	611	0	611
9 Kent	1,275	1,580	686	2,266	920	1,047	12	1,059
10 Sussex	4	5	6	11	0	5	0	5
11 Hampshire	286	404	5	409	191	287	7	294
12 Isle of Wight	0	0	0	0	8	8	0	8
13 Avon and Dorset	4	32	1	33	42	59	3	62
14 Devon	167	393	8	401	46	66	1	67
15 Cornwall	54	55	4	59	0	15	0	15
16 Somerset	80	126	0	126	98	115	0	115
17 Bristol	312	438	1	439	85	101	5	106
18 Severn	1,614	2,539	0	2,539	645	849	0	849
19 Wye	0	4	0	4	6	12	0	12
20 Usk	475	677	0	677	98	180	0	180
21 Glamorgan	2,559	2,639	26	2,665	532	1,789	1,234	3,023
22 South West Wales	799	825	0	825	288	309	0	309
23 Gwynedd	86	98	126	224	5	9	0	9
24 Dee and Clwyd	49	84	114	198	153	167	0	167
25 Mersey and Weaver	4,464	5,695	281	5,976	3,500	4,774	643	5,417
26 Lancashire	760	1,378	289	1,667	406	781	72	853
27 Cumberland	120	128	175	303	29	30	65	95
28 Thames Conservancy	259	1,308	0	1,308	144	440	0	440
29 Lee Conservancy	66	272	0	272	10	308	0	308
30 Port of London Authority (including London Excluded Area)	1,296	16,796	165	16,961	1,140	10,898	374	11,272
Total England and Wales (excluding public sector of the Iron and Steel Industry)	33,472	63,194	2,803	65,997	17,005	36,096	3,124	39,220

Treatment plant operating costs for 1968
(£ thousands)

For discharges to:

Controlled waters excluding contractors services	Controlled waters including contractors services for sludge disposal	Controlled waters and sewers excluding contractors services	Controlled waters and sewers-contractors costs for sludge removal	Controlled waters and sewers including contractors services excluding local authority charges	Controlled waters and sewers including contractors services and local authority charges	Uncontrolled waters and disposal by incineration	Total for all methods of disposal excluding local authority charges	Total for all methods of disposal including local authority charges	
46 (6+12+18)	47 (33+46)	48 (28+46)	49 (32+33)	50 (48+49)	51 (50+29)	52 (22+31+37)	53 (50+52)	54 (51+52)	
373.8	400.2	427.1	155.6	582.7	585.8	215.1	797.8	800.9	1
1,561.1	2,073.5	1,875.0	557.8	2,432.8	2,914.4	24.1	2,456.9	2,938.5	2
2,121.5	2,189.8	2,658.8	154.5	2,813.3	3,840.5	64.2	2,877.5	3,904.7	3
112.7	115.3	123.8	4.4	128.2	134.2	22.5	150.7	156.7	4
30.6	31.9	93.6	23.0	116.6	161.0	1.5	118.1	162.5	5
139.3	143.1	252.2	6.6	258.8	280.4	39.8	298.6	320.2	6
31.1	50.4	47.1	25.4	72.5	122.3	7.9	80.4	130.2	7
32.4	36.8	65.9	13.6	79.5	156.0	16.3	95.8	172.3	8
109.1	109.4	155.4	37.7	193.1	275.5	34.2	227.3	309.7	9
3.3	4.3	5.0	1.5	6.5	8.7	0	6.5	8.7	10
162.6	178.9	177.9	21.4	199.3	225.1	86.7	286.0	311.8	11
0.2	0.2	1.2	0	1.2	1.2	0	1.2	1.2	12
4.0	6.5	5.4	3.3	8.7	61.7	1.9	10.6	63.6	13
41.7	49.3	92.6	372.9	465.5	534.2	5.7	471.2	539.9	14
1.3	1.3	10.3	0.3	10.6	10.6	0	10.6	10.6	15
40.0	40.1	41.9	29.1	71.0	179.2	0	71.0	179.2	16
15.0	17.1	57.8	3.3	61.1	188.6	1.1	62.2	189.7	17
142.0	159.1	245.1	36.7	281.8	341.4	16.2	298.0	357.6	18
0	0	6.8	1.1	7.9	21.1	0	7.9	21.1	19
142.3	177.6	149.6	38.5	188.1	192.1	3.1	191.2	195.2	20
463.0	527.3	541.4	137.3	678.7	682.0	731.0	1,409.7	1,413.3	21
268.4	277.1	274.2	9.7	283.9	283.9	0	283.9	283.9	22
6.0	6.0	6.5	0.1	6.6	6.6	0.1	6.7	6.7	23
179.4	187.6	198.3	10.3	208.6	213.7	117.7	326.3	331.4	24
1,299.3	1,919.4	1,657.2	781.4	2,438.6	3,116.4	158.2	2,596.8	3,274.6	25
171.3	187.9	272.8	38.8	311.6	471.6	138.3	449.9	609.9	26
14.1	14.4	15.5	1.0	16.5	16.5	5.5	22.0	22.0	27
93.2	109.4	331.5	81.1	412.6	674.4	12.5	425.1	686.9	28
4.0	5.4	147.5	10.5	158.0	254.5	22.4	180.4	276.9	29
254.2	282.0	1,011.8	46.6	1,058.4	2,947.7	86.4	1,144.8	3,034.1	30
7,816.9	9,304.3	10,949.2	2,603.5	13,552.7	18,901.3	1,812.4	15,365.1	20,713.7	

RP—D

The CBI Survey
Table 90   Summary (by River Authorities) of daily volumes of effluent discharged in 1968 (x 1,000 gpd)

River authority	To controlled waters (Tidal, non-tidal and canals)			To controlled waters plus sewers			Total effluent to other than controlled waters or sewers	Total effluent
	Total effluent	Cooling water	Process water	Total effluent	Cooling water	Process water		
	55 (1+7+13)	56 (3+9+15)	57 (2+8+14)	58 (23+55)	59 (25+56)	60 (24+57)	61 (19+30+34)	62 (58+61)
1 Northumbrian	580,304	504,670	75,634	588,902	509,513	79,389	920,796	1,509,698
2 Yorkshire	1,403,966	1,337,141	66,825	1,432,734	1,344,844	87,890	4,590	1,437,324
3 Trent	1,751 576	1,676,046	75,530	1,827,244	1,687,579	139,665	3,371	1,830,615
4 Lincolnshire	42,468	22,785	19,683	44,320	22,915	21,405	3,206	47,526
5 Welland and Nene	67,072	63,759	3,313	69,739	63,856	5,883	1	69,740
6 Great Ouse	142,966	132,013	10,953	145,356	132,411	12,945	603	145,959
7 East Suffolk and Norfolk	179,007	168,073	10,934	181,135	168,239	12,896	163,426	344,561
8 Essex	331,367	329,590	1,777	338,037	33,217	5,867	25	338,062
9 Kent	245,408	216,393	29,015	251,615	216,666	34,949	68	251,683
10 Sussex	500	361	139	1,128	439	689	267,017	268,145
11 Hampshire	175,293	161,230	14,063	176,653	161,843	14,810	465,535	642,188
12 Isle of Wight	12	0	12	24	0	24	0	24
13 Avon and Dorset	770	529	241	1,053	537	516	171,270	172,323
14 Devon	30,666	26,581	4,085	31,643	26,607	5,036	111	31,754
15 Cornwall	3,405	30,050	4,009	34,274	30,055	4,219	5,640	39,914
16 Somerset	604,651	595,260	9,391	608,622	596,270	12,352	0	608,622
17 Bristol Avon	57,744	40,749	16,995	79,824	45,979	33,845	51	79,875
18 Severn	947,949	930,245	17,704	954,048	931,067	22,981	7	954,055
19 Wye	0	0	0	14,138	0	14,138	0	14,138
20 Usk	417,637	408,891	8,746	418,079	408,986	9,093	25	418,104
21 Glamorgan	57,171	47,849	9,322	65,004	52,749	12,255	331,907	396,911
22 South West Wales	275,508	262,456	13,052	275,807	262,548	13,259	32	275,839
23 Gwynedd	843,968	843,859	109	844,356	844,218	138	328,085	1,172,441
24 Dee and Clywd	4,371	280	4,091	4,902	548	4,354	60	4,962
25 Mersey and Weaver	1,150,571	1,049,600	100,971	1,227,780	1,062,562	165,218	20,704	1,248,484
26 Lancashire	198,523	140,391	58,132	225,170	151,789	73,381	4,060	229,230
27 Cumberland	37,501	34,734	2,767	39,650	35,323	4,327	15,602	55,252
28 Thames Conservancy	118,393	65,556	52,837	146,012	70,079	75,933	115	146,127
29 Lee Conservancy	93,080	92,689	391	101,620	94,274	7,346	108	101,728
30 Port of London Authority (including London Excluded Area)	3,388,907	3,336,808	52,099	3,458,197	3,338,095	120,102	22,934	3,481,131
Total England and Wales (excluding Public sector of the Iron and Steel Industry)	13,181,408	12,518,588	662,820	13,587,066	12,592,161	994,905	2,729,349	16,316,415

6 Expenditure

Introduction

1 Estimates have been prepared to give an indication of the order of magnitude of the cost of remedial works on those discharges which are unsatisfactory at present and, in the case of sewage effluent and crude sewage discharges, on those which are likely to become unsatisfactory by 1980 because of, for example, increases in population and sewage flow. The estimates were obtained from a number of sources and their dates varied but they can be generally regarded as representing early 1970 prices. The sources of the estimates were as follows:—

Sewage Effluent

In many cases estimates of cost were provided by the sewerage authority on their sub-form B returns. If no estimate was provided and it was likely to be high—of the order of £400,000 or more—the sewerage authority was contacted and, in many cases, was able to assist, either direct or through their consulting engineers. A visit was made to the Upper Tame Main Drainage Authority because of their extensive responsibilities in the Trent River Authority area. Where it was known that a scheme had been submitted to the Department, the estimate on Form K29 was used.

If the estimated costs could not be obtained by any of the above means, they were prepared in the Department by reference to the details provided (on Form B) of the design capacity of the existing treatment works and the details of the forecast conditions in 1980. Estimates were based on an average cost per head for the work of extending from the present design capacity to the likely future capacity. The Welsh Office were consulted in respect of all estimates relating to Wales.

Crude Sewage

Because, in many cases, the remedial measures for crude sewage discharges would involve large intercepting sewers, the costs of which are even more difficult to estimate than extensions to treatment works (a good example of this is the solution proposed for Tyneside), it was decided to contact each individual sewerage authority where costs had not been supplied and were likely to be other than nominal and most of the estimates used were obtained in this way. In many cases detailed estimates were available and in other cases broad estimates were provided. Again the Welsh Office co-operated in the case of discharges in Wales.

Unsatisfactory storm overflows

The wide range of remedial measures shown to be necessary has been described in Chapter 4, but in most cases little or no information was provided to give a guide as to the sizes of sewers or the extent of work required where "re-sewering" was shown to be required or where "segregation of surface water" was suggested—both of these being potentially very costly. However, the Yorkshire River Authority, with 302 unsatisfactory overflows and a wide range of remedial works in a variety of situations, were able to give a great deal of information on costs and this was used as the basis for assessing a national figure. The remedial measures were divided into broad bands—"overflow modification", "new overflow", "local sewerage" and so on and an average cost per overflow for each type of remedial measure was arrived at. These average costs were then applied to all the details provided by the other authorities to give an indication of national costs.

Industrial effluent

In some cases, costs were provided by the river authorities, either from their own estimates or from estimates provided by the discharger. Where no such estimates were given, they were prepared in the Department on the basis of an assessment of the treatment that would be needed to meet the required standards. In all such cases, there was close consultation with the river authorities.

Costs relate only to discharges considered unsatisfactory at present. No attempt was made to forecast future costs because of the difficulty of obtaining reliable information on which to base them. Where a river authority indicated that the remedy should be diversion to the public sewers, the cost of diverting the flow has been included in the estimates but not the cost of any pre-treatment that the sewerage authority might require. The cost of providing the necessary treatment capacity also causes some difficulty. In some cases the local authorities were aware of the intentions and had allowed for the necessary costs in their estimates but there could be cases where extensions beyond those covered by the estimates might be needed to deal with industrial effluent so diverted. The extra costs relative to the costs shown for industrial discharges could be significant, but they are unlikely to be large in relation to the overall total of expenditure. The cost of conveyance and treatment at the sewage disposal works would usually, in any case, be recovered from industry in the form of an annual charge.

The estimates do not include any costs for treating water raised or drained from underground mines.

2 Estimates prepared in this way can at best only be very approximate in the absence of detailed knowledge of each site, the size and condition of existing structures and the extent to which they could be incorporated into extensions. It would therefore be misleading to place reliance on individual estimates but, overall, they give an approximate indication of the magnitude of the costs.

3 It should be noted that the estimates do not include the cost of those works indicated in the returns as being under construction at the time—on the grounds that the finances will already have been allocated. Where, however, a long-term programme of uninterrupted construction was known to be under way, costs were included to cover the amounts still to be sanctioned. Also excluded are the costs of providing sewage treatment exclusively for areas not yet provided with main drainage.

4 No costs are included for sewerage except that, in cases where the flow was to be diverted from one treatment works to another, an attempt was made to estimate the cost of diversion. It is very important to note that the annual capital expenditure on works of sewage treatment is normally less than half of the total annual expenditure on sewerage and sewage treatment. The records of the Departments do not break down expenditure accurately in this way, but it is not unreasonable to assume that for each £1 million shown in the subsequent estimates to be required on sewage treatment, at least a further £1 million will have to be spent on sewerage—and those latter amounts are not included. From the current year Departmental records of costs will show expenditure on sewerage and sewage treatment separately.

5 Because sewage works construction is going on continuously throughout the country, some of the money indicated as requiring to be spent will be associated with remedial works already in hand or perhaps, in some cases, already completed.

Details of the estimates for England and Wales

6 A summary of the estimates is at Table 91. Costs shown under "Remedial works required immediately" usually represent the total cost of remedial measures on discharges considered to be unsatisfactory at present and can be considered as indicating a "backlog".

In actual practice, it would take several years to complete the works which these costs cover.

7 To the figure of £609.352 million has to be added the estimated costs for remedial measures on unsatisfactory storm overflows. As a very approximate guide to the likely cost, the figure derived in the manner already described is about £170 million for the whole of England and Wales. The figure is not sufficiently precise to warrant breaking down.

8 Details of the estimated remedial costs for sewage effluent, crude sewage and industrial effluent by river authorities are shown in Tables 92, 93 and 94 and a further breakdown of costs appears in Tables 218–220 at the end of the Report.

Table 91 Estimates of costs of remedial works in £ millions. England and Wales

Nature of expenditure	Non-tidal rivers		Tidal rivers		All rivers		Canals		All rivers and canals		Total of expenditure
	Remedial works required immediately	Other remedial works to 1980	Remedial works required immediately	Other remedial works to 1980	Remedial works required immediately	Other remedial works to 1980	Remedial works required immediately	Other remedial works to 1980	Remedial works required immediately	Other remedial works to 1980	
1 Discharges of sewage effluent	214.205	117.993	53.102	38.614	267.307	156.607	0.495	1.843	267.802	158.450	426.252
2 Discharges of crude sewage	2.139	0.420	141.325	0.483	143.464	0.903	0	0	143.464	0.903	144.367
3 Discharges of industrial effluent	21.622	—	16.332	—	37.954	—	0.779	—	38.733	—	38.733
Total of estimates	237.966	118.413	210.759	39.097	448.725	157.510	1.274	1.843	449.999	159.353	609.352

Notes:
1 No information was asked for about future costs of treating industrial effluent and none was supplied.
2 In addition to incurring expenditure on direct discharges to rivers and canals, industry also incurs heavy expenditure on pre-treatment of industrial effluents before discharge to sewers, on charges made by sewerage authorities for conveyance and treatment of industrial effluents and on disposal of effluents by other private arrangements. Information about this is given in the part of the report dealing with the CBI survey (Chapter 5).
3 Attention is drawn to that part of paragraph 1 which deals with the estimates of costs of remedial works on discharges of industrial effluent.
4 "Remedial works required immediately" are likely to be phased over several years. "Other remedial works" will be carried out as the need arises.

Table 92 Estimates of costs of remedial works on discharges of sewage effluent in £ millions. England and Wales

River Authority	Non-tidal rivers		Tidal rivers		All rivers		Canals		All rivers and canals		Total of expenditure to 1980
	Remedial works required immediately	Other remedial works to 1980	Remedial works required immediately	Other remedial works to 1980	Remedial works required immediately	Other remedial works to 1980	Remedial works required immediately	Other remedial works to 1980	Remedial works required immediately	Other remedial works to 1980	
Northumbrian	4.321	1.202	1.564	0.850	5.885	2.052	0	0	5.885	2.052	7.937
Yorkshire	35.062	13.954	2.580	1.860	37.642	15.814	0	0.102	37.642	15.916	53.558
Trent	67.124	11.643	0.196	0	67.320	11.643	0.255	1.565	67.575	13.208	80.783
Lincolnshire	0.726	1.933	0	0.018	0.726	1.951	0	0	0.726	1.951	2.677
Welland and Nene	2.469	8.854	0.270	2.747	2.739	11.601	0.010	0.005	2.749	11.606	14.355
Great Ouse	8.366	12.848	0.425	0.210	8.791	13.058	0	0	8.791	13.058	21.849
East Suffolk and Norfolk	0.885	0.573	4.776	1.338	5.661	1.911	0	0	5.661	1.911	7.572
Essex	2.906	2.825	2.427	0.895	5.333	3.720	0	0	5.333	3.720	9.053
Kent	1.765	4.716	0.610	6.978	2.375	11.694	0	0	2.375	11.694	14.069
Sussex	2.365	1.905	0.545	0.070	2.910	1.975	0	0	2.910	1.975	4.885
Hampshire	1.232	0.765	3.315	1.970	4.547	2.735	0	0	4.547	2.735	7.282
Isle of Wight	0.146	0.134	0.300	0	0.446	0.134	0	0	0.446	0.134	0.580
Avon and Dorset	1.151	2.316	0	0.245	1.151	2.561	0	0	1.151	2.561	3.712
Devon	2.386	0.642	2.227	0.018	4.613	0.660	0	0	4.613	0.660	5.273
Cornwall	0.822	0.949	0.045	0.155	0.867	1.104	0	0	0.867	1.104	1.971
Somerset	3.677	1.592	0	1.250	3.677	2.842	0	0	3.677	2.842	6.519
Bristol Avon	3.308	3.926	0.060	0	3.368	3.926	0	0	3.368	3.926	7.294
Severn	7.736	20.668	0.537	2.085	8.273	22.753	0	0.075	8.273	22.828	31.101
Wye	2.093	0.745	0	0	2.093	0.745	0	0	2.093	0.745	2.838
Usk	0.305	0.525	3.200	0	3.505	0.525	0	0	3.505	0.525	4.030
Glamorgan	8.225	0.047	0	6.500	8.225	6.547	0	0	8.225	6.547	14.772
South West Wales	0.635	0.317	0.115	0.835	0.750	1.152	0	0	0.750	1.152	1.902
Gwynedd	0.361	0.598	0.105	0.070	0.466	0.668	0	0	0.466	0.668	1.134
Dee and Clwyd	1.792	0.892	0.210	0.750	2.002	1.642	0	0	2.002	1.642	3.644
Mersey and Weaver	23.841	5.837	2.520	1.500	26.361	7.337	0	0	26.361	7.337	33.698
Lancashire	13.411	2.330	1.827	2.000	15.238	4.330	0	0	15.238	4.330	19.568
Cumberland	1.667	0.210	0.122	0	1.789	0.210	0	0	1.789	0.210	1.999
Thames Conservancy	14.114	10.537	0	0	14.114	10.537	0.230	0.096	14.344	10.633	24.977
Lee Conservancy	1.314	4.510	0	0	1.314	4.510	0	0	1.314	4.510	5.824
Port of London Authority (incl. London Excluded Area)	0	0	25.126	6.270	25.126	6.270	0	0	25.126	6.270	31.396
Total of estimates	214.205	117.993	53.102	38.614	267.307	156.607	0.495	1.843	267.802	158.450	426.252

9 The estimated cost for remedial works on discharges of sewage effluent is about £425 million and nearly two-thirds of this represents the cost of remedial works required immediately. The totals include an allowance for the works necessary to improve the effluent standard where the river authority indicated that higher standards would be called for. Particularly large expenditure is indicated as likely in the Trent RA area (about £80 million), Yorkshire RA area (about £54 million), Mersey and Weaver RA area (about £34 million), Port of London Authority including the London Excluded Area (about £31 million), Severn RA area (about £31 million) and Thames Conservancy Area (about £25 million).

10 These figures relate to only a part of the problem because some authorities are likely to incur much greater expenditure on the treatment of discharges of crude sewage. The estimates indicate an overall cost of about £145 million, almost all of this to deal with discharges at present unsatisfactory. The bulk of this expenditure relates to tidal waters and river authorities which are prominent are Northumbrian (about £68 million) and Mersey and Weaver (about £38 million). Again, it must be remembered that certain costly schemes are excluded from the totals because they relate to discharges to controlled waters outside the scope of the survey—notably, in this context, The Solent and the Bristol Channel.

11 The estimated expenditure on treatment of industrial effluents is spread over 30 different classes of industry with the chemical, paper and board, iron and steel, quarrying, textiles (cotton and synthetic) and petroleum refining accounting for about 75 per cent of all the costs.

12 Unsatisfactory storm overflows may often be remedied as an additional benefit from schemes carried out primarily for another purpose. They are intermittent discharges and although many should clearly receive early attention, it is probable that, given a limited amount of money, there might be advantages in concentrating more on the continuous and unsatisfactory discharges.

13 Table 95 shows the estimated costs per head of the population in each river authority area for dealing with discharges of sewage effluent and crude sewage. The estimates are taken from Tables 92 and 93 and the populations are those given in Table 7 of Volume 1.

14 The costs per head shown on Table 95 can only be approximate because no allowance has been made for increases in population during the period covered by the estimates (ie up to 1980) and the estimates are based on 1970 prices. However, the table indicates the order of expenditure and shows where it is likely to be high and where it is likely to be low.

15 A summary of the situation in each area follows on page 87 but no detailed reference is made to storm-overflow costs. Brief references to the possible extent of the remedial works have been made in Chapter 4.

Table 93 Estimates of costs of remedial works on discharges of crude sewage in £ millions. England and Wales

River Authority	Non-tidal rivers		Tidal rivers		All rivers		
	Remedial works required immediately	Other remedial works to 1980	Remedial works required immediately	Other remedial works to 1980	Remedial works required immediately	Other remedial works to 1980	Total of all estimates
Northumbrian	0	0	67.611	0	67.611	0	67.611
Yorkshire	0.210	0	1.195	0	1.405	0	1.405
Trent	0	0	0	0	0	0	0
Lincolnshire	0	0	1.000	0	1.000	0	1.000
Welland and Nene	0	0	0	0	0	0	0
Great Ouse	0	0	0	0	0	0	0
East Suffolk and Norfolk	0	0	0.200	0	0.200	0	0.200
Essex	0	0	0.354	0	0.354	0	0.354
Kent	0	0	0.765	0	0.765	0	0.765
Sussex	0	0	1.500	0	1.500	0	1.500
Hampshire	0	0	0.100	0	0.100	0	0.100
Isle of Wight	0	0	1.200	0	1.200	0	1.200
Avon and Dorset	0	0	0	0	0	0	0
Devon	0.450	0	4.195	0	4.645	0	4.645
Cornwall	0	0	4.496	0.443	4.496	0.443	4.939
Somerset	0.320	0.040	4.010	0	4.330	0.040	4.370
Bristol Avon	0.793	0	1.715	0	2.508	0	2.508
Severn	0	0	0.850	0	0.850	0	0.850
Wye	0	0	0.110	0	0.110	0	0.110
Usk	0	0	0.762	0	0.762	0	0.762
Glamorgan	0	0	3.110	0	3.110	0	3.110
South West Wales	0.169	0	4.458	0	4.627	0	4.627
Gwynedd	0.197	0.380	1.174	0	1.371	0.380	1.751
Dee and Clwyd	0	0	0.300	0	0.300	0	0.300
Mersey and Weaver	0	0	38.160	0	38.160	0	38.160
Lancashire	0	0	3.930	0.040	3.930	0.040	3.970
Cumberland	0	0	0.130	0	0.130	0	0.130
Thames Conservancy	0	0	0	0	0	0	0
Lee Conservancy	0	0	0	0	0	0	0
Port of London Authority (incl. London Excluded Area)	0	0	0	0	0	0	0
Total of estimates	2.139	0.420	141.325	0.483	143.464	0.903	144.367

Note: There are no discharges of crude sewage to canals.

Table 94 Estimates of costs of remedial works on discharges of industrial effluent in £ millions. England and Wales.

River Authority	Non-tidal rivers	Tidal rivers	All rivers	Canals	Total of expenditure
Northumbrian	0.192	3.615	3.807	0	3.807
Yorkshire	4.404	0.272	4.676	0.063	4.739
Trent	1.074	1.100	2.174	0.361	2.535
Lincolnshire	0.099	0.002	0.101	0	0.101
Welland and Nene	0.038	0.003	0.041	0	0.041
Great Ouse	0.046	0.025	0.071	0	0.071
East Suffolk and Norfolk	0.045	0.079	0.124	0	0.124
Essex	0.143	0.016	0.159	0	0.159
Kent	0.010	1.258	1.268	0	1.268
Sussex	0.024	0.002	0.026	0	0.026
Hampshire	0.009	0.020	0.029	0	0.029
Isle of Wight	0.005	0	0.005	0	0.005
Avon and Dorset	0.013	0	0.013	0	0.013
Devon	0.143	0.020	0.163	0	0.163
Cornwall	3.658	0.002	3.660	0	3.660
Somerset	2.140	1.470	3.610	0	3.610
Bristol Avon	0.152	0	0.152	0	0.152
Severn	0.348	0.002	0.350	0.126	0.476
Wye	0.002	0	0.002	0	0.002
Usk	1.128	0	1.128	0	1.128
Glamorgan	0.369	0.003	0.372	0	0.372
South West Wales	1.388	0.465	1.853	0.004	1.857
Gwynedd	0.012	0	0.012	0	0.012
Dee and Clwyd	0.080	0.203	0.283	0	0.283
Mersey and Weaver	4.833	5.825	10.658	0.200	10.858
Lancashire	0.868	0.329	1.197	0	1.197
Cumberland	0.119	0.200	0.319	0	0.319
Thames Conservancy	0.280	0	0.280	0	0.280
Lee Conservancy	0	0	0	0	0
Port of London Authority (incl. London Excluded Area)	0	1.421	1.421	0.025	1.446
Total of estimates	21.622	16.332	37.954	0.779	38.733

See footnote to Table 91.

Table 95 Estimated costs per head of population for dealing with discharges of sewage effluent and crude sewage. England and Wales.

River Authority	Total estimated cost (£ millions) for dealing with			Approximate population	Approx. cost per head (£)
	Sewage effluent	Crude sewage	Sewage effluent and crude sewage		
Northumbrian	7.937	67.611	75.548	2,708,000	27.9
Yorkshire	53.558	1.405	54.963	4,573,000	12.0
Trent	80.783	0	80.783	5,557,000	14.5
Lincolnshire	2.677	1.000	3.677	658,000	5.6
Welland and Nene	14.355	0	14.355	741,000	19.4
Great Ouse	21.849	0	21.849	1,100,000	19.9
East Suffolk and Norfolk	7.572	0.200	7.772	815,000	9.5
Essex	9.053	0.354	9.407	2,050,000	4.6
Kent	14.069	0.765	14.834	1,770,000	8.4
Sussex	4.885	1.500	6.385	1,138,000	5.6
Hampshire	7.282	0.100	7.382	1,064,000	6.9
Isle of Wight	0.580	1.200	1.780	105,000	17.0
Avon and Dorset	3.712	0	3.712	650,000	5.7
Devon	5.273	4.645	9.918	412,000	24.1
Cornwall	1.971	4.939	6.910	610,000	11.3
Somerset	6.519	4.370	10.889	450,000	24.2
Bristol Avon	7.294	2.508	9.802	925,000	10.6
Severn	31.101	0.850	31.951	2,000,000	16.0
Wye	2.838	0.110	2.948	191,000	15.4
Usk	4.030	0.762	4.792	420,000	11.4
Glamorgan	14.772	3.110	17.882	998,000	17.9
South West Wales	1.902	4.627	6.529	595,000	11.0
Gwynedd	1.134	1.751	2.885	250,000	11.5
Dee and Clwyd	3.644	0.300	3.944	505,000	7.8
Mersey and Weaver	33.698	38.160	71.858	5,250,000	13.7
Lancashire	19.568	3.970	23.538	1,624,000	14.5
Cumberland	1.999	0.130	2.129	304,000	7.0
Thames Conservancy	24.977	0	24.977	3,758,000	6.6
Lee Conservancy	5.824	0	5.824	1,800,000	3.2
Port of London Authority (incl. London Excluded Area)	31.396	0	31.396	5,400,000	5.8
Total of estimates	426.252	144.367	570.619	48,421,000	£11.8 per head average

1 Northumbrian River Authority

Table 96 Estimates of costs of remedial works in £ millions

Nature of expenditure	Non-tidal rivers		Tidal rivers		All rivers		Total of expenditure to 1980
	Remedial works required immediately	Other remedial works to 1980	Remedial works required immediately	Other remedial works to 1980	Remedial works required immediately	Other remedial works to 1980	
1 Discharges of sewage effluent	4.321	1.202	1.564	0.850	5.885	2.052	7.937
2 Discharges of crude sewage	0	0	67.611	0	67.611	0	67.611
3 Discharges of industrial effluent	0.192	—	3.615	—	3.807	—	3.807
Total of estimates	4.513	1.202	72.790	0.850	77.303	2.052	79.355

16 The estimates indicate that about £8 million requires to be spent on discharges of sewage effluent and some £68 million on discharges of crude sewage. All but about £2 million of this expenditure represents the cost of remedial works required urgently. Expenditure on discharges of industrial effluent at present unsatisfactory amounts to nearly £4 million, giving an overall total in the area of about £80 million.

17 The survey shows that the main expenditure on discharges of sewage effluent is likely to be incurred on discharges to the Tyne, the Wear, the Tees, the Lumley Park Burn, the North Burn, the Blyth and the Skerne. This expenditure (much of which has already been committed) is, however, small compared with the great expenditure required on discharges of crude sewage to the tidal estuaries. The estimates show that about £61 million will have to be spent on Tyneside and Teesside to deal with the many discharges of crude sewage to the estuaries there. The balance of about £7 million will be required to deal mainly with discharges of crude sewage to the Tweed, the Blyth and the Wear.

18 The returns indicate that the bulk of the expenditure on unsatisfactory industrial discharges will be incurred on discharges from chemical and allied industries. ☐

2 Yorkshire River Authority

Table 97 Estimates of costs of remedial works in £ millions

Nature of expenditure	Non-tidal rivers		Tidal rivers		All rivers		Canals		All rivers and canals		Total of expenditure to 1980
	Remedial works required immediately	Other remedial works to 1980	Remedial works required immediately	Other remedial works to 1980	Remedial works required immediately	Other remedial works to 1980	Remedial works required immediately	Other remedial works to 1980	Remedial works required immediately	Other remedial works to 1980	
1 Discharges of sewage effluent	35.062	13.954	2.580	1.860	37.642	15.814	0	0.102	37.642	15.916	53.558
2 Discharges of crude sewage	0.210	0	1.195	0	1.405	0	0	0	1.405	0	1.405
3 Discharges of industrial effluent	4.404	—	0.272	—	4.676	—	0.063	—	4.739	—	4.739
Total of estimates	39.676	13.954	4.047	1.860	43.723	15.814	0.063	0.102	43.786	15.916	59.702

19 The estimates indicate that the bulk of the expenditure in the period to 1980 will be incurred on discharges of sewage effluent, mainly to non-tidal rivers. The estimated total of all expenditure approaches £60 million and about three-quarters of this represents remedial works required urgently. Expenditure on discharges of crude sewage will be relatively low at about £1.4 million and, on industrial discharges moderate at £4.7 million, having regard to the very large volume of industrial effluent discharged to rivers in the area.

20 Most of the expenditure on discharges of sewage effluent will be incurred in dealing with discharges to rivers in the Ouse basin, particularly in the South Yorkshire industrial zone. The estimates show a need to spend about £53 million on sewage effluents in the Ouse basin, with particularly heavy expenditure being required on discharges to the Don, the Dearne, the Aire, the Calder, the Wyke Beck, the Wharfe and the Rother which, between them, account for about £36 million. The main expenditure on discharges of crude sewage will be incurred on the tidal stretches of the Ouse.

21 The cost of remedial works on industrial discharges is spread over some 20 different classes of industry, notably chemical and allied industries, electricity generation, engineering of all kinds, food processing, gas and coke production, iron and steel manufacture, paper and board making and textiles.

22 Expenditure on direct discharges to the Humber estuary is not included in the above. ☐

3 Trent River Authority

(see Table 98 over page)

23 A significant point in the area is that no new expenditure will be required on discharges of crude sewage as a scheme is already under construction to deal with the only crude sewage discharges in the area—those from Gainsborough to the lower tidal length of the Trent. However, very heavy expenditure will be required on discharges of sewage effluent—by far the greatest of any river authority in the country. The figure is estimated at just over £80 million, all but about £2 million being applicable to discharges to non-tidal waters and over £67 million represents expenditure required urgently. Most of the £2 million referred to above represents expenditure on discharges to canals.

24 The estimates show a need to spend a sum in the region of £26 million on discharges to the Tame principally on the section which

(3 Trent River Authority continued)

Table 98 Estimates of costs of remedial works in £ millions

Nature of expenditure	Non-tidal rivers		Tidal rivers		All rivers		Canals		All rivers and canals		
	Remedial works required immediately	Other remedial works to 1980	Remedial works required immediately	Other remedial works to 1980	Remedial works required immediately	Other remedial works to 1980	Remedial works required immediately	Other remedial works to 1980	Remedial works required immediately	Other remedial works to 1980	Total of expenditure to 1980
1 Discharges of sewage effluent	67.124	11.643	0.196	0	67.320	11.643	0.255	1.565	67.575	13.208	80.783
2 Discharges of crude sewage	0	0	0	0	0	0	0	0	0	0	0
3 Discharges of industrial effluent	1.074	—	1.100	—	2.174	—	0.361	—	2.535	—	2.535
Total of estimates	68.198	11.643	1.296	0	69.494	11.643	0.616	1.565	70.110	13.208	83.318

passes through West Bromwich, Walsall and Birmingham, and some £9 million on discharges to the Trent, mostly on the upper length between Stoke and Nottingham. Heavy expenditure—in the region of £8 million—is also indicated as being necessary on discharges to the Wolverhampton Arm of the Tame whilst about £9 million will have to be spent on discharges to the Derwent and the Tipton Brook. Substantial expenditure—in the region of about £15 million altogether—is indicated as being necessary on discharges to the Ryton, Oldcoates Dyke, the Erewash, the Soar, Wem Brook, Sketchley Brook, Rough Brook, Saredon Brook, Bottesford Beck, the Wreake and the Blythe. Most of the expenditure on discharges to the Tame, the Wolverhampton Arm of the Tame, Rough Brook, Tipton Brook and the Blythe relates to works owned by Upper Tame Main Drainage Authority who are already engaged upon a major long-term programme of re-construction and remedial works. The expenditure needed on discharges to canals is spread over several, the highest expenditure being most likely on discharges to the Staffordshire and Worcestershire Canal, Nottingham Canal and the Shropshire Union Canal.

25 The expenditure needed on industrial discharges is spread over some 20 different classes of industry. A substantial proportion of the total estimated expenditure relates to the iron and steel manufacturing industry. Others of note are chemical and allied industries, coal-mining process water, electricity generation, engineering, food processing, paper and board making, plating and metal finishing and quarrying, but in relation to the amount of industry in the area, the individual costs against each of these are small. The estimates also show a need to spend a small sum on treating minewater from active coal mines, but this is not included in the total. ☐

4 Lincolnshire River Authority

Table 99 Estimates of costs of remedial works in £ millions

Nature of expenditure	Non-tidal rivers		Tidal rivers		All rivers		
	Remedial works required immediately	Other remedial works to 1980	Remedial works required immediately	Other remedial works to 1980	Remedial works required immediately	Other remedial works to 1980	Total of expenditure to 1980
1 Discharges of sewage effluent	0.726	1.933	0	0.018	0.726	1.951	2.677
2 Discharges of crude sewage	0	0	1.000	0	1.000	0	1.000
3 Discharges of industrial effluent	0.099	—	0.002	—	0.101	—	0.101
Total of estimates	0.825	1.933	1.002	0.018	1.827	1.951	3.778

26 Only moderate expenditure will be required in the sum of just under £4 million, about half of which needs to be spent immediately. About 70 per cent of the total expenditure relates to discharges of sewage effluent to non-tidal waters with the largest expenditure forecast for discharges to the Witham, the Hammond Beck and Hobhole Drain. None of the expenditure on these three rivers, however, is immediately necessary. The urgent expenditure is spread in relatively small amounts over several rivers. Expenditure of £1 million on crude sewage relates to the works necessary to deal with discharges to the tidal section of the Witham.

27 The costs of improving industrial discharges are low and associated mainly with discharges from food processing.

28 Expenditure on direct discharges to the Humber Estuary is not included in the above. ☐

5 Welland and Nene River Authority

Table 100 Estimates of costs of remedial works in £ millions

Nature of expenditure	Non-tidal rivers		Tidal rivers		All rivers		Canals		All rivers and canals		
	Remedial works required immediately	Other remedial works to 1980	Remedial works required immediately	Other remedial works to 1980	Remedial works required immediately	Other remedial works to 1980	Remedial works required immediately	Other remedial works to 1980	Remedial works required immediately	Other remedial works to 1980	Total of expenditure to 1980
1 Discharges of sewage effluent	2.469	8.854	0.270	2.747	2.739	11.601	0.010	0.005	2.749	11.606	14.355
2 Discharges of crude sewage	0	0	0	0	0	0	0	0	0	0	0
3 Discharges of industrial effluent	0.038	—	0.003	—	0.041	—	0	—	0.041	—	0.041
Total of estimates	2.507	8.854	0.273	2.747	2.780	11.601	0.010	0.005	2.790	11.606	14.396

29 Virtually all of the expenditure in the area relates to discharges of sewage effluent to rivers and only about 20 per cent of this is considered to be needed urgently.

30 The biggest expenditure is likely to be on discharges to the Nene, and could be of the order of £9 million, but much of this is associated with town expansion, and will be required in a few years time rather than immediately. Expenditure on discharges to the Welland is likely to exceed £1 million, a good part of this being considered urgent and more than £1 million will have to be spent on the Whilton Arm of the Nene. Expenditure approaching £0.5 million is envisaged on discharges to the Ise, but otherwise the bulk of the estimated costs relate to medium and small amounts on a large number of rivers. The expenditure recorded under "Canals" relates to discharges to the Grand Union Canal and to one of its feeders, the Norton Stream.

31 Expenditure on industrial discharges is insignificant and relates mainly to the leather tanning and fellmongering industry.

6 Great Ouse River Authority

Table 101 Estimates of costs of remedial works in £ millions

Nature of expenditure	Non-tidal rivers		Tidal rivers		All rivers		
	Remedial works required immediately	Other remedial works to 1980	Remedial works required immediately	Other remedial works to 1980	Remedial works required immediately	Other remedial works to 1980	Total of expenditure to 1980
1 Discharges of sewage effluent	8.366	12.848	0.425	0.210	8.791	13.058	21.849
2 Discharges of crude sewage	0	0	0	0	0	0	0
3 Discharges of industrial effluent	0.046	—	0.025	—	0.071	—	0.071
Total of estimates	8.412	12.848	0.450	0.210	8.862	13.058	21.920

32 Most of the expenditure of about £22 million relates to remedial works required for discharges of sewage effluent to non-tidal rivers and some £8.8 million (40 per cent) is considered to be required urgently.

33 Of the expenditure required, about £10 million will be associated with discharges to the upper stretches of the Great Ouse in the earlier part of the period up to 1980 and a big proportion of this relates to the new treatment works for Milton Keynes. Substantial expenditure is also expected on discharges to the River Flit/Ivel Navigation where costs are likely to exceed £2 million, most to be spent in the near future, and on the Middle Level Drainage System where nearly £1.4 million is required immediately. More than £2 million will be required for works discharging to the River Ivel, the Soham Lode and the Cam and most of this is required immediately. There are several other smaller items of expenditure, almost all at works discharging effluent in the Great Ouse basin.

34 Expenditure on industrial discharges is comparatively very low and required for the leather, food processing & laundering industries.

7 East Suffolk and Norfolk River Authority

Table 102 Estimates of costs of remedial works in £ millions

Nature of expenditure	Non-tidal rivers		Tidal rivers		All rivers		
	Remedial works required immediately	Other remedial works to 1980	Remedial works required immediately	Other remedial works to 1980	Remedial works required immediately	Other remedial works to 1980	Total of expenditure to 1980
1 Discharges of sewage effluent	0.885	0.573	4.776	1.338	5.661	1.911	7.572
2 Discharges of crude sewage	0	0	0.200	0	0.200	0	0.200
3 Discharges of industrial effluent	0.045	—	0.079	—	0.124	—	0.124
Total of estimates	0.930	0.573	5.055	1.338	5.985	1.911	7.896

35 Expenditure on effluents from sewage treatment works accounts for nearly all the expenditure in this area and about 75 per cent relates to remedial works required immediately, mostly on discharges to tidal rivers.

36 Nearly £6 million will have to be spent on sewage effluent discharges to the tidal sections of three rivers, the Yare, where about £2.2 million is required, almost all immediately, the Orwell, where about £2.7 million is required immediately, and the Ant, one of the principal Broadland rivers, where about £1 million will be required in the near future. About half of all the expenditure shown to be necessary on non-tidal rivers relates to discharges to the Waveney, the Gipping and the Tiffey. Most of this expenditure is required immediately. A small amount of expenditure will be required for remedial works on crude sewage discharges to the Yare, the Orwell and the Bure.

37 Expenditure on industrial discharges is also small and is required mainly for wastes from leather, food processing, brewing and engineering industries in the area.

8 Essex River Authority

(see Table 103 over page)

38 The estimates indicate that nearly all the necessary expenditure relates to sewage effluent discharges with some £5.3 million of this being required immediately. Rather more expenditure is forecast on non-tidal rivers than on tidal rivers. Only small expenditure is forecast for discharges of crude sewage and industrial effluent.

(8 Essex River Authority continued)

Table 103 Estimates of costs of remedial works in £ millions

Nature of expenditure	Non-tidal rivers		Tidal rivers		All rivers		
	Remedial works required immediately	Other remedial works to 1980	Remedial works required immediately	Other remedial works to 1980	Remedial works required immediately	Other remedial works to 1980	Total of expenditure to 1980
1 Discharges of sewage effluent	2.906	2.825	2.427	0.895	5.333	3.720	9.053
2 Discharges of crude sewage	0	0	0.354	0	0.354	0	0.354
3 Discharges of industrial effluent	0.143	—	0.016	—	0.159	—	0.159
Total of estimates	3.049	2.825	2.797	0.895	5.846	3.720	9.566

39 Expenditure is likely to be particularly high on the Blackwater and the Stour. Remedial works on discharges to the Blackwater are estimated to cost about £1.7 million, £1.3 million of which is required immediately, mainly on tidal stretches. Over the length of the Stour more than £1 million is estimated to be required, most of which will be required immediately. On several other rivers such as the Brain, the Colne, the Roding, the Wid and the Crouch, expenditure on each ranging between £0.5 million and £1 million is estimated to be required with smaller amounts on other rivers. The estimates for dealing with crude sewage discharges relate to the River Stour and to discharges from Tendring Rural District and Harwich.

40 Remedial works are required on some industrial discharges mainly from premises concerned with food processing and engineering. ☐

9 Kent River Authority

Table 104 Estimates of costs of remedial works in £ millions

Nature of expenditure	Non-tidal rivers		Tidal rivers		All rivers		
	Remedial works required immediately	Other remedial works to 1980	Remedial works required immediately	Other remedial works to 1980	Remedial works required immediately	Other remedial works to 1980	Total of expenditure to 1980
1 Discharges of sewage effluent	1.765	4.716	0.610	6.978	2.375	11.694	14.069
2 Discharges of crude sewage	0	0	0.765	0	0.765	0	0.765
3 Discharges of industrial effluent	0.010	—	1.258	—	1.268	—	1.268
Total of estimates	1.775	4.716	2.633	6.978	4.408	11.694	16.102

41 Nearly 90 per cent of the required expenditure of about £16 million relates to remedial works required for discharges of sewage effluent, although only about £2.4 million of this is considered to be required immediately. The major part will be required in the near future for discharges to tidal rivers. The balance of the total expenditure forecast in the area is required for discharges of crude sewage and industrial wastes.

42 Remedial works for discharges on the lower reaches of the River Medway account for about £5 million, most of this being required on discharges to tidal waters in the next few years. Some high expenditure in the near future is also forecast for discharges to the Swale where more than £2.5 million is likely to be spent, and some £1.3 million is likely to be needed on discharges to the Great Stour. On the Wingham, the Grom, the Eden and the Eridge Stream smaller amounts in the region of about £0.5 million each are likely to be needed. The estimate for dealing with crude sewage discharges relates mainly to discharges to the River Medway and the Great Stour.

43 The amount to be spent on industrial discharges relates mainly to the paper and board making and chemical industries. ☐

10 Sussex River Authority

Table 105 Estimates of costs of remedial works in £ millions

Nature of expenditure	Non-tidal rivers		Tidal rivers		All rivers		
	Remedial works required immediately	Other remedial works to 1980	Remedial works required immediately	Other remedial works to 1980	Remedial works required immediately	Other remedial works to 1980	Total of expenditure to 1980
1 Discharges of sewage effluent	2.365	1.905	0.545	0.070	2.910	1.975	4.885
2 Discharges of crude sewage	0	0	1.500	0	1.500	0	1.500
3 Discharges of industrial effluent	0.024	—	0.002	—	0.026	—	0.026
Total of estimates	2.389	1.905	2.047	0.070	4.436	1.975	6.411

44 Most of the expenditure needed in this area relates to improving sewage effluent and crude sewage discharges, with a very small amount being required for industrial effluents. Nearly 70 per cent of the total expenditure will be required immediately.

45 Particularly large expenditure is forecast on the upper part of the River Arun where over £1 million needs to be spent immediately on discharges of sewage effluent. On several other rivers distributed widely throughout the area expenditure of small amounts is forecast. About £1.5 million needs to be spent on the crude sewage discharge from Littlehampton to the tidal part of the River Arun.

46 The small expenditure on industrial discharges relates chiefly to farm wastes. ☐

11 Hampshire River Authority

Table 106 Estimates of costs of remedial works in £ millions

Nature of expenditure	Non-tidal rivers		Tidal rivers		All rivers		
	Remedial works required immediately	Other remedial works to 1980	Remedial works required immediately	Other remedial works to 1980	Remedial works required immediately	Other remedial works to 1980	Total of expenditure to 1980
1 Discharges of sewage effluent	1.232	0.765	3.315	1.970	4.547	2.735	7.282
2 Discharges of crude sewage	0	0	0.100	0	0.100	0	0.100
3 Discharges of industrial effluent	0.009	—	0.020	—	0.029	—	0.029
Total of estimates	1.241	0.765	3.435	1.970	4.676	2.735	7.411

47 Most of the expenditure required in this area relates to sewage effluent discharges with some £4.5 million likely to be required immediately. About three-quarters of the expenditure on sewage effluent discharges is related to tidal rivers, some expenditure being required immediately and some in the near future. Only small expenditure will be required for dealing with crude sewage and industrial effluent discharges.

48 Over £5 million is forecast to be spent on discharges to Langstone Harbour (the tidal reach of the Hermitage Stream), the Wallington, the Itchen and the Test (Southampton Water). The expenditure required immediately for dealing with crude sewage relates to discharges to Southampton Water and the Hamble.

49 Costs for improving industrial effluents are low and are mainly for wastes from the chemical industry.

50 Attention is drawn to paragraph 87 of Chapter 2 and paragraph 39 of Chapter 3 with respect to the future of the South Hampshire area.

51 The costs of remedial works on direct discharges to The Solent are not included. □

12 Isle of Wight River and Water Authority

Table 107 Estimates of costs of remedial works in £ millions

Nature of expenditure	Non-tidal rivers		Tidal rivers		All rivers		
	Remedial works required immediately	Other remedial works to 1980	Remedial works required immediately	Other remedial works to 1980	Remedial works required immediately	Other remedial works to 1980	Total of expenditure to 1980
1 Discharges of sewage effluent	0.146	0.134	0.300	0	0.446	0.134	0.580
2 Discharges of crude sewage	0	0	1.200	0	1.200	0	1.200
3 Discharges of industrial effluent	0.005	—	0	—	0.005	—	0.005
Total of estimates	0.151	0.134	1.500	0	1.651	0.134	1.785

52 Estimated costs for improving discharges to rivers are low and about two-thirds of all the costs relate to crude sewage discharges to the Medina on which remedial works are required immediately. The costs for sewage effluent discharges will be spread over schemes involving discharges to several rivers in the area.

53 The small amount of expenditure forecast for improving industrial discharges relates to wastes from laundering, farming and plating industries.

54 The costs of remedial works on direct discharges to The Solent are not included. □

13 Avon and Dorset River Authority

Table 108 Estimates of costs of remedial works in £ millions

Nature of expenditure	Non-tidal rivers		Tidal rivers		All rivers		
	Remedial works required immediately	Other remedial works to 1980	Remedial works required immediately	Other remedial works to 1980	Remedial works required immediately	Other remedial works to 1980	Total of expenditure to 1980
1 Discharges of sewage effluent	1.151	2.316	0	0.245	1.151	2.561	3.712
2 Discharges of crude sewage	0	0	0	0	0	0	0
3 Discharges of industrial effluent	0.013	—	0	—	0.013	—	0.013
Total of estimates	1.164	2.316	0	0.245	1.164	2.561	3.725

55 Almost all the expenditure needed in the area relates to discharges of sewage effluent, with a small amount relating to industrial discharges. There are no crude sewage discharges in the area.

56 Of the expenditure likely on sewage effluent discharges, about one-third will be required immediately and the balance mainly in the near future. Expenditure required on discharges to the Stour is estimated at about £2 million and about a third of this is needed immediately. The balance of about £1.7 million is spread over several rivers in relatively small amounts.

57 Expenditure on industrial discharges relates to the leather and textile industries. □

14 Devon River Authority

Table 109 Estimates of costs of remedial works in £ millions

Nature of expenditure	Non-tidal rivers		Tidal rivers		All rivers		
	Remedial works required immediately	Other remedial works to 1980	Remedial works required immediately	Other remedial works to 1980	Remedial works required immediately	Other remedial works to 1980	Total of expenditure to 1980
1 Discharges of sewage effluent	2.386	0.642	2.227	0.018	4.613	0.660	5.273
2 Discharges of crude sewage	0.450	0	4.195	0	4.645	0	4.645
3 Discharges of industrial effluent	0.143	—	0.020	—	0.163	—	0.163
Total of estimates	2.979	0.642	6.442	0.018	9.421	0.660	10.081

58 In this area, slightly over half of the total expenditure relates to sewage effluent discharges with the balance to be spent mainly on crude sewage. Only a small sum needs to be spent on industrial discharges.

59 The forecasts for improving discharges of sewage effluent indicate that most of the expenditure will be required immediately and about £2 million of this relates to the River Teign. The remainder is divided between discharges to several rivers throughout the area. Most of the expenditure relating to crude sewage discharges will be incurred on the Torridge, the Exe, the Taw basin, the Teign basin, the Dart and the Kingsbridge Stream system.

60 Costs for industrial discharges relate mainly to metal refining, synthetic and cotton textile industries, food processing, laundering, leather manufacture and wastes from water treatment plants.

15 Cornwall River Authority

Table 110 Estimates of costs of remedial works in £ millions

Nature of expenditure	Non-tidal rivers		Tidal rivers		All rivers		
	Remedial works required immediately	Other remedial works to 1980	Remedial works required immediately	Other remedial works to 1980	Remedial works required immediately	Other remedial works to 1980	Total of expenditure to 1980
1 Discharges of sewage effluent	0.822	0.949	0.045	0.155	0.867	1.104	1.971
2 Discharges of crude sewage	0	0	4.496	0.443	4.496	0.443	4.939
3 Discharges of industrial effluent	3.658	—	0.002	—	3.660	—	3.660
Total of estimates	4.480	0.949	4.543	0.598	9.023	1.547	10.570

61 Expenditure needed in this area is, in the main, divided between crude sewage and industrial effluent discharges, with less than 20 per cent of the total forecast relating to sewage effluent discharges.

62 About a third of the expenditure on sewage effluent discharges is likely to be needed on rivers in the Tamar basin with the balance spread over many rivers throughout the area. About £2.5 million also needs to be spent on discharges of crude sewage to rivers in the Tamar basin, more than half of this for the Tamar itself. Smaller but significant sums for discharges of crude sewage to Newton Creek, the Camel, the Angarrack and the Fowey are also indicated as well as lesser amounts on several other rivers.

63 Almost all of the expenditure for improving discharges of industrial wastes is likely to be needed to deal with discharges from quarries. Small sums are shown to be required on wastes from laundries and from water treatment plants.

16 Somerset River Authority

Table 111 Estimates of costs of remedial works in £ millions

Nature of expenditure	Non-tidal rivers		Tidal rivers		All rivers		
	Remedial works required immediately	Other remedial works to 1980	Remedial works required immediately	Other remedial works to 1980	Remedial works required immediately	Other remedial works to 1980	Total of expenditure to 1980
1 Discharges of sewage effluent	3.677	1.592	0	1.250	3.677	2.842	6.519
2 Discharges of crude sewage	0.320	0.040	4.010	0	4.330	0.040	4.370
3 Discharges of industrial effluent	2.140	—	1.470	—	3.610	—	3.610
Total of estimates	6.137	1.632	5.480	1.250	11.617	2.882	14.499

64 Estimates of costs for remedial works in this area indicate a fairly high level of expenditure with some 80 per cent of the total cost being applicable to remedial works required immediately and divided more or less equally between non-tidal and tidal waters.

65 Most of the expenditure needed on sewage effluent relates to discharges to non-tidal stretches of rivers with particularly large sums forecast to be spent on discharges to the River Tone, the South Drain which flows into the Huntspill River and the Yeovil Yeo. On tidal stretches, the whole of the cost is attributable to the Kingston Seymour works which will discharge to the River Kenn. Smaller costs relate to several other sewage effluent discharges throughout

the area. About £3.5 million out of the total expenditure forecast for crude sewage discharges relates to discharges to the River Parrett. About £0.5 million is likely to be required to deal with discharges to the Congresbury Yeo and small sums will be needed for other discharges in the area including about £0.4 million for crude sewage discharges to non-tidal waters.

66 Substantial expenditure is indicated as being required on industrial discharges mainly from farming and from synthetic and cotton textiles manufacture, with small amounts relating to paper and board manufacture and food processing.

67 The costs of remedial works on direct discharges to the Severn Estuary and the Bristol Channel are not included. ☐

17 Bristol Avon River Authority

Table 112 Estimates of costs of remedial works in £ millions

Nature of expenditure	Non-tidal rivers		Tidal rivers		All rivers		
	Remedial works required immediately	Other remedial works to 1980	Remedial works required immediately	Other remedial works to 1980	Remedial works required immediately	Other remedial works to 1980	Total of expenditure to 1980
1 Discharges of sewage effluent	3.308	3.926	0.060	0	3.368	3.926	7.294
2 Discharges of crude sewage	0.793	0	1.715	0	2.508	0	2.508
3 Discharges of industrial effluent	0.152	—	0	—	0.152	—	0.152
Total of estimates	4.253	3.926	1.775	0	6.028	3.926	9.954

68 About three-quarters of all the expenditure needed in this area relates to sewage effluent discharges and about half of this latter expenditure is likely to be required immediately. Substantial costs for remedial works on crude sewage discharges arise on both non-tidal and tidal stretches.

69 The expenditure relating to sewage effluent will almost all be on discharges to non-tidal waters. Expenditure on discharges to the Bristol Avon itself is likely to exceed £3 million, about a half being required immediately and most of the balance in the near future whilst the River Biss will account for about £1 million, all urgently needed. Smaller costs for remedial works are distributed over many other rivers. The estimates for remedial works on crude sewage discharges to non-tidal waters relate mainly to the Bristol Avon and the Semington Brook whilst the estimate amounting to £1.7 million on tidal waters relates wholly to discharges to the Bristol Avon.

70 Costs for improvements to industrial effluents relate mainly to food processing with small amounts attributable to paper and board manufacture, rubber processing, quarrying, farming and leather industries. ☐

18 Severn River Authority

Table 113 Estimates of costs of remedial works in £ millions

Nature of expenditure	Non-tidal rivers		Tidal rivers		All rivers		Canals		All rivers and canals		
	Remedial works required immediately	Other remedial works to 1980	Remedial works required immediately	Other remedial works to 1980	Remedial works required immediately	Other remedial works to 1980	Remedial works required immediately	Other remedial works to 1980	Remedial works required immediately	Other remedial works to 1980	Total of expenditure to 1980
1 Discharges of sewage effluent	7.736	20.668	0.537	2.085	8.273	22.753	0	0.075	8.273	22.828	31.101
2 Discharges of crude sewage	0	0	0.850	0	0.850	0	0	0	0.850	0	0.850
3 Discharges of industrial effluent	0.348	—	0.002	—	0.350	—	0.126	—	0.476	—	0.476
Total of estimates	8.084	20.668	1.389	2.085	9.473	22.753	0.126	0.075	9.599	22.828	32.427

71 The forecasts of expenditure indicate that almost all of it is likely to be required for sewage effluent discharges and the level of expenditure promises to be high with the larger part being required in the near future. Only relatively small expenditure is forecast for dealing with discharges of crude sewage and industrial effluent.

72 A large proportion of the required expenditure will be concentrated on discharges to a few rivers. The cost of remedial works on sewage effluent discharges to the Worcestershire Stour which drains the south-western fringes of the West Midlands conurbation will approach £5 million. On the Severn itself, sewage effluent discharges to the river over its whole length are estimated to require the expenditure of nearly £4 million. On discharges to the Sowe, flowing south from Coventry, over £3 million is forecast to be spent and a further £3 million will be needed on discharges to the Holesmouth Pill, a short watercourse north of Bristol. More than £2 million will have to be spent on discharges to the Warwickshire Avon and more than £1 million each on the Arrow and the Tern. Small expenditure will be required on the one discharge to the Grand Union Canal. Only a fairly small proportion of the heavy expenditure referred to above is considered to be urgently needed. The expenditure forecast for discharges of crude sewage relates to discharges to the tidal Severn, several of which are to be diverted to treatment works.

73 Expenditure required on industrial discharges is spread over some 15 different classes of industry, with engineering, water treatment, food processing, and plating and finishing works accounting for about £350,000. The balance relates mainly to wastes from chemical works, paper and board manufacture, electricity generation and the wool textile and iron and steel industries. ☐

19 Wye River Authority

Table 114 Estimates of costs of remedial works in £ millions

Nature of expenditure	Non-tidal rivers		Tidal rivers		All rivers		
	Remedial works required immediately	Other remedial works to 1980	Remedial works required immediately	Other remedial works to 1980	Remedial works required immediately	Other remedial works to 1980	Total of expenditure to 1980
1 Discharges of sewage effluent	2.093	0.745	0	0	2.093	0.745	2.838
2 Discharges of crude sewage	0	0	0.110	0	0.110	0	0.110
3 Discharges of industrial effluent	0.002	—	0	—	0.002	—	0.002
Total of estimates	2.095	0.745	0.110	0	2.205	0.745	2.950

74 Only a relatively small level of expenditure is forecast, mainly relating to sewage effluent discharges, with about two-thirds likely to be required immediately and the balance in the near future. No expenditure is foreseen on sewage effluent discharges to tidal waters, but small expenditure is required in respect of discharges of crude sewage to tidal waters and also industrial effluent to non-tidal waters.

75 Nearly £2 million out of the expenditure forecast for improving sewage effluent discharges relates to those to the River Wye itself and most of this is required immediately. Similarly the expenditure required for remedial works on crude sewage discharges also relates to the River Wye on the tidal stretches and it is all required immediately.

76 The small expenditure needed on industrial discharges relates to pottery making.

20 Usk River Authority

Table 115 Estimates of costs of remedial works in £ millions

Nature of expenditure	Non-tidal rivers		Tidal rivers		All rivers		
	Remedial works required immediately	Other remedial works to 1980	Remedial works required immediately	Other remedial works to 1980	Remedial works required immediately	Other remedial works to 1980	Total of expenditure to 1980
1 Discharges of sewage effluent	0.305	0.525	3.200	0	3.505	0.525	4.030
2 Discharges of crude sewage	0	0	0.762	0	0.762	0	0.762
3 Discharges of industrial effluent	1.128	—	0	—	1.128	—	1.128
Total of estimates	2.433	0.525	3.962	0	5.395	0.525	5.920

77 In this area about two-thirds of all the expenditure is expected to relate to remedial works on sewage effluent discharges and the remainder to discharges of crude sewage and industrial wastes.

78 Sewage effluent discharges to the River Usk are likely to account for about £3.6 million, and about £3.2 million of this is needed immediately for those discharges which are to the tidal sections. Costs forecast for improving other discharges are generally small. Costs forecast for dealing with crude sewage discharges all relate to the River Usk. They are required immediately and the intention is that the works should follow on after completion of the present authorised works which are under construction.

79 Most of the costs for remedial works on industrial discharges relate to the iron and steel industry, where the forecast is about £1 million. Small sums relate to discharges from processes allied to coal mining, the plating and finishing industry and food processing.

80 The costs of remedial works on direct discharges to the Severn Estuary are not included.

21 Glamorgan River Authority

Table 116 Estimates of costs of remedial works in £ millions

Nature of expenditure	Non-tidal rivers		Tidal rivers		All rivers		
	Remedial works required immediately	Other remedial works to 1980	Remedial works required immediately	Other remedial works to 1980	Remedial works required immediately	Other remedial works to 1980	Total of expenditure to 1980
1 Discharges of sewage effluent	8.225	0.047	0	6.500	8.225	6.547	14.772
2 Discharges of crude sewage	0	0	3.110	0	3.110	0	3.110
3 Discharges of industrial effluent	0.369	—	0.003	—	0.372	—	0.372
Total of estimates	8.594	0.047	3.113	6.500	11.707	6.547	18.254

81 The estimates indicate that the major part of the cost relates to remedial works on sewage effluent discharges. Almost all the expenditure on discharges to non-tidal waters is required immediately whereas on tidal waters it will, in the main, be required in the

near future. The estimated costs for dealing with discharges of crude sewage and industrial effluents represent about one-fifth of the total for the area.

82 Expenditure on regional schemes for several discharges to non-tidal stretches of the Taff, the Ely and other rivers in these river basins is estimated at over £7 million and a further £6.5 million is estimated to be required for a regional scheme to deal with several discharges to rivers in the Ogmore basin. The effluent will discharge to the tidal part of the River Ogmore. Together, these three river basins account for almost all of the total estimated cost for dealing with sewage effluent discharged in the area. The estimates for dealing with crude sewage discharges relate to the tidal waters of the Ely, the Afan and the Neath, and the whole of the expenditure is required immediately.

83 Forecasts of costs for industrial discharges relate mainly to chemical, glue and gelatine manufacture and some coal by-product processes. The estimate also shows a need to spend a small sum on treating mine water from active coal mines, but this is not included in the total.

84 The costs of remedial works on direct discharges to the Severn Estuary and Swansea Bay are not included. □

22 South West Wales River Authority

Table 117 Estimates of costs of remedial works in £ millions

Nature of expenditure	Non-tidal rivers		Tidal rivers		All rivers		Canals		All rivers and canals		
	Remedial works required immediately	Other remedial works to 1980	Remedial works required immediately	Other remedial works to 1980	Remedial works required immediately	Other remedial works to 1980	Remedial works required immediately	Other remedial works to 1980	Remedial works required immediately	Other remedial works to 1980	Total of expenditure to 1980
1 Discharges of sewage effluent	0.635	0.317	0.115	0.835	0.750	1.152	0	0	0.750	1.152	1.902
2 Discharges of crude sewage	0.169	0	4.458	0	4.627	0	0	0	4.627	0	4.627
3 Discharges of industrial effluent	1.388	—	0.465	—	1.853	—	0.004	—	1.857	—	1.857
Total of estimates	2.192	0.317	5.038	0.835	7.230	1.152	0.004	0	7.234	1.152	8.386

85 In this area more than half of the estimated expenditure for improvement relates to discharges of crude sewage, mostly to tidal waters and the balance is divided about equally between sewage effluent and industrial effluent.

86 The costs relating to sewage effluent discharges are generally made up of small amounts distributed over many rivers, the largest single cost being about £0.4 million on discharges to the River Towy. Much larger costs are forecast for crude sewage discharges, with about £3 million to be spent on a scheme for dealing with several discharges from Llanelli to rivers in the adjacent area.

87 Expenditure on industrial effluents is likely to be fairly high and spread over some 12 different types of industry with metal refining being prominent. Other significant amounts relate to chemical works, process water from coal mining, food processing and the iron and steel industry with lesser amounts relating to engineering, gas and coke production, plating and finishing and farming. There are also indications of substantial expenditure needed to deal with effluents from derelict mines but this is not included in the total.

88 The costs of remedial works on direct discharges to the Bristol Channel are not included. □

23 Gwynedd River Authority

Table 118 Estimates of costs of remedial works in £ millions

Nature of expenditure	Non-tidal rivers		Tidal rivers		All rivers		
	Remedial works required immediately	Other remedial works to 1980	Remedial works required immediately	Other remedial works to 1980	Remedial works required immediately	Other remedial works to 1980	Total of expenditure to 1980
1 Discharges of sewage effluent	0.361	0.598	0.105	0.070	0.466	0.668	1.134
2 Discharges of crude sewage	0.197	0.380	1.174	0	1.371	0.380	1.751
3 Discharges of industrial effluent	0.012	—	0	—	0.012	—	0.012
Total of estimates	0.570	0.978	1.279	0.070	1.849	1.048	2.897

89 About 60 per cent of all the estimated expenditure is likely to be required for discharges of crude sewage and most of the balance for discharges of sewage effluent. Expenditure will be divided fairly evenly between non-tidal and tidal stretches. Overall, the level of expenditure is low.

90 Expenditure forecast for improving sewage effluent discharges will be distributed in fairly small sums over many rivers throughout the area but most of the costs for dealing with crude sewage discharges relate to four rivers, the Mawddach, the Artro, the Dwyryd and the Conway. The largest sum, nearly £0.7 million, is expected to be spent on dealing with discharges to the Conway.

91 The small expenditure needed to improve industrial effluent relates to an effluent from a laundry.

92 The costs of remedial works on direct discharges to the Menai Straits are not included. □

24 Dee and Clwyd River Authority

Table 119 Estimates of costs of remedial works in £ millions

Nature of expenditure	Non-tidal rivers		Tidal rivers		All rivers		
	Remedial works required immediately	Other remedial works to 1980	Remedial works required immediately	Other remedial works to 1980	Remedial works required immediately	Other remedial works to 1980	Total of expenditure to 1980
1 Discharges of sewage effluent	1.792	0.892	0.210	0.750	2.002	1.642	3.644
2 Discharges of crude sewage	0	0	0.300	0	0.300	0	0.300
3 Discharges of industrial effluent	0.080	—	0.203	—	0.283	—	0.283
Total of estimates	1.872	0.892	0.713	0.750	2.585	1.642	4.227

93 Most of the forecast expenditure of just over £4 million relates to improving sewage effluent discharges, part of it needed immediately and part in the near future. Small sums are required to deal with discharges of crude sewage and industrial effluent.

94 The largest items of expenditure on sewage effluent are expected to be incurred on discharges to the River Dee, the Alyn, the Black Brook and the Pulford Brook, and about half of this expenditure is immediately necessary. Expenditure of £0.3 million on crude sewage relates to the works necessary to deal with a discharge at Holywell to a small stream near the mouth of the Dee.

95 Costs for improving industrial effluents are relatively low and associated primarily with the manufacture of man-made fibres, the paper and board industry, chemical works and farming. □

25 Mersey and Weaver River Authority

Table 120 Estimates of costs of remedial works in £ millions

Nature of expenditure	Non-tidal rivers		Tidal rivers		All rivers		Canals*		All rivers and canals		
	Remedial works required immediately	Other remedial works to 1980	Remedial works required immediately	Other remedial works to 1980	Remedial works required immediately	Other remedial works to 1980	Remedial works required immediately	Other remedial works to 1980	Remedial works required immediately	Other remedial works to 1980	Total of expenditure to 1980
1 Discharges of sewage effluent	23.841	5.837	2.520	1.500	26.361	7.337	0	0	26.361	7.337	33.698
2 Discharges of crude sewage	0	0	38.160	0	38.160	0	0	0	38.160	0	38.160
3 Discharges of industrial effluent	4.833	—	5.825	—	10.658	—	0.200	—	10.858	—	10.858
Total of estimates	28.674	5.837	46.505	1.500	75.179	7.337	0.200	0	75.379	7.337	82.716

*For the purpose of the survey, the Manchester Ship Canal is considered to be a part of the river systems of the Mersey and the Irwell and the costs for dealing with discharges to it are recorded under the costs for rivers.

96 Heavy expenditure is forecast in this area and of the total of nearly £83 million, all but £7.3 million is considered to be required immediately. About £38 million is estimated to be needed immediately for dealing with the crude sewage discharges and about £34 million for sewage effluent discharges, most of this needed urgently. Expenditure on industrial discharges is also high and is estimated at about £11 million.

97 The costs for dealing with crude sewage discharges, which are second only to the corresponding costs for the Northumbrian River Authority area, relate entirely to discharges to the tidal waters of the Mersey from Warrington down to the sea. Expenditure on sewage effluent discharges to the Weaver basin is estimated at about £4 million, nearly £3 million of which relates to discharges to the Weaver and the Wheelock, and over £2 million will have to be spent on the Alt, but the bulk of the estimated expenditure—about £27 million—will be needed to deal with discharges to rivers in the Mersey basin. Expenditure on the Mersey, the Irwell and the Manchester Ship Canal alone would run to about £15 million and expenditure on discharges to the Sankey Brook, the Roch, the Tame and the Bollin is likely to reach about £7 million. Most of the heavy expenditure in the Mersey basin will be required on discharges in the South East Lancashire conurbation area.

98 The cost of remedial works on industrial discharges is spread over more than 20 different classes of industry, those prominent being the chemical and petroleum refining industries and, to a lesser extent, paper and board, cotton and synthetic textiles, soap and detergent and food processing industries. The costs for dealing satisfactorily with wastes from these particular industries represent more than 90 per cent of the total.

99 No account has been taken of the possible South Lancashire and North Cheshire sludge pipeline by which sludge from a wide area might be disposed of from a sludge terminal to sea. □

26 Lancashire River Authority
(see Table 121)

100 Almost 80 per cent of the total estimated cost of remedial works in the area will be for works on sewage effluent discharges and about three-quarters of this is considered to be required immediately. Expenditure on discharges of crude sewage will be nearly £4 million and on industrial discharges rather lower at about £1.2 million.

Table 121 Estimates of costs of remedial works in £ millions

Nature of expenditure	Non-tidal rivers		Tidal rivers		All rivers		
	Remedial works required immediately	Other remedial works to 1980	Remedial works required immediately	Other remedial works to 1980	Remedial works required immediately	Other remedial works to 1980	Total of expenditure to 1980
1 Discharges of sewage effluent	13.411	2.330	1.827	2.000	15.238	4.330	19.568
2 Discharges of crude sewage	0	0	3.930	0.040	3.930	0.040	3.970
3 Discharges of industrial effluent	0.868	—	0.329	—	1.197	—	1.197
Total of estimates	14.279	2.330	6.086	2.040	20.365	4.370	24.735

101 About half of all the expenditure forecast for sewage effluent relates to only two rivers, the Douglas flowing north through Wigan to the Ribble, and the Calder, another tributary of the Ribble, which drains the area around Accrington and Burnley. Many existing discharges to these rivers are to be abandoned and diverted to new works. A further amount of about £5 million or so is likely to be required on discharges to the Ribble, the Hole Brook, the Walney Channel at Barrow-in-Furness and the Boathouse Sluice near Ormskirk. Smaller sums have been estimated for many other discharges to rivers in the area. The expenditure on crude sewage discharges is almost all considered to be urgently required. A big proportion of the total relates to discharges to the Ribble, the Lune, the Walney Channel and the adjacent Salthouse Pool. Several other discharges are likely to be the subject of smaller expenditure in the immediate future.

102 Estimated expenditure needed to improve industrial discharges relates chiefly to wastes from paper and board manufacture and cotton and synthetic textiles. Other smaller sums relate to quarrying wastes, disposal tips, chemical plants, food processing, water treatment plants and engineering concerns.

103 Costs of remedial works on a few discharges to the seaward end of Morecambe Bay are not included.

27 Cumberland River Authority

Table 122 Estimates of costs of remedial works in £ millions

Nature of expenditure	Non-tidal rivers		Tidal rivers		All rivers		
	Remedial works required immediately	Other remedial works to 1980	Remedial works required immediately	Other remedial works to 1980	Remedial works required immediately	Other remedial works to 1980	Total of expenditure to 1980
1 Discharges of sewage effluent	1.667	0.210	0.122	0	1.789	0.210	1.999
2 Discharges of crude sewage	0	0	0.130	0	0.130	0	0.130
3 Discharges of industrial effluent	0.119	—	0.200	—	0.319	—	0.319
Total of estimates	1.786	0.210	0.452	0	2.238	0.210	2.448

104 The expenditure forecast for this area is relatively low at about £2.5 million with the sum to be spent on sewage effluent discharges accounting for about 80 per cent of the total. Small expenditure is forecast for discharges of crude sewage and industrial wastes.

105 Most of the expenditure on sewage effluent discharges is spread in small amounts over many rivers but the remedial works on discharges to the River Eden are likely to cost some £0.8 million. The bulk of the expenditure is required immediately. The small expenditure shown to be needed on crude sewage discharges relates to discharges to the Southern Esk and the Ehen and it is all required immediately.

106 Estimates of cost for improving industrial discharges are relatively low and principally relate to chemical works and, to a lesser extent, to the engineering and textile industries.

107 The costs of remedial works on direct discharges to the Solway Firth are not included.

28 Thames Conservancy Catchment Board

Table 123 Estimates of costs of remedial works in £ millions

Nature of expenditure	Non-tidal rivers		Tidal rivers		All rivers		Canals		All rivers and canals		
	Remedial works required immediately	Other remedial works to 1980	Remedial works required immediately	Other remedial works to 1980	Remedial works required immediately	Other remedial works to 1980	Remedial works required immediately	Other remedial works to 1980	Remedial works required immediately	Other remedial works to 1980	Total of expenditure to 1980.
1 Discharges of sewage effluent	14.114	10.537	0	0	14.114	10.537	0.230	0.096	14.344	10.633	24.977
2 Discharges of crude sewage	0	0	0	0	0	0	0	0	0	0	0
3 Discharges of industrial effluent	0.280	—	0	—	0.280	—	0	—	0.280	—	0.280
Total of estimates	14.394	10.537	0	0	14.394	10.537	0.230	0.096	14.624	10.633	25.257

108 In the non-tidal Thames basin, the forecast of expenditure amounting to some £25 million almost all concerns remedial works on sewage effluent discharges. Numerous works are involved, many at relatively low cost, but substantial expenditure will be needed on several rivers. Over £3 million is forecast to be spent on discharges to the River Wey and expenditure in excess of £1 million each is likely on the Thames, the Hogsmill River, the Mole, the Earlswood Brook, the Wye, the Blackwater, the Thame and the Cherwell, all but the last two tributaries being in the area to the immediate west and south of London. Some expenditure will also be needed on discharges to the Grand Union, Oxford South, and Kennet and Avon canals.

109 The small costs estimated for improving industrial discharges relate mainly to effluents from paper and board making, general manufacturing and chemical industries.

29 Lee Conservancy Catchment Board

Table 124 Estimates of costs of remedial works in £ millions

Nature of expenditure	Non-tidal rivers		Tidal rivers		All rivers		Total of expenditure to 1980
	Remedial works required immediately	Other remedial works to 1980	Remedial works required immediately	Other remedial works to 1980	Remedial works required immediately	Other remedial works to 1980	
1 Discharges of sewage effluent	1.314	4.510	0	0	1.314	4.510	5.824
2 Discharges of crude sewage	0	0	0	0	0	0	0
3 Discharges of industrial effluent	0	0	0	0	0	0	0
Total of estimates	1.314	4.510	0	0	1.314	4.510	5.824

110 Expenditure in this area relates entirely to sewage effluent discharges and the sum of about £5.8 million has been estimated for the necessary remedial works with about 25 per cent of this being required immediately. The largest expenditure—in the region of £4.5 million—is likely to be required on discharges to the River Lee between Luton and Hoddesdon and about £1 million will be required for improving discharges to the Great Hallingbury Brook at Bishop's Stortford. There are no discharges of crude sewage in the area and all the discharges of industrial effluent are classed as satisfactory so that no remedial costs arise.

30 Port of London Authority (including the London Excluded Area)

Table 125 Estimates of costs of remedial works in £ millions

Nature of expenditure	Non-tidal rivers		Tidal rivers		All rivers		Canals		All rivers and canals		Total of expenditure to 1980
	Remedial works required immediately	Other remedial works to 1980	Remedial works required immediately	Other remedial works to 1980	Remedial works required immediately	Other remedial works to 1980	Remedial works required immediately	Other remedial works to 1980	Remedial works required immediately	Other remedial works to 1980	
1 Discharges of sewage effluent	0	0	25.126	6.270	25.126	6.270	0	0	25.126	6.270	31.396
2 Discharges of crude sewage	0	0	0	0	0	0	0	0	0	0	0
3 Discharges of industrial effluent	0	—	1.421	—	1.421	—	0.025	—	1.446	—	1.446
Total of estimates	0	0	26.547	6.270	26.547	6.270	0.025	0	26.572	6.270	32.842

111 The forecasts of expenditure in this area all relate to tidal waters and mainly concern sewage effluent discharges. Large expenditure will be required on discharges direct to the Thames, especially from those works under the control of the Greater London Council and the West Kent Main Sewerage Board. Some £25 million out of about £31 million estimated for discharges to the Thames is considered to be required immediately. Only a small amount of expenditure (2 per cent of the total) is forecast on discharges to tributaries of the Thames since schemes affecting several of these discharges are under construction now and their costs are not included in the estimates.

112 The principal expenditure on industrial effluent relates to paper and board making with other smaller amounts being needed mainly on discharges from quarrying, food processing, electricity generation and chemical works.

7 Forecasts of improvement

Introduction

1 The river authorities were asked to provide information about the effect on the rivers of carrying out the remedial measures on discharges at present unsatisfactory. An indication of what the river authorities think should be done has already been given in Chapters 2–5 and costs have been dealt with in Chapter 6.

2 The forecasts of upgrading can only be regarded as a general guide to possible improvement because other unforeseen factors such as increased abstractions of water could affect the forecasts. It must also be remembered that possible improvements within a class—say from a "poor" Class 2 to a "good" Class 2 are not recorded and so, to that extent, the forecasts could under-estimate the changes. It is also relevant that three river authorities* expressed the view that, whatever remedial works were carried out, some of their rivers because of natural characteristics could never be better than Class 2.

3 The speed with which a river would recover and improve, consequent upon remedial measures on discharges would vary widely between rivers of different physical characteristics and, within a given river, between the water, the bed and the fauna and flora. It cannot therefore be assumed that overall improvement would always follow immediately upon completion of all the necessary works and further expenditure on remedial works on the river itself might even be required to remove deposits.

4 Forecasts of river improvement were of two kinds. Firstly an indication was given of any upgrading in chemical classification that would occur and secondly an indication of any additional lengths of river which might become suitable for public water supply.

River upgrading. England and Wales

5 Table 126 summarises the data provided by the returns in terms of miles likely to be upgraded from one class to another. The total mileage of rivers likely to be upgraded (nearly 2,800 miles) represents some 45 per cent of the total length of Class 2, 3 and 4 river included in the survey, namely 6,242 miles. For canals the proportion is about 20 per cent. Altogether, a total of 992 miles of Class 4 waters are expected to be upgraded, 953 miles of Class 3 and 1,013 miles of Class 2. The Class 3 and 4 mileage likely to be upgraded represents about 70 per cent of all the Class 3 and 4 mileage in the survey.

6 From the information in Table 126 it is possible to compare the situation after upgrading with the 1970 situation which was presented in Tables 5 and 6 of Volume 1 and this is done in Tables 127 and 128. The figures in the 'Net change' column are obtained from Table 126. It is significant to note that, on the basis of the forecasts of the improvement following necessary remedial works, the mileage of non-tidal rivers in Class 1 and Class 2 should increase from 90.9 per cent of the total non-tidal to 96.3 per cent of the total and that the mileage in the worst non-tidal class should be reduced by about 80 per cent. In the case of tidal waters, the proportion in Class 1 and Class 2 should increase from 71.5 per cent of the total to 84.2 per cent whilst the mileage in the worst class should be reduced by over 70 per cent. The corresponding figures for canals are 84.5 per cent up to 90.9 per cent for the two best classes and the mileage in the worst class reduced by about 86 per cent.

Table 126 Upgrading in chemical classification following remedial works on discharges at present unsatisfactory. England and Wales

Extent of upgrading	Miles upgraded				
	Non-tidal rivers	Tidal rivers	All rivers	Canals	All rivers and canals
From Class 4 to Class 3	330	42	372	44	416
From Class 4 to Class 2	370	96	466	45	511
From Class 4 to Class 1	53	12	65	0	65
From Class 3 to Class 2	637	110	747	53	800
From Class 3 to Class 1	144	9	153	0	153
From Class 2 to Class 1	872	116	988	25	1,013
Total miles upgraded	2,406	385	2,791	167	2,958

* See paragraphs 28, 30 and 33 of Chapter 7.

Table 127 Comparison of mileages by chemical classification following remedial works on discharges at present unsatisfactory. Rivers

	Chemical classification	1970 situation (miles)	%	Net change (miles)	Situation after remedial works (miles)	%
Non-tidal rivers	1	17,000	76.1	+1,069	18,069	81.0
	2	3,290	14.8	+ 135	3,425	15.3
	3	1,071	4.8	− 451	620	2.8
	4	952	4.3	− 753	199	0.9
Total		22,313			22,313	
Tidal rivers	1	862	48.1	+ 137	999	55.8
	2	419	23.4	+ 90	509	28.4
	3	301	16.8	− 77	224	12.5
	4	209	11.7	− 150	59	3.3
Total		1,791			1,791	
All rivers	1	17,862	74.1	+1,206	19,068	79.1
	2	3,709	15.4	+ 225	3,934	16.3
	3	1,372	5.7	− 528	844	3.5
	4	1,161	4.8	− 903	258	1.1
Total		24,104			24,104	

Table 128 Comparison of mileages by chemical classification following remedial works on discharges at present unsatisfactory. Canals

Chemical classification	1970 situation (miles)	%	Net change (miles)	Situation after remedial works (miles)	%
1	700	45.4	+25	725	47.1
2	601	39.1	+73	674	43.8
3	136	8.8	− 9	127	8.2
4	103	6.7	−89	14	0.9
Total	1,540			1,540	

7 Tables 129 and 130 give a river authority by river authority breakdown of the figures and show the forecast situation in the individual areas following remedial works, as compared with the 1970 situation represented by Tables 5 and 6 of Volume 1.

Table 129 Estimates of miles of river by chemical classification following remedial works. England and Wales

Authority		Non-tidal						Tidal				
		Total Length	Total	Class 1	Class 2	Class 3	Class 4	Total	Class 1	Class 2	Class 3	Class 4
Northumbrian	Miles	1,817	1,734	1,468	249	16	1	83	22	61	0	0
	%			84.6	14.4	0.9	0.1		26.5	73.5		
Yorkshire	Miles	3,556	3,432	2,965	231	144	92	124	49	3	72	0
	%			86.4	6.7	4.2	2.7		39.5	2.4	58.1	
Trent	Miles	1,511	1,459	644	709	97	9	52	0	26	0	26
	%			44.1	48.6	6.7	0.6			50.0		50.0
Lincolnshire	Miles	521	510	482	20	8	0	11	4	7	0	0
	%			94.5	3.9	1.6			38.1	61.9		
Welland and Nene	Miles	361	319	224	84	10	1	42	0	42	0	0
	%			70.2	26.4	3.1	0.3			100		
Great Ouse	Miles	950	911	679	197	31	4	39	39	0	0	0
	%			74.6	21.6	3.4	0.4		100			
East Suffolk & Norfolk	Miles	541	376	376	0	0	0	165	86	79	0	0
	%			100					52.1	47.9		
Essex	Miles	460	372	260	94	18	0	88	67	21	0	0
	%			69.9	25.3	4.8			76.1	23.9		
Kent	Miles	605	529	517	12	0	0	76	58	7	8	3
	%			97.7	2.3				76.3	9.2	10.5	4.0
Sussex	Miles	343	269	226	43	0	0	74	74	0	0	0
	%			84.0	16.0				100			
Hampshire	Miles	424	369	254	111	4	0	55	39	15	1	0
	%			68.8	30.1	1.1			70.9	27.3	1.8	
Isle of Wight	Miles	89	73	66	6	1	0	16	16	0	0	0
	%			90.4	8.2	1.4			100			
Avon and Dorset	Miles	433	419	411	8	0	0	14	14	0	0	0
	%			98.1	1.9				100			
Devon	Miles	1,002	924	878	46	0	0	78	46	32	0	0
	%			95.0	5.0				59.0	41.0		
Cornwall	Miles	1,232	1,093	979	100	0	14	139	131	8	0	0
	%			89.6	9.1		1.3		94.2	5.8		
Somerset	Miles	440	406	400	6	—a	0	34	1	30	1	2
	%			98.5	1.4	0.1			2.9	88.3	2.9	5.9
Bristol Avon	Miles	378	363	280	77	6	0	15	8	7	0	0
	%			77.1	21.2	1.7			53.3	46.7		
Severn	Miles	1,163	1,115	871	188	46	10	48	48	0	0	0
	%			78.1	16.9	4.1	0.9		100			
Wye	Miles	1,065	1,049	1,019	29	1	0	16	15	1	0	0
	%			97.1	2.8	0.1			93.7	6.3		
Usk	Miles	256	237	148	69	20	—b	19	6	13	0	0
	%			62.3	29.1	8.4	0.2		31.6	68.4		
Glamorgan	Miles	329	293	148	107	32	6	36	2	12	11	11
	%			50.5	36.5	10.9	2.1		5.6	33.2	30.6	30.6
South West Wales	Miles	703	608	531	40	29	8	95	57	29	9	—c
	%			87.3	6.6	4.8	1.3		59.8	30.4	9.4	0.4
Gwynedd	Miles	361	292	279	13	—d	0	69	69	0	—d	0
	%			95.5	4.4	0.1			99.4		0.6	
Dee and Clwyd	Miles	432	403	382	21	0	—c	29	29	0	0	0
	%			94.8	5.1		0.1		100			
Mersey and Weaver	Miles	942	875	354	367	114	40	67	1	6	45	15
	%			40.5	41.9	13.0	4.6		1.5	8.9	67.2	22.4
Lancashire	Miles	1,187	1,000	662	305	28	5	187	98	81	8	0
	%			66.2	30.5	2.8	0.5		52.4	43.3	4.3	
Cumberland	Miles	1,223	1,188	1,170	18	0	0	35	20	15	0	0
	%			98.5	1.5				57.2	42.8		
Thames Conservancy	Miles	1,444	1,444	1,254	188	1	1	0	0	0	0	0
	%			86.8	13.0	0.1	0.1					
Lee Conservancy	Miles	181	174	142	16	14	2	7	0	0	5	2
	%			81.7	9.2	8.0	1.1				71.5	28.5
Port of London Authority (incl. London Excluded Area)	Miles	155	77	0	71	0	6	78	0	14	64	—e
	%				92.2		7.8			17.9	81.8	0.3
Total England & Wales	Miles	24,104	22,313	18,069	3,425	620	199	1,791	999	509	224	59
	%			81.0	15.3	2.8	0.9		55.8	28.4	12.5	3.3

a Actual mileage 0.2, Class 3
b Actual mileage 0.5, Class 4
c Actual mileage 0.4, Class 4
d Actual mileage 0.4, Class 3
e Actual mileage 0.3, Class 4

Table 130 Estimates of miles of canal by chemical classification following remedial works. England and Wales

Authority		Total Length	Class 1	Class 2	Class 3	Class 4
Yorkshire	Miles	170	86	12	64	8
	%		50.6	7.1	37.6	4.7
Trent	Miles	439	90	291	52	6
	%		20.5	66.3	11.8	1.4
Lincolnshire	Miles	5	5	0	0	0
	%		100			
Welland and Nene	Miles	41	41	0	0	0
	%		100			
Great Ouse	Miles	31	31	0	0	0
	%		100			
Avon and Dorset	Miles	16	16	0	0	0
	%		100			
Somerset	Miles	14	14	0	0	0
	%		100			
Bristol Avon	Miles	20	20	0	0	0
	%		100			
Severn	Miles	245	87	158	0	0
	%		35.5	64.5		
Usk	Miles	46	33	13	0	0
	%		71.7	28.3		
Glamorgan	Miles	14	14	0	0	0
	%		100			
South West Wales	Miles	13	12	—*	1	0
	%		89.1	3.1	7.8	
Dee and Clwyd	Miles	26	26	0	0	0
	%		100			
Mersey and Weaver	Miles	246	145	91	10	0
	%		58.9	37.0	4.1	
Lancashire	Miles	122	44	78	0	0
	%		36.1	63.9		
Thames Conservancy	Miles	61	61	0	0	0
	%		100			
Lee Conservancy	Miles	1	0	1	0	0
	%			100		
Port of London Authority (incl. London Excluded Area)	Miles	30	0	30	0	0
	%			100		
Total England & Wales	Miles	1,540	725	674	127	14
	%		47.1	43.8	8.2	0.9

*Actual mileage 0.4, Class 2.
The following authorities have no canals in the survey: Northumbrian, East Suffolk and Norfolk, Essex, Kent, Sussex, Hampshire, Isle of Wight, Devon, Cornwall, Wye, Gwynedd and Cumberland.

Table 131 Suitability for public water supply. Situation after treatment of unsatisfactory discharges. Non-tidal rivers, England and Wales

	1970 situation			Additional miles becoming suitable	Situation after remedial works		
	Total miles in class	Miles at present suitable for further supply	% of total		Total miles in class	Miles suitable after treatment of effluents	% of total
Class 1	17,000	4,940	29.1	13	18,069	4,953	27.4
Class 2	3,290	485	14.7	183	3,425	668	19.5
Class 1 and Class 2	20,290	5,425	26.8	196	21,494	5,621	26.1

Suitability of improved rivers for further public water supplies

8 In Chapter 5 of Volume 1 it was explained that river authorities had been asked to indicate those stretches of Class 1 and Class 2 rivers which might be used as sources for further public water supplies. At the same time the authorities were also asked to indicate the additional lengths (if any) that might become suitable following the carrying out of the remedial measures they thought to be necessary. Details of these additional lengths are given in Table 131.

9 These additional lengths becoming suitable would occur not only within the upgraded miles stated in Table 126 but also within stretches of river where no change in chemical class is forecast. This means that the improvement within the class would be sufficient to make the river suitable, as, for example, where a "poor" class 2 river became a "good" Class 2 river. As with the information provided in Volume 1 about rivers suitable for public water supplies, these additional mileages can only be a broad guide and the degree of usage could be established only after detailed examination of any proposal put forward.

10 The change shown by Table 131 is not as great as might have been expected because it can be seen that, although the length in Class 1 and Class 2 may be increased by some 1,200 miles, the additional lengths becoming suitable will amount to only about 200 miles. This can largely be explained by the fact that the suitability or otherwise of Class 1 or Class 2 waters for public water supplies is not so much governed by quality criteria as by available quantity and the interests of other users (see Table 71, Volume 1) and these same criteria influence the upgraded mileage. The proportions of available mileage actually suitable or becoming suitable are very similar at 26.8 per cent in 1970 and 26.1 per cent (of the increased mileage) in the future.

Details of the survey for England and Wales

11 The report follows with a description of the changes occurring in each of the river authority areas.

1 Northumbrian River Authority

Table 132 Comparison of mileages by chemical classification following remedial works on discharges at present unsatisfactory. Rivers

	Non-tidal river. Chemical classes				Tidal river. Chemical classes				All rivers. Chemical classes			
	1	2	3	4	1	2	3	4	1	2	3	4
Present mileage in class	1,441	214	66	13	17	12	22	32	1,458	226	88	45
% of total in class	83.1	12.3	3.8	0.8	20.4	14.5	26.5	38.6	80.3	12.4	4.8	2.5
Net increase or decrease	+27	+35	−50	−12	+5	+49	−22	−32	+32	+84	−72	−44
Future mileage in class	1,468	249	16	1	22	61	0	0	1,490	310	16	1
% of total in class	84.6	14.4	0.9	0.1	26.5	73.5	—	—	82.1	17.0	0.9	0.1
Total miles	1,734				83				1,817			

Upgrading

12 The survey shows that a total of 137 miles of river are likely to be upgraded following the completion of remedial measures to give a net increase of 116 miles in Classes 1 and 2 and an equal net decrease in Classes 3 and 4. Of the 133 miles of Class 3 and 4 existing in 1970, only 17 miles are likely to remain after remedial measures have had their effect and only about 1 mile of this will be Class 4. All of the tidal waters should be Class 1 or Class 2.

13 Many rivers will be improved over short lengths but the Skerne (about 30 miles long) is expected to be improved over a length of some 22 miles, the tidal Tees over a length of about 19 miles and the tidal Tyne over some 16 miles.

14 The remaining Class 3 mileage is likely to be spread over some 8 rivers and streams, notably the Brierdene Burn (3.6 miles), the Skerne (3.4 miles), the Twizell Burn (3.0 miles) and the Team (2.7 miles). No significant lengths of Class 4 river will remain.

Future suitability for additional public water supply

15 A length of only 0.2 miles of the River Wear becomes available for public water supplies following the remedial measures.

2 Yorkshire River Authority

Table 133 Comparison of mileages by chemical classification following remedial works on discharges at present unsatisfactory. Rivers

	Non-tidal river. Chemical classes				Tidal river. Chemical classes				All rivers. Chemical classes			
	1	2	3	4	1	2	3	4	1	2	3	4
Present mileage in class	2,874	212	111	235	48	4	56	16	2,922	216	167	251
% of total in class	83.8	6.1	3.2	6.9	38.7	3.2	45.2	12.9	82.1	6.1	4.7	7.1
Net increase or decrease	+91	+19	+33	−143	+1	−1	+16	−16	+92	+18	+49	−159
Future mileage in class	2,965	231	144	92	49	3	72	0	3,014	234	216	92
% of total in class	86.4	6.7	4.2	2.7	39.5	2.4	58.1	—	84.8	6.5	6.1	2.6
Total miles	3,432				124				3,556			

Table 134 Comparison of mileages by chemical classification following remedial works on discharges at present unsatisfactory. Canals

	Chemical classes			
	1	2	3	4
Present mileage in class	73	24	29	44
% of total in class	42.9	14.1	17.1	25.9
Net increase or decrease	+13	−12	+35	−36
Future mileage in class	86	12	64	8
% of total in class	50.6	7.1	37.6	4.7
Total miles	170			

Upgrading

16 The survey shows that a total of 279 miles of river are likely to be upgraded following the completion of remedial works. The net effect will be a decrease of 159 miles in the length of Class 4, balanced by increases in the lengths in Classes 1, 2 and 3. The largest net increase (92 miles) is forecast on the length in Class 1. A total of 92 miles are likely to remain in Class 4, all on non-tidal waters, and the remaining Class 3 mileages will be divided in the proportion of two-thirds non-tidal to one-third tidal.

17 Many rivers will be improved over short lengths but the biggest improvements are likely to occur on the Aire and the Calder, where more than 100 miles will be upgraded, including all Class 4 waters and on the Don in the region of Conisbrough, over the length from Penistone and through Sheffield, where improvement is forecast over some 18 miles. Many improvements to small lengths of Class 2 rivers account for the large overall increase in Class 1 rivers.

18 The 92 miles of non-tidal Class 4 river remaining will be found in the Don (19 miles), in the Rother (26 miles), and spread in short lengths over some 25 small streams mostly in the West Riding. The Class 3 non-tidal miles will occur in the Aire (35 miles), the Calder (35 miles) and the Don (10 miles) and again, over short lengths, in 30 or more small streams and becks mostly in the West Riding. The Class 3 tidal miles remaining will be in the Ouse, the Don, the Aire and the Hull River.

19 On canals, 50 miles should be upgraded, principally in the Aire and Calder Navigation and the Leeds and Liverpool Canal. Shorter improvements are foreseen in the Calder and Hebble Navigation and in the Huddersfield Broad. The net effect will be a reduction in Class 4 and Class 2 mileages balanced by nearly equal increases in Class 3 and Class 1. The remaining Class 4 miles will be principally in the Sheffield and South Yorkshire Canal and the Class 3 miles in the Aire and Calder Navigation (improved from Class 4) and in the Sheffield and South Yorkshire Canal.

Future suitability for additional public water supply

20 On only two rivers do additional lengths become available following remedial measures. These are on the Lambwath Stream north of Hull and the River Foss at York. The totals are 4 miles on Class 1 river and 10 miles on Class 2 river.

3 Trent River Authority

Table 135 Comparison of mileages by chemical classification following remedial works on discharges at present unsatisfactory. Rivers

	Non-tidal river. Chemical classes				Tidal river. Chemical classes				All rivers. Chemical classes			
	1	2	3	4	1	2	3	4	1	2	3	4
Present mileage in class	607	503	171	178	0	26	0	26	607	529	171	204
% of total in class	41.6	34.5	11.7	12.2	—	50	—	50	40.0	35.2	11.3	13.5
Net increase or decrease	+37	+206	−74	−169	0	0	0	0	+37	+206	−74	−169
Future mileage in class	644	709	97	9	0	26	0	26	644	735	97	35
% of total in class	44.1	48.6	6.7	0.6	—	50	—	50	42.6	48.7	6.4	2.3
Total miles	1,459				52				1,511			

Table 136 Comparison of mileages by chemical classification following remedial works on discharges at present unsatisfactory. Canals

	Chemical classes			
	1	2	3	4
Present mileage in class	88	210	91	50
% of total in class	20.1	47.8	20.7	11.4
Net increase or decrease	+2	+81	−39	−44
Future mileage in class	90	291	52	6
% of total in class	20.5	66.3	11.8	1.4
Total miles	439			

Upgrading

21 Some 350 miles on rivers are likely to be upgraded following remedial measures, all of the forecast upgrading being on the non-tidal sections. There should be a very big increase in the Class 2 mileage accompanied by decreases in the lower classes, particularly non-tidal Class 4, where the present length of 178 miles is expected to be reduced to 9 miles only.

22 The longest length of improvement is forecast for the Trent with some 68 miles in all upgraded. This should bring virtually all of the non-tidal river up to Class 2, the main exception being a length of about 9 miles below Stoke where improvement from Class 4 to Class 3 is forecast. A small increase in the Class 1 length right at the head of the river is also expected. The grossly polluted River Tame which is wholly Class 4 at present is expected to be improved to either Class 2 or Class 3 over its entire length of about 46 miles. Other badly polluted rivers where significant lengths are expected to be improved are the Bottesford Beck, the Maun, the Erewash, the Churnet, the Anker, the Cole, the Wolverhampton Arm of the Tame and Ford Brook.

23 On the non-tidal waters the few remaining Class 4 miles will be found in the upper Churnet, Hooborough Brook, Minworth Conduit and the Pyford Brook. The Class 3 miles will be found mainly on the Tame (31 miles), the Trent (9 miles), the Wolverhampton Arm of the Tame (6 miles), Ford Brook (6 miles), Repton Brook (5 miles) with the balance spread in short lengths over some 20 or so brooks. Much of this Class 3 mileage will have been improved from Class 4. No change is expected on the tidal waters so that 26 Class 4 miles will remain, all in the lower Trent. It is the river authority's view that this Class 4 rating is primarily the result of the natural phenomenon of high silt content.

24 On canals more than 90 miles will be upgraded principally in the Birmingham area with smaller stretches on the Trent and Mersey Canal, the Erewash Canal, the Chesterfield Canal and the Coventry Canal. The net effect will be a reduction in Class 3 and Class 4 mileages and an equal increase mainly in Class 2. The remaining Class 3 and Class 4 miles will be found principally in the Sheffield and South Yorkshire Canal, Birmingham Canal Navigation, the Birmingham and Fazeley Canal and the Grand Union Canal.

Future suitability for additional public water supply

25 A further 100 miles of river should become suitable for public water supply in addition to the 376 miles already suitable, 44 miles of this extra length occurring on the River Trent. Other rivers where significant additional lengths are likely to become available are the Wreake (13 miles), the Anker (12 miles), the Derwent (10 miles) and the Penk (15 miles). ☐

4 Lincolnshire River Authority

Table 137 Comparison of mileages by chemical classification following remedial works on discharges at present unsatisfactory. Rivers

	Non-tidal river. Chemical classes				Tidal river. Chemical classes				All rivers. Chemical classes			
	1	2	3	4	1	2	3	4	1	2	3	4
Present mileage in class	445	42	22	1	4	0	7	0	449	42	29	1
% of total in class	87.3	8.2	4.3	0.2	38.1	—	61.9	—	86.1	8.0	5.7	0.2
Net increase or decrease	+37	−22	−14	−1	0	+7	−7	0	+37	−15	−21	−1
Future mileage in class	482	20	8	0	4	7	0	0	486	27	8	0
% of total in class	94.5	3.9	1.6	—	38.1	61.9	—	—	93.3	5.2	1.5	—
Total miles	510				11				521			

Upgrading

26 The survey shows that a total of 56 miles of river are likely to be upgraded following completion of remedial measures, the net effect being an increase of 37 miles in the length in Class 1 balanced by decreases in the other 3 classes. All existing Class 4 waters are likely to be improved and only 8 miles are likely to remain in Class 3. Substantial improvements are expected in the Witham and the Till. The Witham should become mostly Class 1 and the Till wholly Class 1. The 8 miles remaining in Class 3 will be spread over the non-tidal waters of the New Cut Drain, the Old River Slea and the Duckpool/Reeds Beck. All the canals are Class 1 at present.

Future suitability for additional public water supply

27 No additional lengths of river will become available for public water supply following completion of the remedial measures. ☐

5 Welland and Nene River Authority

Table 138 Comparison of mileages by chemical classification following remedial works on discharges at present unsatisfactory. Rivers

	Non-tidal river. Chemical classes				Tidal river. Chemical classes				All rivers. Chemical classes			
	1	2	3	4	1	2	3	4	1	2	3	4
Present mileage in class	224	84	10	1	0	42	0	0	224	126	10	1
% of total in class	70.2	26.4	3.1	0.3	—	100	—	—	62.0	34.9	2.8	0.3
Net increase or decrease	0	0	0	0	0	0	0	0	0	0	0	0
Future mileage in class	224	84	10	1	0	42	0	0	224	126	10	1
% of total in class	70.2	26.4	3.1	0.3	—	100	—	—	62.0	34.9	2.8	0.3
Total miles	319				42				361			

Upgrading

28 No changes are forecast in the area, but attention is drawn to the observations of the Authority (as recorded in paragraph 7 on page 12 of Volume 1) concerning the impracticability of improving lowland Class 2 rivers. The short lengths of Class 3 and 4 are likely to remain in the Corby area and in the area between Peterborough and Market Deeping. All the canals in the area are Class 1 at present.

Future suitability for additional public water supply

29 No additional lengths of river will become available for public water supply following remedial measures.

6 Great Ouse River Authority

Table 139 Comparison of mileages by chemical classification following remedial works on discharges at present unsatisfactory. Rivers

	Non-tidal river. Chemical classes				Tidal river. Chemical classes				All rivers. Chemical classes			
	1	2	3	4	1	2	3	4	1	2	3	4
Present mileage in class	540	309	58	4	39	0	0	0	579	309	58	4
% of total in class	59.3	33.9	6.4	0.4	100	—	—	—	61.0	32.5	6.1	0.4
Net increase or decrease	+139	−112	−27	0	0	0	0	0	+139	−112	−27	0
Future mileage in class	679	197	31	4	39	0	0	0	718	197	31	4
% of total in class	74.6	21.6	3.4	0.4	100	—	—	—	75.5	20.8	3.3	0.4
Total miles	911				39				950			

Upgrading

30 The survey shows that 163 miles of rivers are likely to be upgraded. All the tidal waters are already Class 1 and the whole of the improvements are therefore on non-tidal waters, the net effect being an expected increase of 139 miles of Class 1 with the corresponding decrease being largely in the Class 2 lengths. Attention is, however, drawn to the observations of the Authority (as recorded in paragraph 7 on page 12 of Volume 1) concerning the way in which natural characteristics can often prevent improvement from Class 2 to Class 1.

31 The biggest improvements should occur on the Ouse where almost all of the Class 2 lengths are expected to be improved to Class 1. Substantial lengths of Class 3 are likely to be improved on the Middle Level System and on the River Cam. The residual Class 4 mileage will be in the Pix Brook and the Class 3 mileage spread mainly over the Middleton Stop Drain, the upper end of the Middle Level System and the High Lode on the Sixteen Foot Drain System. All the canal waters are Class 1 at present.

Future suitability for additional public water supply

32 No additional lengths of river will become available for public water supply following remedial measures.

7 East Suffolk and Norfolk River Authority

Table 140 Comparison of mileages by chemical classification following remedial works on discharges at present unsatisfactory. Rivers

	Non-tidal river. Chemical classes				Tidal river. Chemical classes				All rivers. Chemical classes			
	1	2	3	4	1	2	3	4	1	2	3	4
Present mileage in class	287	89	0	0	78	87	0	0	365	176	0	0
% of total in class	76.3	23.7	—	—	47.3	52.7	—	—	67.5	32.5	—	—
Net increase or decrease	+89	−89	0	0	+8	−8	0	0	+97	−97	0	0
Future mileage in class	376	0	0	0	86	79	0	0	462	79	0	0
% of total in class	100	—	—	—	52.1	47.9	—	—	85.4	14.6	—	—
Total miles	376				165				541			

Upgrading

33 In this area, 97 miles are likely to be upgraded following the completion of remedial measures, the net effect being that, on the non-tidal lengths, all the rivers in the area should become Class 1. On the tidal lengths small improvements from Class 2 to Class 1 are envisaged on the Glaven and the Orwell. Attention is, however, drawn to the observations of the Authority (as recorded in paragraph

7 of page 12 of Volume 1) concerning the way in which natural characteristics can often prevent improvement from Class 2 to Class 1, particularly in the lowland rivers. There are no rivers classified lower than Class 2 at present.

Future suitability for additional public water supply
34 Rivers where additional mileages are likely to become suitable for public water supplies are the Ant (5 miles), the Waveney (9 miles), the Tas (7 miles) and the Tiffey (6 miles). ☐

8 Essex River Authority

Table 141 Comparison of mileages by chemical classification following remedial works on discharges at present unsatisfactory. Rivers

	Non-tidal river. Chemical classes				Tidal river. Chemical classes				All rivers. Chemical classes			
	1	2	3	4	1	2	3	4	1	2	3	4
Present mileage in class	214	117	32	9	63	14	6	5	277	131	38	14
% of total in class	57.5	31.5	8.6	2.4	71.6	15.9	6.8	5.7	60.2	28.5	8.3	3.0
Net increase or decrease	+46	−23	−14	−9	+4	+7	−6	−5	+50	−16	−20	−14
Future mileage in class	260	94	18	0	67	21	0	0	327	115	18	0
% of total in class	69.9	25.3	4.8	—	76.1	23.9	—	—	71.1	25.0	3.9	—
Total miles	372				88				460			

Upgrading
35 The survey shows that 80 miles should be upgraded following the completion of necessary remedial works. The net effect on non-tidal waters is that the length of Class 1 should be increased by 46 miles, balanced by reductions of 23 miles, 14 miles and 9 miles respectively in Classes 2, 3 and 4; 18 miles of Class 3 are likely to remain. The improvements on tidal lengths will mean that all tidal waters will become either Class 1 or Class 2.

36 The biggest improvements should occur on the Colne where a total of 13 miles are likely to be upgraded mainly to Class 1 and where all Class 3 and 4 river will be eliminated. Other sizeable lengths of improvement should occur on the Stour (11 miles), the Roding (11 miles) and the Brain (8 miles). The balance of improved mileage should be spread in short lengths over some 20 rivers and streams. Short Class 3 lengths are likely to remain on the non-tidal waters of the Ramsey, the Wid, the Crouch, Mar Dyke, the Beam, Wantz Stream and the Roding.

Future suitability for additional public water supply
37 No additional lengths of river will become available for public water supply following completion of remedial measures. ☐

9 Kent River Authority

Table 142 Comparison of mileages by chemical classification following remedial works on discharges at present unsatisfactory. Rivers

	Non-tidal river. Chemical classes				Tidal river. Chemical classes				All rivers. Chemical classes			
	1	2	3	4	1	2	3	4	1	2	3	4
Present mileage in class	506	20	3	0	51	7	15	3	557	27	18	3
% of total in class	95.6	3.8	0.6	—	67.1	9.2	19.7	4.0	92.0	4.5	3.0	0.5
Net increase or decrease	+11	−8	−3	0	+7	0	−7	0	+18	−8	−10	0
Future mileage in class	517	12	0	0	58	7	8	3	575	19	8	3
% of total in class	97.7	2.3	—	—	76.3	9.2	10.5	4.0	95.1	3.1	1.3	0.5
Total miles	529				76				605			

Upgrading
38 The survey shows that 23 miles of river are likely to be upgraded. The net effect on non-tidal waters will be an increase of 11 miles in Class 1 balanced by reductions in Classes 2 and 3 after which all non-tidal waters should be in either Class 1 or Class 2. The net effect on tidal waters should be a moderate increase in Class 1 with an equal decrease in Class 3. The remaining Class 3 and 4 tidal waters are likely to occur on the Medway and Middle Cut and on the Milton Creek respectively.

Future suitability for additional public water supply
39 No additional lengths of river will become available for public water supplies following completion of remedial measures. ☐

10 Sussex River Authority

Table 143 Comparison of mileages by chemical classification following remedial works on discharges at present unsatisfactory. Rivers

	Non-tidal river. Chemical classes				Tidal river. Chemical classes				All rivers. Chemical classes			
	1	2	3	4	1	2	3	4	1	2	3	4
Present mileage in class	168	93	5	3	54	20	0	0	222	113	5	3
% of total in class	62.6	34.5	1.8	1.1	73.0	27.0	—	—	64.7	32.9	1.5	0.9
Net increase or decrease	+58	−50	−5	−3	+20	−20	0	0	+78	−70	−5	−3
Future mileage in class	226	43	0	0	74	0	0	0	300	43	0	0
% of total in class	84.0	16.0	—	—	100	—	—	—	88.0	12.0	—	—
Total miles	269				74				343			

Upgrading

40 In this area, some 85 miles of river are forecast to be upgraded. The net effect on non-tidal waters will be a substantial increase in Class 1 balanced by reductions mainly in Class 2 and the complete elimination of all Class 3 and Class 4 lengths. On tidal waters, the forecast of upgrading indicates that the improvements should bring all tidal waters up to Class 1. The rivers where longest lengths of improvement are likely to occur are the Arun (20 miles) and the Adur (13.5 miles) with some 20 other rivers likely to be improved over shorter lengths.

Future suitability for additional public water supply

41 No additional lengths of river will become available for public water supply following completion of remedial measures. ☐

11 Hampshire River Authority

Table 144 Comparison of mileages by chemical classification following remedial works on discharges at present unsatisfactory. Rivers

	Non-tidal river. Chemical classes				Tidal river. Chemical classes				All rivers. Chemical classes			
	1	2	3	4	1	2	3	4	1	2	3	4
Present mileage in class	239	114	14	2	35	15	2	3	274	129	16	5
% of total in class	64.8	30.9	3.8	0.5	63.7	27.2	3.6	5.5	64.7	30.4	3.8	1.1
Net increase or decrease	+15	−3	−10	−2	+4	0	−1	−3	+19	−3	−11	−5
Future mileage in class	254	111	4	0	39	15	1	0	293	126	5	0
% of total in class	68.8	30.1	1.1	—	70.9	27.3	1.8	—	69.1	29.7	1.2	—
Total miles		369				55				424		

Upgrading

42 In this area 21 miles of river are likely to be upgraded following the necessary remedial measures. Class 4 is likely to be eliminated altogether, and apart from 5 miles (Class 3) all rivers should be in Class 1 or Class 2. The longest improved stretch is forecast in the Beaulieu (about 9 miles) and the other improvements will occur in short lengths over 10 or so other rivers. The stretches in Class 3 likely to remain will be found in short lengths mainly in the Denmead Stream, the Farlington Stream, the Hamble and the Badnam Stream.

Future suitability for additional public water supply

43 No additional lengths of river will become available for public water supply following completion of remedial measures. ☐

12 Isle of Wight River and Water Authority

Table 145 Comparison of mileages by chemical classification following remedial works on discharges at present unsatisfactory. Rivers

	Non-tidal river. Chemical classes				Tidal river. Chemical classes				All rivers. Chemical classes			
	1	2	3	4	1	2	3	4	1	2	3	4
Present mileage in class	55	12	5	1	11	4	1	0	66	16	6	1
% of total in class	75.4	16.4	6.8	1.4	68.7	25.0	6.3	—	74.2	18.0	6.7	1.1
Net increase or decrease	+11	−6	−4	−1	+5	−4	−1	0	+16	−10	−5	−1
Future mileage in class	66	6	1	0	16	0	0	0	82	6	1	0
% of total in class	90.4	8.2	1.4	—	100	—	—	—	92.1	6.7	1.2	—
Total miles		73				16				89		

Upgrading

44 In the Isle of Wight 17 miles of river are likely to be upgraded following completion of the remedial measures. The improvements will be spread over 14 rivers and the net result will be to bring most of the rivers to Class 1. Only 6 miles will remain in Class 2 with a short length of just under a mile of Class 3 in the Whale Chine.

Future suitability for additional public water supply

45 No additional lengths of river will become available for public water supply following completion of remedial measures. ☐

13 Avon and Dorset River Authority

Table 146 Comparison of mileages by chemical classification following remedial works on discharges at present unsatisfactory. Rivers

	Non-tidal river. Chemical classes				Tidal river. Chemical classes				All rivers. Chemical classes			
	1	2	3	4	1	2	3	4	1	2	3	4
Present mileage in class	401	7	3	8	14	0	0	0	415	7	3	8
% of total in class	95.7	1.7	0.7	1.9	100	—	—	—	95.8	1.6	0.7	1.9
Net increase or decrease	+10	+1	−3	−8	0	0	0	0	+10	+1	−3	−8
Future mileage in class	411	8	0	0	14	0	0	0	425	8	0	0
% of total in class	98.1	1.9	—	—	100	—	—	—	98.3	1.7	—	—
Total miles		419				14				433		

Upgrading

46 The 1970 survey showed 415 miles of river to be in Class 1 out of a total of 433 miles, with 11 miles in Classes 3 and 4. The forecast is that 18 miles are likely to be improved including all Class 3 mileage upgraded to Class 1 and all Class 4 to Class 2. This should bring all the rivers up to the two highest grades, with 425 miles in Class 1 and 8 miles in Class 2. All the canal mileage is already Class 1.

Future suitability for additional public water supply

47 No additional lengths of river will become available for public water supply following the completion of remedial measures. ☐

14 Devon River Authority

Table 147 Comparison of mileages by chemical classification following remedial works on discharges at present unsatisfactory. Rivers

	Non-tidal river. Chemical classes				Tidal river. Chemical classes				All rivers. Chemical classes			
	1	2	3	4	1	2	3	4	1	2	3	4
Present mileage in class	841	58	25	0	32	25	21	0	873	83	46	0
% of total in class	91.0	6.3	2.7	—	41.0	32.1	26.9	—	87.2	8.2	4.6	—
Net increase or decrease	+37	−12	−25	0	+14	+7	−21	0	+51	−5	−46	0
Future mileage in class	878	46	0	0	46	32	0	0	924	78	0	0
% of total in class	95.0	5.0	—	—	59.0	41.0	—	—	92.2	7.8	—	—
Total miles	924				78				1,002			

Upgrading

48 The forecast is that 95 miles are likely to be upgraded. The length of non-tidal river in Class 1 should go up by 37 miles with an equal reduction in Class 2 and Class 3. On tidal rivers, the length in Class 1 and Class 2 is likely to increase by 21 miles with a corresponding reduction in Class 3. After these improvements, no Class 3 or 4 waters will remain and most will be in Class 1.

49 The river likely to be improved over the greatest length is the Culm where 17 miles should be upgraded from Class 3 to Class 2. Other rivers where substantial improvements are expected are the Axe where about 13 miles should be improved, the Taw where 9 miles should be improved and the Exe where 8 miles should be improved. The balance of the upgraded lengths will be found in some 14 other rivers.

Future suitability for additional public water supply

50 No additional lengths of river will become available for public water supply following completion of remedial measures. ☐

15 Cornwall River Authority

Table 148 Comparison of mileages by chemical classification following remedial works on discharges at present unsatisfactory. Rivers

	Non-tidal river. Chemical classes				Tidal river. Chemical classes				All rivers. Chemical classes			
	1	2	3	4	1	2	3	4	1	2	3	4
Present mileage in class	941	75	27	50	99	39	0	1	1,040	114	27	51
% of total in class	86.1	6.8	2.5	4.6	71.2	28.1	—	0.7	84.4	9.2	2.2	4.2
Net increase or decrease	+38	+25	−27	−36	+32	−31	0	−1	+70	−6	−27	−37
Future mileage in class	979	100	0	14	131	8	0	0	1,110	108	0	14
% of total in class	89.6	9.1	—	1.3	94.2	5.8	—	—	90.1	8.8	—	1.1
Total miles	1,093				139				1,232			

Upgrading

51 The forecast is that about 125 miles of river are likely to be improved as a result of remedial measures, with the net result that the length in Class 1 should go up by 70 miles with a corresponding decrease almost wholly in the two worst classes. Apart from 14 miles of non-tidal river which will remain in Class 4, the whole of the 1,200 and more miles should be in the two top classes, and mostly in Class 1.

52 The improvements will be spread over about 60 rivers and streams, with the greatest lengths being in the Fal (15 miles), St. Austell White River (8 miles), Charlestown Leat (8 miles), Luxulyan River (6 miles), Tory Brook (6 miles), Red River (6 miles) and the Portreath Stream (5 miles). The 14 miles of Class 4 remaining will be concentrated in the Carnon and its tributaries in the Fal basin, south-west of Truro.

Future suitability for additional public water supply

53 No additional lengths of river will become available for public water supply following the completion of remedial measures. ☐

16 Somerset River Authority

(see Table 149 over page)

54 The survey shows that 104 miles of river should be upgraded as a result of remedial measures, the improvements being largely on non-tidal rivers. The net effect should be an overall increase in Class 1 waters balanced by a decrease in the other 3 classes. A large proportion of the improvements are expected to be in Class 2 rivers, and in the more polluted Class 3 and Class 4 lengths 43 miles should be reduced to 3 miles.

55 The river where most improvement is likely to occur is the Parrett and the survey records a probable improvement over about

(16 Somerset River Authority continued)

Table 149 Comparison of mileages by chemical classification following remedial works on discharges at present unsatisfactory. Rivers

	Non-tidal river. Chemical classes				Tidal river. Chemical classes				All rivers. Chemical classes			
	1	2	3	4	1	2	3	4	1	2	3	4
Present mileage in class	319	64	20	3	0	14	6	14	319	78	26	17
% of total in class	78.6	15.8	4.9	0.7	—	41.2	17.6	41.2	72.5	17.7	5.9	3.9
Net increase or decrease	+81	−58	−20	−3	+1	+16	−5	−12	+82	−42	−25	−15
Future mileage in class	400	6	—*	0	1	30	1	2	401	36	1	2
% of total in class	98.5	1.4	0.1	—	2.9	88.3	2.9	5.9	91.2	8.2	0.2	0.4
Total miles		406				34				440		

*Actual mileage 0.2, Class 3

18 miles, mostly stretches at present in Class 3 and 4, all expected to be improved to Class 2. Other rivers likely to be improved over substantial lengths are the Yeovil Yeo, the Tone, the Cary, the Axe and the Doniford Stream. Improvements over shorter lengths are expected in about 30 other rivers and streams. Short stretches in Class 3 and 4 are likely to remain on the Axe, mainly on the tidal waters at the mouth of the river. The canals in the area are all Class 1 at present.

Future suitability for additional public water supply
56 On the River Avill, a further 0.5 mile of Class 2 river is likely to become suitable for public water supply.

17 Bristol Avon River Authority

Table 150 Comparison of mileages by chemical classification following remedial works on discharges at present unsatisfactory. Rivers

	Non-tidal river. Chemical classes				Tidal river. Chemical classes				All rivers. Chemical classes			
	1	2	3	4	1	2	3	4	1	2	3	4
Present mileage in class	198	138	21	6	0	0	0	15	198	138	21	21
% of total in class	54.6	38.0	5.8	1.6	—	—	—	100	52.4	36.4	5.6	5.6
Net increase or decrease	+82	−61	−15	−6	+8	+7	0	−15	+90	−54	−15	−21
Future mileage in class	280	77	6	0	8	7	0	0	288	84	6	0
% of total in class	77.1	21.2	1.7	—	53.3	46.7	—	—	76.2	22.2	1.6	—
Total miles		363				15				378		

Upgrading
57 On the rivers in this area, about 108 miles should be upgraded following the remedial measures, the net effect being an increase of 90 miles in Class 1 balanced by reductions in the other 3 classes. The biggest improvement will be in Class 2 but significant reductions are expected in the two worst classes, where the present length of 42 miles is expected to be reduced to 6 miles with all Class 4 waters improved to Class 1 or Class 2.

58 The Bristol Avon is the principal river in the area and 47 miles of its length of 86 miles are forecast to be improved, with all the Class 3 and 4 sections likely to be improved to Class 1 or Class 2. The Somerset Frome is expected to improve over 12 miles, such that the whole river should be Class 1 and the Brinkworth Brook is also likely to improve over about 12 miles to either Class 1 or Class 2. Improvements over shorter lengths are expected in about 15 other rivers and streams. About 6 miles of river are likely to remain in Class 3, principally in the Whatley Brook and the Poulshot Stream. The canal lengths are all Class 1 at present.

Future suitability for additional public water supply
59 The survey indicates that no additional lengths of river are likely to become available for public water supply following the completion of remedial measures.

18 Severn River Authority

Table 151 Comparison of mileages by chemical classification following remedial works on discharges at present unsatisfactory. Rivers

	Non-tidal river. Chemical classes				Tidal river. Chemical classes				All rivers. Chemical classes			
	1	2	3	4	1	2	3	4	1	2	3	4
Present mileage in class	857	173	17	68	48	0	0	0	905	173	17	68
% of total in class	76.9	15.5	1.5	6.1	100	—	—	—	77.8	14.8	1.5	5.9
Net increase or decrease	+14	+15	+29	−58	0	0	0	0	+14	+15	+29	−58
Future mileage in class	871	188	46	10	48	0	0	0	919	188	46	10
% of total in class	78.1	16.9	4.1	0.9	100	—	—	—	79.0	16.2	3.9	0.9
Total miles		1,115				48				1,163		

Upgrading
60 The survey shows that 73 miles of river, all on non-tidal stretches, are forecast to be upgraded following the completion of the remedial measures, the effect being an increase in each of Class 1, 2 and 3, with a decrease in Class 4 of 58 miles.

61 Improvement over the greatest length is likely to occur in the Worcestershire Stour where some 22 miles of Class 4 river should become Class 3. About 11 miles of the Arrow should be improved from Class 4 to Class 2 with a similar improvement over 9 miles of the Warwickshire Avon. Improvements over shorter lengths are expected

14 other rivers. There are likely to remain about 46 miles in Class 3 and 10 miles in Class 4. The Worcestershire Stour will account for 5 miles of Class 3 (mostly upgraded from Class 4), the Salwarpe for about 6 miles and the Chelt for about 5 miles, with shorter lengths in the Frome, Hatherley Brook, Hyde Brook, Cow Honeybourne Brook, the Sowe and the Mouseweet Brook. Short lengths of Class 4 are likely to remain on the Salwarpe, Strine Brook, Tach Brook and the Sowe. No change is forecast in the classification of canal mileage where 35.5 per cent is in Class 1 and 64.5 per cent in Class 2.

Future suitability for additional public water supply
62 A further length of about 5 miles of the River Morda near Oswestry should become suitable for public water supply following completion of remedial measures. ☐

19 Wye River Authority

Table 152 Comparison of mileages by chemical classification following remedial works on discharges at present unsatisfactory. Rivers

	Non-tidal river. Chemical classes				Tidal river. Chemical classes				All rivers. Chemical classes			
	1	2	3	4	1	2	3	4	1	2	3	4
Present mileage in class	1,012	28	9	—*	15	1	0	0	1,027	29	9	—*
% of total in class	96.4	2.6	0.9	0.1	93.7	6.3	—	—	96.5	2.6	0.8	0.1
Net increase or decrease	+7	+1	−8	(less 0.4)	0	0	0	0	+7	+1	−8	(less 0.4)
Future mileage in class	1,019	29	1	0	15	1	0	0	1,034	30	1	0
% of total in class	97.1	2.8	0.1	—	93.7	6.3	—	—	97.1	2.8	0.1	—
Total miles	1,049				16				1,065			

*Actual mileage 0.4, Class 4

Upgrading
63 The Wye River Authority already has 99 per cent of its rivers in the two top classes and a further 12 miles are forecast to be upgraded, all on non-tidal stretches, with a net effect of increasing the mileage in Class 1 and Class 2. The River Wye would be improved over about 4 miles and the remaining improvements would be spread over 6 other rivers. Short lengths of Class 3 water are likely to remain in the Lewstone Brook, Wellington Brook and, over a very short stretch, in the Wye, but no Class 4 will remain.

Future suitability for additional public water supply
64 No additional lengths of river are likely to become available for public water supply following the completion of remedial works. ☐

20 Usk River Authority

Table 153. Comparison of mileages by chemical classification following remedial works on discharges at present unsatisfactory. Rivers

	Non-tidal river. Chemical classes				Tidal river. Chemical classes				All rivers. Chemical classes			
	1	2	3	4	1	2	3	4	1	2	3	4
Present mileage in class	148	48	21	20	6	4	6	3	154	52	27	23
% of total in class	62.3	20.3	8.9	8.5	31.6	21.0	31.6	15.8	59.8	20.2	10.5	9.5
Net increase or decrease	0	+21	−1	−20	0	+9	−6	−3	0	+30	−7	−23
Future mileage in class	148	69	20	0*	6	13	0	0	154	82	20	0*
% of total in class	62.3	29.1	8.4	0.2	31.6	68.4	—	—	60.0	32.0	7.8	0.2
Total miles	237				19				256			

*Actual length 0.5m, Class 4

Upgrading
65 On the rivers in the area, 29 miles are expected to be upgraded with the effect of increasing the mileage in Class 2 and decreasing the mileages in Classes 3 and 4. All tidal waters should become Class 1 or Class 2 but 20 miles of non-tidal waters are likely to remain in Class 3, these being mostly on the penned reens where the Class 3 condition is due more to stagnant conditions than to polluting discharges.

66 The Ebbw is expected to improve over a length of about 23 miles from Class 4 to Class 2. The tidal section of the Usk should be improved from Class 3 to Class 2 over about 6 miles and a small improvement is expected in the Clydach. A short section of an unnamed tributary of the Ebbw Fach will remain in Class 4. No change is forecast in the classification of canal mileage where 71.7 per cent is in Class 1 and 28.3 per cent in Class 2.

Future suitability for additional public water supply
67 No additional lengths are likely to become available for public water supply following the completion of remedial measures. ☐

21 Glamorgan River Authority

(see Table 154 over page)

Upgrading
68 A total of 109 miles of river are likely to be upgraded following completion of remedial measures with the effect of reducing the lengths in the two worst classes from 135 miles to 60 miles accompanied by a corresponding increase of 75 miles in the two highest classes.

69 Improvements over some 23 miles are forecast on the Rhymney where all the Class 4 length should be improved to Class 3 and much of the Class 3 should be improved to Class 2. The Taff is forecast to be improved over a length of about 29 miles and again all Class 4 should be eliminated and there should be a big reduction in Class 3. The Rhondda Fach should have its 7.5 miles Class 3 improved to Class 2.

(21 Glamorgan River Authority continued)

Table 154. Comparison of mileages by chemical classification following remedial works on discharges at present unsatisfactory. Rivers

	Non-tidal river. Chemical classes				Tidal river. Chemical classes				All rivers. Chemical classes			
	1	2	3	4	1	2	3	4	1	2	3	4
Present mileage in class	120	68	87	18	2	4	16	14	122	72	103	32
% of total in class	41.0	23.2	29.7	6.1	5.6	11.1	44.4	38.9	36.6	22.0	31.6	9.8
Net increase or decrease	+28	+39	−55	−12	0	+8	−5	−3	+28	+47	−60	−15
Future mileage in class	148	107	32	6	2	12	11	11	150	119	43	17
% of total in class	50.5	36.5	10.9	2.1	5.6	33.2	30.6	30.6	45.5	36.2	13.1	5.2
Total miles	293				36				329			

Improvements over shorter lengths are forecast in about 20 other rivers including the Ely (6 miles), the Cynon (5.7 miles), the Neath (5.3 miles) and the Afan (4.6 miles).

70 On non-tidal rivers, there are likely to remain some 32 miles of Class 3 and 6 miles of Class 4. The former will be spread over 18 rivers, notably the Rhymney (3 miles), the Ely (5.6 miles), the Llynfi (3.4 miles), the Kenfig (3.8 miles), the Afan (4 miles), the Pelena (3.2 miles) and the Dulais (2.7 miles). The Class 4 non-tidal waters will be mainly on the Ely and the Clun. The Class 3 tidal waters will be mainly on the Taff and the Rhymney and the Class 4 tidal waters on the Ely. The canals in the area are already Class 1.

Future suitability for additional public water supply

71 No additional lengths of river are likely to become available for public water supply following completion of remedial measures.

22 South West Wales River Authority

Table 155 Comparison of mileages by chemical classification following remedial works on discharges at present unsatisfactory. Rivers

	Non-tidal river. Chemical classes				Tidal river. Chemical classes				All rivers. Chemical classes			
	1	2	3	4	1	2	3	4	1	2	3	4
Present mileage in class	526	28	43	11	51	18	11	15	577	46	54	26
% of total in class	86.5	4.6	7.1	1.8	53.7	18.9	11.6	15.8	82.1	6.5	7.7	3.7
Net increase or decrease	+5	+12	−14	−3	+6	+11	−2	−15	+11	+23	−16	−18
Future mileage in class	531	40	29	8	57	29	9	—*	588	69	38	8
% of total in class	87.3	6.6	4.8	1.3	59.8	30.4	9.4	0.4	83.6	9.8	5.5	1.1
Total miles	608				95				703			

*Actual mileage 0.4 miles, Class 4

Upgrading

72 A total of 52 miles of river should be upgraded as a result of the completion of remedial measures, resulting in a net decrease in the Class 3 and 4 lengths of 34 miles, but still leaving some 46 miles altogether in the two lowest classes.

73 The main improvements in the area are likely to be to the rivers in the Loughor basin, with about 13 miles of the Loughor improved including about 12 miles of Class 3 and Class 4 improved to Class 2 and some 7 miles of the Lliedi improved. The Tawe is expected to be upgraded over about 7 miles and the Gwendraeth Fawr over 5 miles. The balance of the improvements would be spread over 9 other rivers in shorter lengths.

74 About 29 miles of non-tidal river will still be in Class 3, some having been upgraded from Class 4. The Rheidol will be Class 3 over about 12 miles and the Afon Yswyth over about 8 miles and six other rivers will have Class 3 stretches over short distances, namely the Tawe, the Dafen, the Melindwr, the Mynach, the Castell and the Clarach. Eight miles of Class 4 non-tidal river remaining will be divided mostly between the Yswyth and the Magwr. Certain tidal waters will still be in Class 3, mainly the Tawe and the Lleidi over about 4 miles each and a very short tidal length on the Nant-y-Fendrod will remain Class 4. No improvement is expected in the rivers flowing in the valleys above Aberswyth. No changes are forecast in the canals in the area, most of the mileage being already Class 1.

Future suitability for additional public water supply

75 No additional lengths of river are likely to become available for public water supply as a result of the remedial measures.

23 Gwynedd River Authority

Table 156 Comparison of mileages by chemical classification following remedial works on discharges at present unsatisfactory. Rivers

	Non-tidal river. Chemical classes				Tidal river. Chemical classes				All rivers. Chemical classes			
	1	2	3	4	1	2	3	4	1	2	3	4
Present mileage in class	279	12	—*	1	69	0	—*	0	348	12	—*	1
% of total in class	95.5	4.1	0.1	0.3	99.4	—	0.6	—	96.0	3.4	0.3	0.3
Net increase or decrease	0	+1	0	−1	0	0	0	0	0	+1	0	−1
Future mileage in class	279	13	—*	0	69	0	—*	0	348	13	—**	0
% of total in class	95.5	4.4	0.1	—	99.4	—	0.6	—	96.0	3.7	0.3	—
Total miles	292				69				361			

*Actual mileage 0.4, Class 3   **Actual mileage 0.8, Class 3

Upgrading

76 The only expected change relates to 0.7 miles improved from Class 4 to Class 2 in the Garsiwn Ditch near Machynlleth. A short length at the lower end of the Afon Wen near Pwllheli is likely to remain Class 3, but all other rivers are or will be in the top grades, with 96 per cent in Class 1.

Future suitability for additional public water supply

77 No additional lengths of river are likely to become available for public water supply following the remedial measures. ☐

24 Dee and Clwyd River Authority

Table 157 Comparison of mileages by chemical classification following remedial works on discharges at present unsatisfactory. Rivers

	Non-tidal river. Chemical classes				Tidal river. Chemical classes				All rivers. Chemical classes			
	1	2	3	4	1	2	3	4	1	2	3	4
Present mileage in class	361	39	3	—*	14	15	0	0	375	54	3	—*
% of total in class	89.5	9.7	0.7	0.1	48.3	51.7	—	—	86.7	12.5	0.7	0.1
Net increase or decrease	+21	−18	−3	0	+15	−15	0	0	+36	−33	−3	0
Future mileage in class	382	21	0	—*	29	0	0	0	411	21	0	—*
% of total in class	94.8	5.1	—	0.1	100	—	—	—	95.1	4.8	—	0.1
Total miles	403				29				432			

*Actual mileage 0.4, Class 4

Table 158 Comparison of mileages of chemical classification following remedial works on discharges at present unsatisfactory. Canals

	Chemical classes			
	1	2	3	4
Present mileage in class	18	8	0	0
% of total in class	69.2	30.8	—	—
Net increase or decrease	+8	−8	0	0
Future mileage in class	26	0	0	0
% of total in class	100	—	—	—
Total miles	26			

Upgrading

78 The forecast is that 39 miles of rivers will be upgraded following the carrying out of remedial measures with the net effect of increasing the mileage in Class 1 by 36 miles, balanced by a reduction of 33 miles in Class 2 and 3 miles in Class 3 to bring more than 95 per cent of the total mileage into the Class 1 category. The tidal Dee should be further improved to Class 1 over about 13 miles, to bring the whole river up to Class 1 and an improvement to Class 1 over about 11 miles is also expected in the Worthenbury and Wych Brooks. Other improvements over short lengths are expected in a further 8 rivers. Two short Class 3 lengths in the Finchett's Gutter and the Pulford Brook should improve to Class 2, to eliminate all Class 3, but two very short lengths of Class 4 are likely to remain in the Queensferry Drain and Sandycroft Drain. On the canals, 8 miles of the Shropshire Union Canal in the vicinity of Chester should be upgraded from Class 2 to Class 1 so that all canals in the area should become Class 1.

Future suitability for additional public water supply

79 No additional lengths of river are likely to become available for public water supply following the remedial measures. ☐

25 Mersey and Weaver River Authority

Table 159 Comparison of mileages by chemical classification following remedial works on discharges at present unsatisfactory. Rivers

	Non-tidal river. Chemical classes				Tidal river. Chemical classes				All rivers. Chemical classes			
	1	2	3	4	1	2	3	4	1	2	3	4
Present mileage in class	329	269	100	177	1	3	40	23	330	272	140	200
% of total in class	37.6	30.8	11.4	20.2	1.5	4.5	59.7	34.3	35.0	28.9	14.9	21.2
Net increase of decrease	+25	+98	+14	−137	0	+3	+5	−8	+25	+101	+19	−145
Future mileage in class	354	367	114	40	1	6	45	15	355	373	159	55
% of total in class	40.5	41.9	13.0	4.6	1.5	8.9	67.2	22.4	37.7	39.6	16.9	5.8
Total miles	875				67				942			

Table 160 Comparison of mileages by chemical classification following remedial works on discharges at present unsatisfactory. Canals

	Chemical classes			
	1	2	3	4
Present mileage in class	143	83	14	6
% of total in class	58.1	33.7	5.7	2.5
Net increase or decrease	+2	+8	−4	−6
Future mileage in class	145	91	10	0
% of total in class	58.9	37.0	4.1	—
Total miles	246			

Upgrading

80 The survey indicates that 216 miles of river are likely to be upgraded and 172 miles of this total represents improvements to rivers in the Mersey basin. The net effect of all the improvements would be a reduction of 145 miles in the Class 4 lengths with an increase in the other 3 classes, principally in Class 2. However, even after remedial works, 159 miles are still expected to be in Class 3 and 55 miles in Class 4.

81 The greatest improvements are expected on the Irwell and on the Manchester Ship Canal on the Irwell system where a total of about 37 miles are likely to be upgraded with big reductions in the lengths in Class 3 and Class 4 throughout most of the length above Davyhulme. The Tame is expected to be improved over some 19 miles, all of this being an upgrading from Class 4 to Class 3. Improvements over some 15 miles should occur in the Roch, resulting in the elimination of all Class 4 mileage and a good length improved to Class 2. Other rivers showing likely improvements over significant

lengths are the Alt (13 miles with all Class 4 eliminated), the Irk (12 miles with all Class 4 eliminated), the Sankey and Hardshaw Brooks (11 miles with all Class 4 eliminated), the Wheelock (about 9 miles with all Class 3 and 4 eliminated), the Tonge and Eagley Brook (about 9 miles with all Class 3 and 4 eliminated) and the Medlock (about 9 miles with all Class 4 eliminated). The remaining improvements would be spread over some 30 other rivers. The only improvement in class expected on the Mersey itself relates to an upgrading from Class 4 to Class 3 over 2.7 miles on the tidal section below Ram's Brook.

82 After completion of remedial works, over 150 miles of non-tidal river, mostly in the Mersey basin, will still be in Class 3 or Class 4. This mileage will be spread over about 50 rivers and a good deal of the Class 3 will have been upgraded from Class 4. Conspicuous on the list are the Mersey (18 miles Class 3, 15 miles Class 4), the Tame (19 miles Class 3), the Irwell and Manchester Ship Canal (9 miles Class 3, 2 miles Class 4), the Irk (9 miles Class 3), the Prescott and Ditton Brooks (6.7 miles Class 4), the Medlock (6.7 miles Class 3), the Alt (6.6 miles Class 3) and the Weaver (6.5 miles Class 3). Of the tidal waters likely to remain in Class 3 or 4, the Mersey should account for 35 miles Class 3 and 13 miles Class 4 and the Weaver for 9 miles Class 3, with the balance spread over 5 other rivers and brooks in short lengths.

83 Improvements from Class 3 to Class 2 are expected over 3.8 miles on the Shropshire Union Canal in the Ellesmere Port area, and on the St. Helens Canal improvements are expected over 4.2 miles in the Warrington area. All the Class 4 waters in the Leeds and Liverpool Canal are expected to improve to Class 2 and in the Bridgewater Canal to Class 3. Stretches in Class 3 are, however, still likely to remain in the Weaver Navigation, the Trent and Mersey Canal and the Bridgewater Canal.

Future suitability for additional public water supply

84 Altogether 18 miles of Class 2 river should become suitable for further public water supply. The lengths becoming suitable are on the Valley Brook (1 mile) near Crewe with the balance on the River Bollin, subject to the diversion of certain industrial wastes at present discharged into it.

26 Lancashire River Authority

Table 161 Comparison of mileages by chemical classification following remedial works on discharges at present unsatisfactory. Rivers

	Non-tidal river. Chemical classes				Tidal river. Chemical classes				All rivers. Chemical classes			
	1	2	3	4	1	2	3	4	1	2	3	4
Present mileage in class	615	162	119	104	91	42	27	27	706	204	146	131
% of total in class	61.5	16.2	11.9	10.4	48.6	22.8	14.3	14.3	59.5	17.2	12.3	11.0
Net increase of decrease	+47	+143	−91	−99	+7	+39	19	−27	+54	+182	−110	−126
Future mileage in class	662	305	28	5	98	81	8	0	760	386	36	5
% of total in class	66.2	30.5	2.8	0.5	52.4	43.3	4.3	—	64.1	32.5	3.0	0.4
Total miles	1,000				187				1,187			

Table 162 Comparison of mileages by chemical classification following remedial works on discharges at present unsatisfactory. Canals

	Chemical classes			
	1	2	3	4
Present mileage in class	44	74	1	3
% of total in class	36.1	60.6	0.8	2.5
Net increase or decrease	0	+4	−1	−3
Future mileage in class	44	78	0	0
% of total in class	36.1	63.9	—	—
Total miles	122			

Upgrading

85 The forecast is that 301 miles of river are likely to be upgraded with the net effect of a reduction of 236 miles of Class 3 and 4 balanced by an increase in Class 1 and Class 2, mostly in the latter. Only 5 miles are likely to remain in Class 4 out of a present length of 131 miles.

86 Improvements over considerable lengths are expected in the Ribble (34 miles), the Douglas (31 miles, all Class 4), the Darwen (17 miles, all Class 4), the Calder (16 miles), the Yarrow (13 miles), the Lune (11 miles) and the Lostock (10 miles). Several of these, heavily polluted at present, will have most or all of Class 3 and Class 4 eliminated. Other rivers likely to be improved over lengths of between about 5 and 10 miles are the Three Pools Waterway, Freckleton Pool, Hyndburn Brook, Pendle Water, the Wyre, Main Dyke and Broad Fleet. The balance of improvements will be spread in lengths up to about 5 miles over some 75 other rivers.

87 Following remedial works, there are likely to remain lengths of non-tidal waters in Class 3 totalling about 28 miles, and in Class 4 about 5 miles. These will be found spread throughout some 29 rivers and brooks mostly in very short isolated lengths, but the Calder will have about 6 miles in Class 3 (partly improved from Class 4), the Douglas 2.4 miles in Class 3 and a short length in Class 4, Freckleton Pool 2.4 miles in Class 3 and Church Beck 2 miles in Class 3. The Brun will still have about 2 miles in Class 4. The 8 Class 3 miles remaining on tidal waters will be mainly on the Broad Fleet at the mouth of the Lune, this length having been improved from Class 4.

88 Improvement of about 3 miles from Class 4 to Class 3 is expected on the Leeds and Liverpool Canal in the Wigan area. Improvement of about 1 mile from Class 3 to Class 2 is expected in the Lancaster Canal at Lancaster to bring all canals up to Class 1 or Class 2.

Future suitability for additional public water supply

89 Nearly 21 additional miles of river should become suitable for public water supply, 14 miles occurring on the Ribble over its upstream lengths. Other rivers where stretches should become suitable are the Stock Beck (2.5 miles), Eller Brook (2.4 miles), the River Wyre (1.6 miles) and the River Kent (0.4 miles).

27 Cumberland River Authority

(see Table 163)

Upgrading

90 The survey shows that 62 miles of river in the area are likely to be upgraded as a result of the remedial measures. The net result should be that all rivers will be either Class 1 or Class 2 with over 97 per cent of the total of 1,223 miles in Class 1.

91 The Keekle which is heavily polluted over its entire length should improve from Class 3 to Class 2 and the River Ellen should have all its Class 3 waters upgraded in the same way. In the Waver about 5 miles of Class 4 should be improved to Class 2 and in the Leith,

some 9 miles of Class 3 are likely to be upgraded to Class 1. Improvements on shorter lengths are expected in 15 or so other rivers and streams.

Future suitability for additional public water supply

92 No additional lengths of river become available for further public water supply as a result of the remedial measures in this area. ☐

Table 163 Comparison of mileages by chemical classification following remedial works on discharges at present unsatisfactory. Rivers

	Non-tidal river. Chemical classes				Tidal river. Chemical classes				All rivers. Chemical classes			
	1	2	3	4	1	2	3	4	1	2	3	4
Present mileage in class	1,125	39	24	0	20	9	1	5	1,145	48	25	5
% of total in class	94.7	3.3	2.0	—	57.2	25.7	2.8	14.3	93.6	3.9	2.1	0.4
Net increase or decrease	+45	−21	−24	0	0	+6	−1	−5	+45	−15	−25	−5
Future mileage in class	1,170	18	0	0	20	15	0	0	1,190	33	0	0
% of total in class	98.5	1.5	—	—	57.2	42.8	—	—	97.3	2.7	—	—
Total miles		1,188				35				1,223		

28 Thames Conservancy Catchment Board

Table 164 Comparison of mileages by chemical classification following remedial works on discharges at present unsatisfactory. Rivers

	Non-tidal river. Chemical classes			
	1	2	3	4
Present mileage in class	1,213	186	37	8
% of total in class	84.0	12.9	2.6	0.5
Net increase or decrease	+41	+2	−36	−7
Future mileage in class	1,254	188	1	1
% of total in class	86.8	13.0	0.1	0.1
Total miles		1,444		

Note: There are no tidal waters in the Thames Conservancy area

Upgrading

93 The Board's forecast is that 83 miles of river in the area should be improved with the net effect that virtually all of the total of 1,444 miles should be in the top two grades, the mileage in the heavily polluted Classes 3 and 4 having been reduced from 45 miles to 2 miles.

94 The Thames, which has a short length of about 4.5 miles Class 2 near Swindon, should be upgraded to become Class 1 over its entire length. Other substantial improvements should occur in the Blackwater (some 18 miles of Class 3 improved to Class 2), in the Mole (about 12 miles improved, with Class 3 eliminated), in the Cherwell (7.5 miles with Class 3 eliminated) and in the Wye (7.3 miles with Class 3 and Class 4 eliminated). The balance of improvement would be spread over some 14 other rivers, in lengths of up to about 5 miles. The short polluted lengths likely to remain will be in the Roundmoor Ditch between Windsor and Slough, where there will be about a mile of Class 3 (upgraded from Class 4), and in the Earlswood Brook at Reigate where about a mile of Class 4 will remain. All the canals in the area are already in Class 1.

Future suitability for additional public water supply

95 A further 10 miles of the River Mole should become suitable for public water supply following the remedial measures. ☐

29 Lee Conservancy Catchment Board

Table 165 Comparison of mileages by chemical classification following remedial works on discharges at present unsatisfactory. Rivers

	Non-tidal river. Chemical classes				Tidal river. Chemical classes				All rivers. Chemical classes			
	1	2	3	4	1	2	3	4	1	2	3	4
Present mileage in class	115	27	18	14	0	0	0	7	115	27	18	21
% of total in class	66.1	15.6	10.3	8.0	—	—	—	100	63.5	14.9	10.0	11.6
Net increase or decrease	+27	−11	−4	−12	0	0	+5	−5	+27	−11	+1	−17
Future mileage in class	142	16	14	2	0	0	5	2	142	16	19	4
% of total in class	81.7	9.2	8.0	1.1	—	—	71.5	28.5	78.5	8.8	10.5	2.2
Total miles		174				7				181		

Upgrading

96 The forecast is that 43 miles of river are likely to be upgraded as a result of remedial measures. The net effect will be an increase in the Class 1 mileage, a small increase in Class 3, a decrease in Class 2 and a significant decrease in Class 4. The present total of 39 miles in the two lowest classes should be reduced to 23 miles.

97 Some 22 miles of the Lee should be upgraded including all the Class 4 mileage through London which would be improved to Class 3. Improvements to Class 1 in some of those stretches at present in Classes 2 and 3 are also expected, to bring the Class 1 length in the River Lee up to about 53 miles. The Stort should be upgraded over about 10 miles to bring the whole river to Class 1. Other improvements over short lengths of up to about 2 miles are forecast in 7 other rivers including Pymmes Brook, Moselle Brook, Salmons Brook, Canon's Brook and Stevenage Brook.

98 Some mileage of Class 3 and Class 4 will remain in 7 rivers, most of the Class 3 mileage having been upgraded from Class 4. The Lee will have about 10 miles in Class 3, all in the London area, and other shorter lengths will be found in Pymmes Brook, Moselle Brook and Salmons Brook in the North London area and in Cobbins Brook and Spital Brook at Waltham Abbey and Hoddesdon respectively. The Class 4 mileage remaining will all be in the Dagenham Brook in the Waltham Forest area. No change in the condition of the Class 2 Grand Union Canal is expected.

Future suitability for additional public water supply

99 No additional lengths of the rivers in this area will become available for public water supply. ☐

30 Port of London Authority (including the London Excluded Area)

Table 166 Comparison of mileages by chemical classification following remedial works on discharges at present unsatisfactory. Rivers

	Non-tidal river. Chemical classes				Tidal river. Chemical classes				All rivers. Chemical classes			
	1	2	3	4	1	2	3	4	1	2	3	4
Present mileage in class	0	60	0	17	0	14	64	—*	0	74	64	17
% of total in class	—	77.9	—	22.1	—	17.9	81.8	0.3	—	47.7	41.3	11.0
Net increase or decrease	0	+11	0	−11	0	0	0	0	0	+11	0	−11
Future mileage in class	0	71	0	6	0	14	64	—*	0	85	64	6
% of total in class	—	92.2	—	7.8	—	17.9	81.8	0.3	—	54.8	41.3	3.9
Total miles		77				78				155		

*Actual mileage 0.3, Class 4

Upgrading

100 The survey shows an expected upgrading of 11 non-tidal miles from Class 4 to Class 2, but no change of chemical class on the tidal waters. The rivers likely to be improved are the Wandle (5 miles upgraded) and the Beverley Brook (6 miles upgraded).

101 On non-tidal waters, 6 miles will remain in Class 4—about 1 mile at the upper end of the Wandle and on the Beddington Branch, about 2 miles at the upper end of the Beverley Brook and about 3 miles on the Pyl Brook. On tidal waters, the Thames will account for some 55 miles of Class 3 and the Darenth for 2.5 miles, the balance being spread over 10 short tributaries in Essex and Kent and in the London area. The very short tidal stretches of the Wandle and the Beverley Brook are likely to remain Class 4. No changes are forecast in the Class 2 Grand Union Canal.

Future suitability for additional public water supply

102 No additional lengths of river become available for public water supply following completion of remedial measures in the area.

8 River water quality and discharges in England and Wales

Population and River Pollution

1 The survey shows that badly polluted rivers tend to be associated with the most densely populated areas. More than half the total population of 43½ million people recorded as discharging sewage effluent or crude sewage to rivers and canals, discharge to the Class 3 (poor quality) or Class 4 (grossly polluted) waters. The length of rivers and canals in these two classes is 2,772 miles or just over one tenth of the total length recorded so that on the average considerably more people discharge to each mile of the badly polluted rivers than to each mile of the cleaner rivers.

2 Table 167 presents the data for non-tidal rivers, showing that 30 per cent of the total population are served by treatment works which discharge their sewage effluent to grossly polluted waters. The population served by the crude sewage discharges is negligible in comparison. The population per mile discharging to Class 4 (grossly polluted) non-tidal river averages 8,700, to Class 3 (poor) 3,200, to Class 2 (doubtful) 1,750, and to Class 1 (clean) 600.

3 The data on tidal rivers (Table 168) shows that nearly 80 per cent of the population served by treatment works was discharged to Class 3 (poor) waters, while most of the balance was discharged to the cleaner rivers. The average population per mile discharging to tidal rivers was notably higher than for non-tidal rivers. The large population discharging to Class 3 tidal rivers is almost entirely attributable to the London area where the sewage effluent from some 7½ million people discharged into the tidal River Thames, where it is Class 3.

4 The population discharging to canals is very small but again a greater proportion of the population discharged to Class 3 and Class 4 canals (Table 169).

Discharges and River Pollution

(i) Non-tidal rivers

5 There was a tendency for discharges to the cleaner rivers to be more satisfactory than those to rivers which were badly polluted, with the highest proportion of satisfactory effluent being discharged to Class 2 waters. About a third of all the sewage effluent discharged to non-tidal rivers was discharged to grossly polluted waters and only 36 per cent of this was satisfactory (Table 167).

6 The few discharges of crude sewage were generally small in volume. All but 1 of the discharges were made to Class 1 and Class 2 waters and none of the discharges were considered to be satisfactory.

7 More than 60 per cent of the number and over 50 per cent by volume of all discharges of industrial effluent to non-tidal rivers were made to rivers in Class 1 and Class 2. Again there was a tendency for discharges to the cleaner rivers to be more satisfactory.

Table 167 Summary of selected information about non-tidal rivers. England and Wales

Detail	Class 1	Class 2	Class 3	Class 4	Total
(a) *River quality*					
1 Miles of river	17,000	3,290	1,071	952	22,313
(b) *Sewage effluent*					
2 (i) Number of discharges	2,938	655	227	227	4,047
(ii) Percentage satisfactory	68	60	46	40	64
3 (i) Population served by treatment works discharging to	10,182,726	5,774,522	3,432,762	8,302,253	27,692,263
(ii) Percentage served by treatment works considered satisfactory	57	62	36	38	50
4 (i) DWF of sewage effluent (x 1,000 gpd) discharged to	527,639	319,752	202,531	526,171	1,576,093
(ii) Percentage DWF discharged to a satisfactory standard	52	61	32	36	46
(c) *Crude sewage*					
5 (i) Number of discharges	34	12	0	1	47
(ii) Percentage satisfactory	0	0	—	0	0
6 (i) Population served by crude sewage discharges to	25,699	12,670	0	620	38,989
(ii) Percentage served by crude sewage discharges considered satisfactory	0	0	—	0	0
7 (i) DWF of crude sewage (x 1,000 gpd) discharged to	652	555	0	19	1,226
(ii) Percentage DWF discharged to a satisfactory standard	0	0	—	0	0
(d) *Industrial effluent excluding solely cooling water and mine water*					
8 (i) Number of discharges	845	415	310	457	2,027
(ii) Percentage satisfactory	47	55	36	33	44
9 (i) DWF of industrial effluent (x 1,000 gpd) discharged to	389,322	165,481	278,664	234,041	1,067,508
(ii) Percentage satisfactory	72	75	58	37	62
(e) *Unsatisfactory storm overflows*					
10 Number of discharges	430	520	236	840	2,026
(f) *Estimates of costs of remedial works (£ millions)*					
11 Discharges of sewage effluent	124.975	63.567	40.737	102.919	332.198
12 Discharges of crude sewage	1.641	0.888	0	0.030	2.559
13 Discharges of industrial effluent	5.692	2.406	4.638	8.886	21.622
14 Total expenditure	132.308	66.861	45.375	111.835	356.379
(g) *Miles expected to be upgraded*					
15 Miles expected to be improved from present Class to a higher Class	—	872	781	753	2,406

(ii) Tidal rivers

8 Over three-quarters of the sewage effluent or crude sewage discharged to tidal rivers was discharged to badly polluted waters (Table 168). This high proportion is partly due to the large quantity of sewage effluent discharged to the tidal River Thames. Less than a quarter of this large quantity of effluent was satisfactory, while the proportion of satisfactory effluent discharged to the cleaner rivers was higher.

9 About 70 per cent of the population discharging crude sewage to tidal rivers discharged to Class 3 or Class 4 waters, with Class 4 taking rather more than Class 3. Virtually all of the crude sewage discharged to Class 2, 3 or 4 waters was unsatisfactory but a fairly high proportion of the crude sewage discharged to Class 1 tidal waters was considered to be satisfactory, most of this in Hampshire.

10 The greater part of the industrial effluent discharged to tidal rivers was discharged to waters in Class 1, 3 or 4 with Class 1 receiving about 23 per cent of the volume, Class 3 about 43 per cent and Class 4 about 28 per cent. Only in discharges to Class 4 waters was there a high incidence of unsatisfactory effluent.

(iii) Canals

11 Table 169 gives information about canals and shows a tendency for sewage effluent discharged to the poorer quality waters to be more satisfactory and this trend shows up also in the case of discharges of industrial effluent.

Costs of Improvements related to River Water Quality

12 For non-tidal rivers nearly 70 per cent of the estimated cost of remedial works will relate to Class 1 and Class 4 waters, with Class 1 accounting for rather more than Class 4 and the expenditure on discharges to the latter representing about £115,000 per mile of Class 4 river.

13 The 2,406 miles upgraded will be fairly evenly divided between Class 2, Class 3 and Class 4.

14 For tidal rivers costs of remedial works tend to be greater on discharges to the badly polluted classes of water, with discharges to Class 4 accounting for about 36 per cent of the costs and discharges to Class 3 for about 27 per cent. The cost of remedial works on discharges to Class 4 tidal waters represents about £430,000 per mile.

15 Nearly 40 per cent of the miles likely to be upgraded will be Class 4 lengths, with about 30 per cent of the total each being improvements in lengths at present in Class 2 and Class 3.

16 For canals the bulk of the expenditure will be on discharges to Class 2 and Class 3 canals and the most significant improvements will be in Class 3 and Class 4 waters.

Summary of the situation in the river authority areas

17 A study of all the information suggests that the river authorities can be grouped according to the extent and seriousness of pollution in the area. The groups are not intended to be precise and the dividing lines are not clear-cut but it was not too difficult to decide the group into which each authority should be placed. The summary which follows deals with the authorities in each group together, so that a better appreciation can be obtained of the areas where there are similar problems, whether the problems be great or small. The groups are:

Group A. Eight authorities where problems are not generally serious and the principal consideration is or will be the maintaining of present standards.
Welland and Nene, East Suffolk and Norfolk, Sussex, Isle of Wight, Avon and Dorset, Wye, Gwynedd, Dee and Clwyd.

Group B. Eight authorities where there is fairly serious pollution but generally concentrated in a few areas.
Lincolnshire, Great Ouse, Essex, Kent, Hampshire, Devon, Cumberland, Thames Conservancy.

Group C. Eight authorities where there is very serious pollution but generally concentrated in a few areas.
Northumbrian, Cornwall, Somerset, Bristol Avon, Severn, Usk, South West Wales, Lee Conservancy.

Group D. Six authorities where there is very serious and widespread pollution.
Yorkshire, Trent, Glamorgan, Mersey and Weaver, Lancashire, Port of London Authority and London Excluded Area.

18 It should be noted that the summary which follows relates to the situation as it was reported at the time of the survey. The situation in some places will have changed since then and these changes will be made known when the survey is updated. An updating to January 1972 is being done at present. It should also be noted that, when comments are made about costs in each river authority area, they do not take account of costs associated with storm overflows or with works of sewerage generally. An indication of the national cost of dealing with unsatisfactory storm overflows has been given in chapter 6, paragraph 7.

Authorities in Group A

Welland and Nene River Authority

19 Nearly 97 per cent of the river miles are in Class 1 or Class 2. Many of the discharges of sewage effluent are unsatisfactory but most of these are from relatively small treatment works and a large proportion of them could be rectified by better maintenance. There are no discharges of crude sewage and no unsatisfactory storm overflows. A high proportion of all the industrial effluent is cooling water and more than 90 per cent of all the industrial effluent is discharged to a satisfactory standard.

20 There is not much of a back-log of urgently needed expenditure but the average cost per head will be high. Most of the expenditure in the future will be associated with major town expansion. No changes of river classification are expected, but the authority have drawn attention to the natural characteristics of the rivers in the area and to the impracticability of improving Class 2 to Class 1.

21 Both the Welland and the Nene are significant because of discharges of effluent upstream of public water supply intakes.

East Suffolk and Norfolk River Authority

22 All of the rivers are in either Class 1 or Class 2. Most of the sewage treatment works are small and most of the unsatisfactory sewage effluent arises from 2 discharges to the tidal waters of 2 rivers, the Yare and the Orwell. Crude sewage discharges are made from 2 coastal towns, but only some relatively inexpensive preliminary treatment is likely to be needed on them. Works to deal with the main unsatisfactory storm overflows are in hand. About 5 mgd of industrial effluent is discharged to an unsatisfactory standard.

23 Most of the expenditure in the area will be directed towards dealing with discharges to the tidal waters of the Yare, the Orwell and the Ant. It is expected that all the non-tidal Class 2 waters will be upgraded to Class 1 but only a small improvement is expected on tidal waters. The river authority have drawn attention to the natural characteristics of the tidal rivers in the area which largely prevent their ever being placed in Class 1.

24 The River Gipping is significant because of discharges of effluent upstream of public water supply intakes.

Sussex River Authority

25 Nearly 98 per cent of the river miles in the area are Class 1 or Class 2. A fairly large proportion of the sewage effluent discharges are unsatisfactory but with a few exceptions these are small discharges and several could be improved by better maintenance. There are 2 discharges of crude sewage to tidal waters from towns on or near to the coast, one of which was considered satisfactory at the time of the survey, but water conservation considerations may require some remedial works. A sea outfall scheme is proposed for the other discharge. Schemes are in hand or imminent for dealing with several unsatisfactory storm overflows, and there are no particular problems with industrial effluent, the amount discharged being small.

Table 168 Summary of selected information about tidal rivers. England and Wales

Detail		Class 1	Class 2	Class 3	Class 4	Total
(a)	*River quality*					
1	Miles of river	862	419	301	209	1,791
(b)	*Sewage effluent*					
2	(i) Number of discharges	141	65	70	30	306
	(ii) Percentage satisfactory	68	58	49	63	61
3	(i) Population served by treatment works discharging to	953,185	1,190,831	8,455,088	337,490	10,936,594
	(ii) Percentage served by treatment works considered satisfactory	68	40	23	62	30
4	(i) DWF of sewage effluent (x 1,000 gpd) discharged to	46,126	61,584	514,351	15,668	637,729
	(ii) Percentage DWF discharged to a satisfactory standard	62	38	22	61	27
(c)	*Crude sewage*					
5	(i) Number of discharges	109	123	129	76	437
	(ii) Percentage satisfactory	10	1	1	1	3
6	(i) Population served by crude sewage discharges to	628,798	793,324	1,278,301	1,726,988	4,427,591
	(ii) Percentage served by crude sewage discharges considered satisfactory	45	1	0.1	0.1	7
7	(i) DWF of crude sewage (x 1,000 gpd) discharged to	43,359	49,981	70,810	98,633	262,783
	(ii) Percentage DWF discharged to a satisfactory standard	37	1	0.3	0.1	6
(d)	*Industrial effluent excluding solely cooling water and mine water*					
8	(i) Number of discharges	51	50	107	123	331
	(ii) Percentage satisfactory	60	24	42	26	36
9	(i) DWF of industrial effluent (x 1,000 gpd) discharged to	297,532	76,068	543,136	357,731	1,274,467
	(ii) Percentage satisfactory	70	73	86	14	61
(e)	*Unsatisfactory storm overflows*					
10	Number of discharges	26	48	41	9	124
(f)	*Estimates of costs of remedial works (£ millions)*					
11	Discharges of sewage effluent	23.546	19.628	43.609	4.933	91.716
12	Discharges of crude sewage	10.683	36.139	20.387	74.599	141.808
13	Discharges of industrial effluent	0.806	1.486	2.659	11.381	16.332
14	Total expenditure	35.035	57.253	66.655	90.913	249.856
(g)	*Miles expected to be upgraded*					
15	Miles expected to be improved from present Class to a higher Class	—	116	119	150	385

Table 169 Summary of selected information about canals. England and Wales

Detail		Class 1	Class 2	Class 3	Class 4	Total
(a)	*River quality*					
1	Miles of canal	700	601	136	103	1,540
(b)	*Sewage effluent*					
2	(i) Number of discharges	9	7	8	2	26
	(ii) Percentage satisfactory	67	57	38	100	58
3	(i) Population served by treatment works discharging to	27,551	77,831	169,986	26,874	302,242
	(ii) Percentage served by treatment works considered satisfactory	57	58	93	100	81
4	(i) DWF of sewage effluent (x 1,000 gpd) discharged to	1,056	5,487	10,981	841	18,365
	(ii) Percentage DWF discharged to a satisfactory standard	51	44	95	100	78
(c)	*Crude sewage*					
5	(i) Number of discharges	0	0	0	0	0
	(ii) Percentage satisfactory	—	—	—	—	—
6	(i) Population served by crude sewage discharges to	0	0	0	0	0
	(ii) Percentage served by crude sewage discharges considered satisfactory	—	—	—	—	—
7	(i) DWF of crude sewage (x 1,000 gpd) discharged to	0	0	0	0	0
	(ii) Percentage DWF discharged to a satisfactory standard	—	—	—	—	—
(d)	*Industrial effluent excluding solely cooling water and mine water*					
8	(i) Number of discharges	9	14	33	29	85
	(ii) Percentage satisfactory	23	50	28	41	39
9	(i) DWF of industrial effluent (x 1,000 gpd) discharged to	5,904	1,558	11,579	11,144	30,185
	(ii) Percentage satisfactory	16	25	49	28	35
(e)	*Unsatisfactory storm overflows*					
10	Number of discharges	6	2	0	4	12
(f)	*Estimates of costs of remedial works (£ millions)*					
11	Discharges of sewage effluent	0.341	0.770	1.125	0.102	2.338
12	Discharges of crude sewage	0	0	0	0	0
13	Discharges of industrial effluent	0.048	0.191	0.242	0.298	0.779
14	Total expenditure	0.389	0.961	1.367	0.400	3.117
(g)	*Miles expected to be upgraded*					
15	Miles expected to be improved from present Class to a higher Class	—	25	53	89	167

26 About three-quarters of the expenditure will be on discharges of sewage effluent and most of the remainder on the sea outfall scheme for Littlehampton. The average cost per head is likely to be low. Substantial improvements are expected, to bring all the rivers up to Class 1 or Class 2, with most in Class 1.

Isle of Wight River and Water Authority

27 More than 90 per cent of the river miles are Class 1 or Class 2 with a few short polluted lengths to be found mainly on the rivers discharging to The Solent. Most of the treatment works are small and a high proportion of the unsatisfactory effluent is discharged to the tidal section of the River Medina. The Medina also receives several crude sewage discharges on its tidal length and there are also several crude sewage discharges to The Solent, which are outside the scope of the survey. There are problems with a number of storm overflows, but little trouble with industrial effluent. The amount discharged is small, and mostly cooling water and, with minor exceptions, it is considered satisfactory.

28 Most of the expenditure will be incurred in dealing with the discharges of crude sewage to the Medina and The Solent and the average cost per head will be fairly high. Significant improvements are expected, to bring most of the rivers to Class 1.

Avon and Dorset River Authority

29 Nearly 96 per cent of the river miles are already Class 1 and serious pollution occurs in only 2 small areas. A high proportion of sewage effluent is discharged to a satisfactory standard and there are no discharges of crude sewage. There are some problems with storm overflows but a number of schemes are in hand to improve this situation. Most of the industrial effluent is water used for driving turbo-generators and 98 per cent of it is satisfactory.

30 The expenditure envisaged is relatively low and the average cost per head will also be fairly low. Most of the expenditure will be associated with sewage effluent discharges, and it is expected that almost all of the river mileage will be brought to Class 1.

31 Some of the rivers in the area are of special significance because of discharges of effluent at points upstream of public water supply intakes, the Avon and the Stour being prominent examples.

Wye River Authority

32 The rivers in the area are already of an exceptionally high standard with 99 per cent in the two top classes and 96.5 per cent in Class 1. On the whole, treatment works throughout the area are small and although a fairly high proportion of sewage effluent is unsatisfactory, most of this arises from only 2 treatment works discharging to the Wye. There are 4 discharges of crude sewage, one of which is satisfactory, and there are proposals for dealing with the others. The few unsatisfactory storm overflows should be remedied by a scheme which is due to start soon. There are no serious problems with industrial effluent, the amount discharged being small and almost all satisfactory.

33 The expenditure envisaged is low with most of it relating to sewage effluent but the average cost per head will be fairly high. The improvements forecast would increase still further the lengths in Class 1 and Class 2, leaving only 1 mile in Class 3.

34 The Wye is significant as a river where there are discharges of effluent upstream of public water supply intakes.

Gwynedd River Authority

35 This authority also has rivers of an exceptionally high standard with all but about 1 of 361 miles already in Class 1 or Class 2. Sewage treatment works are generally small, all serving under 5,000 persons, and about half of the effluent is discharged to a satisfactory standard. There are several discharges of crude sewage to be dealt with and a scheme has been prepared to remedy the few unsatisfactory storm overflows. There is only 1 small discharge of industrial effluent.

36 The expenditure envisaged is low, with more than half relating to crude sewage discharges, and a small improvement is expected by way of the upgrading of a short Class 4 stretch of river to Class 2.

Dee and Clwyd River Authority

37 This authority has more than 99 per cent of its river miles in Class 1 or Class 2. There are several large discharges of sewage effluent, but a high proportion of all the effluent is discharged to a satisfactory standard. There are a few discharges of crude sewage, mostly to the Dee estuary, but schemes are under construction for dealing with most of this. There are several unsatisfactory storm overflows and although some are likely to be improved or eliminated by a scheme at present under consideration, other major schemes are likely to be needed. About 55 mgd of industrial effluent is discharged, with a very high proportion of process water. Nearly 80 per cent of the industrial effluent is considered to be satisfactory.

38 As major schemes for dealing with crude discharges are under construction, most of the future expenditure will relate to sewage effluent and the average cost per head is likely to be below average. Further improvements are expected with an increase in the mileage of Class 1 river and the virtual elimination of Class 3 and Class 4 throughout the area.

39 The Dee is a prominent example of a river where there are many discharges of effluent upstream of public water supply intakes.

Authorities in Group B

Lincolnshire River Authority

40 About 94 per cent of the river miles are in the two top classes but there are 29 miles in Class 3. There are a number of fairly large treatment works in the area and a big proportion of the effluent is considered satisfactory. There are 3 discharges of crude sewage—all to the tidal Witham—which are to be given partial treatment. There are also several discharges of crude sewage to the Humber which are outside the scope of the survey. There are a number of storm overflows requiring attention and although the amount of industrial effluent discharged is not great, a big proportion of it is considered to be unsatisfactory.

41 Expenditure on improvements is likely to be relatively low, and about three-quarters of it will relate to sewage effluent and most of the balance to crude sewage discharges. The length in Class 1 is expected to be increased by some 37 miles following remedial works, with the length in Class 3 substantially reduced and Class 4 eliminated.

Great Ouse River Authority

42 Over 93 per cent of river miles are in Class 1 or Class 2, but there are 58 miles of Class 3 and 4 miles of Class 4 in a few small but clearly defined areas. There are several large treatment works in the area, and more than 200 small ones serving up to 5,000 persons. Just over half of all the sewage effluent is considered to be unsatisfactory, but the effluent at several treatment works could probably be made satisfactory by better maintenance. More than two-thirds of all the unsatisfactory effluent is discharged to 3 rivers—the Ouse, the Cam and the Ouzel. There is a considerable amount of crude sewage discharged from King's Lynn but schemes are already under construction for diverting the crude sewage to treatment works. There are no unsatisfactory storm overflows. Cooling water accounts for a very big proportion of the industrial effluent. Discharges of solely cooling water are all considered to be satisfactory but about one-third of the balance of the industrial effluent is considered to be unsatisfactory.

43 Future expenditure will be almost wholly on sewage effluent and a big proportion of this will relate to the new town of Milton Keynes. The average cost per head in the area is likely to be high. Big increases are expected in the Class 1 mileage and significant reductions in the Class 3 mileage, but a substantial length of Class 3 will remain. The river authority have drawn attention to the natural characteristics of some of their rivers which largely prevent their ever being placed in Class 1.

44 Some rivers in the area, particularly the Ouse, are of special significance because of discharges of effluent at points upstream of public water supply intakes.

Essex River Authority

45 There are several small areas of fairly heavy pollution particularly near to the larger towns, the Thames Estuary and Greater London, with 52 miles of river in Class 3 and Class 4. The sizes of treatment works cover a wide range with about 60 per cent of the effluent discharged to a satisfactory standard. There are 5 discharges of crude sewage with a high proportion of the flow through the outfalls being industrial effluent (cooling water). Schemes are in preparation for improvements and the reduction of numbers of outfalls. A good deal of work is indicated as being necessary on storm overflows and major sewerage schemes may be required in places. Cooling water accounts for a very big proportion of the industrial effluent and most of the discharges of solely cooling water are satisfactory, but about half of the balance of industrial effluent is considered to be unsatisfactory.

46 Expenditure is likely to be largely on sewage treatment works but, because of high population density, the average cost per head is likely to be low. There is a forecast of a substantial reduction in the mileage of rivers in Class 3 and Class 4 and an increase in Class 1, but short lengths of Class 3 river totalling some 18 miles are likely to remain.

47 Some rivers in the area, particularly the Colne, the Stour, the Blackwater and the Chelmer are of special significance because of discharges of effluent upstream of public water supply intakes.

Kent River Authority

48 More than 96 per cent of the river miles are in Class 1 or Class 2 and apart from the pollution on the tidal length of the Medway there are only isolated cases of pollution on a very restricted scale, but there are altogether 21 miles of river in Class 3 and Class 4. A large proportion of the sewage treatment works are in the smaller ranges and about three-quarters of all the sewage effluent is considered to be satisfactory. A good number of the unsatisfactory discharges could be improved by better maintenance. There are 8 discharges of crude sewage, one of which is satisfactory, and the returns indicate that steps are being taken or will be taken to deal with the others. There are 11 unsatisfactory storm overflows which should soon be put right as the necessary works for improving them are either under construction or are likely to be built in the near future. The volume of industrial effluent discharged in the area is large and includes a significant amount of process water. More than 100 mgd of industrial effluent is discharged to an unsatisfactory standard, mostly to tidal waters.

49 Expenditure is likely to be fairly high at about £16 million with some £14 million on sewage effluent and the balance distributed between crude sewage and industrial effluent but because of fairly high population density, the average cost per head is likely to be below average. A moderate improvement is visualised with an increase in the Class 1 mileage, but some 11 miles of tidal rivers are likely to remain in Class 3 and Class 4.

50 The River Medway is significant because of discharges of effluent upstream of public water supply intakes.

Hampshire River Authority

51 About 95 per cent of the river miles are in the two top grades but there are several areas around Southampton and Portsmouth where pollution is encountered on a limited scale. The number of sewage effluent discharges is not great and the treatment works mostly serve populations of under 10,000. Only about 30 per cent of the sewage effluent is discharged to a satisfactory standard and most of the unsatisfactory effluent is discharged to tidal waters. Of 5 crude sewage discharges to tidal waters, 3 are considered not to require further treatment, but lengthening of outfalls may be needed. Discharges direct to The Solent are not included in these figures. There are a few unsatisfactory storm overflows, some of which could require construction of relief sewers. The volume of industrial effluent discharged is very high, being nearly 1,800 mgd, but all of this, apart from some 21 mgd, is cooling water. Almost all the industrial effluent is considered to be satisfactory.

52 Long-term proposals associated with major expansion in South Hampshire could affect many aspects of future works in the area, but so far as could be estimated at the time of the survey, expenditure is likely to be about £7.5 million, almost entirely related to discharges of sewage effluent and the average cost per head is likely to be fairly low. It is possible, however, that major works not covered by the estimates might be carried out before 1980. Improvements are expected, with an increase of the mileage in Class 1, a decrease of the mileage in Class 3 and the elimination of Class 4 altogether.

53 The Test and the Itchen are significant because of discharges of effluent upstream of public water supply intakes.

Devon River Authority

54 Although there is a very high proportion of the river miles in Class 1 and Class 2 there are some quite serious areas of pollution, particularly on the Culm, and in the lower reaches of the Exe and other rivers. Most of the treatment works are small and a high proportion of all the sewage effluent is considered to be unsatisfactory, including almost all of the effluent discharged to tidal waters. The number of crude discharges is recorded as 29 but some of these represent multiple outlets, so the number of outfalls is greater. Schemes are either under construction or in preparation for dealing with them. Some extensive works are likely to be needed to deal with 33 unsatisfactory storm overflows and several schemes are already under construction or imminent. Excluding discharges of solely cooling water, which are mostly satisfactory, only about 60 per cent of the industrial effluent in the area is discharged to a satisfactory standard.

55 Expenditure is likely to be of the order of £10 million, fairly evenly divided between sewage effluent and crude sewage with a small sum needed on industrial effluents. Because of the low population density, the average cost per head is likely to be very high. Substantial improvements are expected with the complete elimination of Class 3 and a big increase in the lengths in Class 1, and all rivers should then be in the two top classes.

56 Some rivers in the area, notably the Exe, are of special significance because of discharges of effluent upstream of public water supply intakes.

Cumberland River Authority

57 More than 97 per cent of river miles are in Class 1 or Class 2, but there are a few lengths of river showing signs of fairly serious pollution with a total of 30 miles in Class 3 or Class 4. There is a very high proportion of small treatment works and, similarly, a high proportion of unsatisfactory effluent. The indications are that many of the discharges could be improved by better maintenance. Excluding direct discharges to the Solway Firth, which are outside the scope of the survey, there are 4 discharges of crude sewage. Remedial works are under consideration. There are 20 unsatisfactory storm overflows, with some re-sewering needed, but several could probably be improved at small cost. The amount of industrial effluent discharged is not great and is, for the most part, cooling water, but apart from discharges of solely cooling water, which are all satisfactory, only about 40 per cent of the industrial effluent is discharged to a satisfactory standard.

58 Expenditure on improvements in the area is likely to be low—in the region of about £2.5 million—with about £2 million being required on sewage effluent and most of the balance being required for remedial works on industrial effluent. Improvements are expected to result in the elimination altogether of Class 3 and Class 4 and in a substantial increase in the Class 1 mileage.

59 The Eden and the Derwent are conspicuous examples of rivers where there are discharges of effluent upstream of public water supply intakes.

Thames Conservancy Catchment Board

60 This very extensive area has nearly 97 per cent of its river miles in Class 1 or Class 2, but there are 45 miles in Class 3 and Class 4 mainly in the Home Counties. The number of sewage effluent discharges is large, there being 350 in all, and several of these are from large-sized treatment works. About half of all the sewage effluent is discharged to a satisfactory standard. There are no discharges of crude sewage in the area, but there are 10 unsatisfactory storm overflows. Remedial works to deal with many of these have been prepared and some are under construction. Considering the

size of the area, the quantity of industrial effluent discharged is not excessive, but a fairly high proportion is process water and the discharges from several industries represent high proportions of the national totals for those industries. Excluding discharges of solely cooling water, most of which are satisfactory, about one-third of the industrial effluent is unsatisfactory.

61 As to be expected in an area so large, expenditure is likely to be fairly high, the estimates showing a total requirement of over £25 million, almost all associated with sewage effluent discharges but the cost per head is likely to be below average. Substantial river improvements are expected, with the elimination of Class 3 and Class 4 apart from minor streams that function as effluent carriers, and a significant increase in the Class 1 mileage.

62 Some rivers in the area, particularly the Thames and the Kennet, are significant because of discharges of effluent upstream of public water supply intakes.

Authorities in Group C

Northumbrian River Authority
63 With 133 river miles in Class 3 and Class 4 this area is borderline between Group C and Group D but the worst of the pollution is concentrated in the lower parts of the basins of the Tyne, the Wear and the Tees in a wide strip running from Newcastle and Tynemouth down to Teesside. Sewage treatment works cover a wide range of sizes, but there is a high proportion of small works. Just under a half of all the sewage effluent is discharged to a satisfactory standard but more than a dozen discharges could probably be much improved by better maintenance. There are very serious problems caused by the discharge of crude sewage to the estuaries and the authority shares with the Mersey and Weaver River Authority the distinction of having the greatest volume of crude sewage discharged to its rivers. Each authority receives about one quarter of all the crude sewage recorded in the survey and in this authority's area, almost all of it is discharged to the estuaries of the Tyne, the Tees and, to a smaller extent, the Wear. Major regional schemes are being prepared for the estuaries of the Tyne and the Tees and schemes are also being prepared to deal with most of the other discharges. Problems with storm overflows are not encountered on a wide scale, but in some cases the necessary remedial works could be very costly. Large volumes of industrial effluent are discharged in the area and a significant proportion is process water. Discharges of solely cooling water are all satisfactory, but only about 10 per cent of the balance of the industrial effluent is discharged to a satisfactory standard. Minewater is significant in the area.

64 Expenditure over the next 10 years is likely to be high, with the bulk of it related to discharges of crude sewage and the average cost per head is likely to be higher than in any other area in the country. The sewers proposed under the regional schemes for crude sewage discharges would also intercept much of the industrial effluent discharging at present. Big improvements are expected with substantial reductions in the mileages of Class 3 and Class 4 but short lengths of heavily polluted rivers are likely to remain.

65 Some rivers in the area, particularly the Tyne, the Tees and the Coquet, are significant because of discharges of effluent upstream of public water supply intakes.

Cornwall River Authority
66 Although the authority has a high proportion of river miles in Class 1 and Class 2, there are nearly 80 miles of Class 3 and Class 4 river concentrated in several fairly small areas mainly around Plymouth, St. Austell and Camborne-Redruth and in the Fal basin on either side of Truro. Sewage treatment works are mainly in the smaller categories, mostly serving under 5,000 population and a high proportion of the sewage effluent is discharged to a satisfactory standard. About half of all the unsatisfactory discharges could probably be substantially improved by better maintenance. Of the 37 discharges of crude sewage, 5 are considered to be satisfactory. The greater part of the crude sewage comes from Plymouth and Falmouth and schemes are being planned to deal with most of the discharges. There are many unsatisfactory storm overflows and expensive schemes are likely to be required if they are to be put right. There are many problems with industrial effluent, particularly from process water associated with quarrying and mining. A very high proportion of all the effluent is process water and apart from discharges of solely cooling water, all of which are satisfactory, nearly a half is considered to be unsatisfactory. Water pumped from mines is of some significance.

67 Having regard to the size of the area, future expenditure will be moderate and largely associated with works to deal with crude sewage and industrial effluent. Substantial improvements are expected in the rivers, with the elimination of Class 3 and Class 4 except for the Carnon and its tributaries south west of Truro.

68 Some rivers in the area, notably the Tavy, are significant because of discharges of effluent upstream of public water supply intakes.

Somerset River Authority
69 About 90 per cent of the river miles are in Class 1 or Class 2, but there is heavy pollution in several areas, particularly around Bridgwater. A big proportion of the treatment works are in the smaller ranges and a feature of the area is that nearly 80 per cent of all the sewage effluent is required to comply with standards more stringent than the Royal Commission's. Only about 20 per cent is discharged to a satisfactory standard. There are many discharges of crude sewage, mainly to the tidal length of the Parrett, with smaller amounts discharged to the Brue and to the Congresbury Yeo. Several schemes are in preparation for dealing with the principal discharges. There are a number of discharges of crude sewage to the Bristol Channel but these are not included in the survey. A number of storm overflows give trouble and some new sewers are likely to be needed in places. Two schemes are already in hand. Apart from one fairly large discharge, industrial effluent does not generally give rise to serious problems although many of the discharges are in need of improvement and farm effluent is troublesome.

70 Future expenditure is likely to be in the region of £15 million with just under a half being needed for works on sewage effluent and the remainder divided fairly evenly between crude sewage and industrial discharges. Because of the low population density, the cost per head is likely to be high. Big improvements in the rivers are expected with the length of Class 1 river likely to be increased by about 80 miles accompanied by big reductions in the mileages in Class 3 and Class 4.

Bristol Avon River Authority
71 There are 42 miles of river in Class 3 and Class 4 with particularly heavy pollution in the Bristol area and pollution over short lengths in a number of inland rivers. The area has a fair proportion of medium sized treatment works. Only about 40 per cent of the sewage is discharged to a satisfactory standard, but it is thought that about one-third of all the unsatisfactory sewage effluent discharges could be improved by better maintenance. The area has the third highest total discharge of crude sewage reported in the survey, nearly all of it being discharged from Bristol to the Avon, but major schemes are planned for diverting this and the other crude sewage discharges to treatment works. There are many unsatisfactory storm overflows, some of which will require major works of sewerage; several schemes are under construction and others have been prepared. Although the volume of industrial effluent discharged is not excessive, it contains a substantial proportion of process water and nearly one-third of all the industrial effluent is considered to be unsatisfactory.

72 The greater part of the future expenditure will be on discharges of sewage effluent but a considerable sum is likely to be spent on discharges of crude sewage. Big improvements are expected with the length of Class 1 river increased by 90 miles and substantial reductions in the mileage of Class 3 and Class 4. Short lengths of Class 3 will remain but all Class 4 waters will be upgraded.

Severn River Authority
73 Despite great lengths of unpolluted rivers there are 85 miles of river in Class 3 and Class 4, much of this mileage occurring on the

fringes of the West Midlands around Stourbridge, Kidderminster, Bromsgrove and Rugby, with shorter lengths near some of the other larger towns in the Severn basin. There are nearly 350 sewage treatment works covering a wide range of sizes and just over a half of all the sewage effluent is discharged to a satisfactory standard. There are a few discharges of crude sewage and schemes are being prepared for diverting most of this to treatment works. The survey records 123 unsatisfactory storm overflows and it is evident that major works of sewerage will be required in a number of places if they are to be improved. Very large quantities of industrial effluent are discharged, but they mainly consist of cooling water. Discharges of solely cooling water are mostly satisfactory and about 75 per cent of the balance is discharged to a satisfactory standard. Water raised from derelict coal mines is significant in the area.

74 Heavy expenditure is expected, with more than £31 million being required on sewage effluent discharges. Expenditure on crude sewage and industrial effluent, by comparison, is likely to be small. Overall, the cost per head is likely to be somewhat above average. The most noticeable improvements are expected to be on the Class 4 rivers, with a reduction in length of nearly 60 miles balanced by increases in Classes 1, 2 and 3. Some 46 of Class 3 and 10 miles of Class 4 are likely to remain, much of this being around the fringes of the West Midlands.

75 The River Severn is significant because of discharges of effluent upstream of public water supply intakes.

Usk River Authority

76 A high proportion of the river mileage is in Class 3 and Class 4. These polluted lengths are mainly in the Ebbw which is Class 4 for its entire length, in the lower tidal part of the Usk at Newport and in the many reens discharging to the Bristol Channel. There are 31 sewage treatment works recorded in the survey, most of them in the smaller range; only 7 are regarded as unsatisfactory, but they account for about 70 per cent of all the effluent. Nearly all the unsatisfactory effluent is discharged to tidal waters and a big proportion of it is in a single discharge to the tidal waters of the Usk, but attention is drawn to a note on this particular discharge in paragraph 140 of Chapter 2. There are 18 discharges of crude sewage (excluding direct discharges to the Severn Estuary which are outside the scope of the survey), all from Newport, and most of them discharge to the Usk. A scheme is under construction for intercepting these by stages and for diverting the flow to a new treatment works. Although there are 176 unsatisfactory storm overflows recorded, the indications are that some of them could be improved at relatively low cost and others are likely to be improved consequent upon major sewerage works already under construction. A big proportion of the industrial effluent discharged is cooling water. Discharges of solely cooling water are mostly satisfactory, but only about 10 per cent of the balance is discharged to a satisfactory standard.

77 Expenditure is likely to be moderate with about £4 million on sewage effluent, just over £1 million on industrial effluent and just under £1 million on crude sewage. Improvements are likely on the more polluted rivers with almost all of the Class 3 and Class 4 waters being improved to Class 2, apart from the reens where special conditions prevail which tend to inhibit improvement (see Volume 1, page 20, paragraph 99).

78 The River Usk is significant on account of discharges of effluent upstream of public water supply intakes.

79 It should be noted that the effluent from a substantial proportion of the population and industry of the area is not included in the survey as it is discharged to the Severn Estuary.

South West Wales River Authority

80 This area has 80 miles of river in Class 3 and Class 4 with heavy pollution mainly around Swansea, Llanelli, and in the valleys near Aberystwyth. The sewage treatment works are mostly in the smaller ranges, serving under 5,000 persons, and about 70 per cent of all the sewage effluent is discharged to a satisfactory standard. A third of all the unsatisfactory discharges could probably be improved by better maintenance. There are 34 discharges of crude sewage and proposals are in preparation or under construction for dealing with many of them. Discharges to the Bristol Channel are not included in the above. There are 23 unsatisfactory storm overflows, but the indications are that attention to the overflows and extra capacity at treatment works will be required rather than re-sewering. The quantity of industrial effluent discharged is not great, but a very substantial proportion is process water. A high percentage of all the industrial effluent is discharged to a satisfactory standard. Water drained from derelict mines is significant.

81 Considering the size of the area, expenditure is likely to be moderate with much of the money being required to deal with discharges of crude sewage. Some reduction in the Class 3 and Class 4 mileage is expected but there are likely to remain some 46 miles of badly polluted river largely around and above Aberystwyth.

82 Some rivers in the area, notably the Tavy and the Teify, are significant on account of discharges of sewage effluent upstream of public water supply intakes.

Lee Conservancy Catchment Board

83 The authority has more than 20 per cent of its river miles in Class 3 and Class 4, the main polluted lengths being in the Lee and its tributaries in the London area, with serious pollution also occurring in the Lee between Luton and Welwyn Garden City and in shorter lengths on rivers and streams in the vicinity of Stevenage, Bishop's Stortford, Harlow, Hoddesdon and Waltham Abbey. Although a high proportion of the treatment works are considered to be satisfactory, the unsatisfactory effluents come mainly from the larger works, so that only about 40 per cent of all the effluent is discharged to a satisfactory standard. Most of the unsatisfactory effluent comes from two works in the London area. A feature of the area is that much of the effluent is required to comply with standards more stringent than Royal Commission and 96 per cent of this effluent is satisfactory. There are no discharges of crude sewage in the area and no unsatisfactory storm overflows are recorded. Industrial effluent consists almost entirely of cooling water and all the discharges are considered to be satisfactory.

84 All expenditure will be on discharges of sewage effluent and a major extension is already under way at the principal GLC works in the area. Because of the high population density, the cost per head is likely to be well below average. A big increase in Class 1 mileage is expected, together with a substantial reduction in the Class 4 mileage, but some waters will remain in Class 3 or Class 4.

85 The River Lee is conspicuous on account of several discharges of effluent upstream of public water supply intakes.

Authorities in Group D

Yorkshire River Authority

86 The longest lengths of Class 3 and 4 river are found in this area with widespread and heavy pollution throughout an area of the West Riding on and to the south of a line through Keighley, Leeds and Selby and there is occasional heavy pollution in other areas. Long lengths of Class 4 river are to be found, particularly in the vicinity of Leeds, Bradford, Huddersfield, Wakefield, Sheffield, Rotherham and Chesterfield and many miles of canal in the same general areas are also heavily polluted. There are over 430 discharges of sewage effluent with a big proportion in the medium-sized and large categories. Nearly two-thirds of all the discharges are satisfactory, but in terms of volume, less than 30 per cent of all the effluent is discharged to a satisfactory standard. A significant point shown up is that the authority are likely to be seeking standards more stringent than Royal Commission for many of the discharges in the future. By comparison, crude sewage discharges are much less significant. There are only 12 such discharges and they carry the sewage from less than 30,000 population (excluding discharges to the Humber which are outside of the survey). Steps are being taken to deal with most of these and several schemes are under construction. There are over 300 unsatisfactory storm overflows many, again, discharging to rivers in the West Riding. Many expensive schemes involving

re-sewering are likely to be needed. Nearly 1,500 mgd of industrial effluent is discharged in the area and a significant proportion is process water. More than 90 per cent of effluent discharged as solely cooling water is satisfactory, but only about 60 per cent of the remainder is discharged to a satisfactory standard. A substantial quantity of mine water is discharged to rivers and canals in the area.

87 Expenditure is likely to be very high with most of the costs being related to discharges of sewage effluent, but, because of fairly high population density, the cost per head is not likely to be much above average. A very big reduction in the Class 4 river mileage is expected with increases in the mileage in the other 3 classes. Improvements are also expected on canals. However, long lengths of both rivers and canals, particularly in the West Riding, will still remain heavily polluted.

88 A number of rivers in the area, notably the Ouse and the Esk, are significant on account of discharges of effluent upstream of public water supply intakes.

Trent River Authority
89 The authority has 375 river miles and 141 miles of canals in Class 3 and Class 4. The worst of the pollution occurs in the West Midlands towards the upper end of the Trent basin, but serious pollution is also to be found in many other parts of the area notably on the Trent and on the rivers in the vicinity of and downstream of such places as Stoke-on-Trent, Leek, Bedworth, Scunthorpe, Ilkeston, Mansfield and Nottingham. The largest number of discharges of sewage effluent in the survey is recorded here—more than 480 in all with many in the medium-sized and large-sized category. Just over half of the discharges are satisfactory and about 65 per cent of the total volume is discharged to a satisfactory standard. The returns suggest that a considerable number of the discharges could be improved by better maintenance. It is evident that major grouping of discharges is contemplated as the survey indicates that, by 1980, the number of treatment works should be reduced to under 400. There are only a few discharges of crude sewage and these are being dealt with by a scheme which is under construction at present. There are 80 storm overflows requiring attention and a good deal of re-sewering is likely to be needed. More than 1,700 mgd of industrial effluent is discharged in the area and a significant proportion is process water. About 97 per cent of the effluent discharged as solely cooling water is satisfactory and about 83 per cent of the balance is discharged to a satisfactory standard. Fairly large quantities of mine water are discharged in the area.

90 Most of the expenditure is likely to be related to discharges of sewage effluent and the estimates indicate a need to spend over £80 million on this type of discharge. The bulk of this is urgently needed. A big proportion of the total cost relates to expenditure on treatment works owned by the Upper Tame Main Drainage Authority who are already engaged on a long-term programme of remedial works. Despite high population density, the cost per head is likely to be above average. A big reduction in the mileage of Class 3 and Class 4 rivers and canals is expected, but about 190 miles of river and canal are likely to remain heavily polluted, much of this mileage, again, in the West Midlands.

91 The Derwent and the Dove are conspicuous as rivers where there are discharges of effluent upstream of public water supply intakes. The suitability of the River Trent as a source for future water supplies is being studied.

Glamorgan River Authority
92 Pollution is widespread throughout the area, with 135 miles of river in Class 3 or Class 4 and, in terms of miles of polluted river per unit of area, one of the worst in the survey. The heaviest pollution occurs in the river basins of the Taff and the Rhymney and there is further pollution, but over shorter lengths, in a number of other river basins, particularly the Neath, the Afan and the Afon Kenfig. There are 40 discharges of sewage effluent including several large discharges but only 13 are satisfactory and only about 8 per cent of the total volume of effluent is discharged to a satisfactory standard. Most of the unsatisfactory effluent is discharged to the Taff and the Ogmore. The survey records 29 discharges of crude sewage, but several very large discharges to the Bristol Channel are excluded. Major schemes are under construction or are planned for dealing with the crude discharges to the rivers. The 5 unsatisfactory storm overflows should be improved or eliminated under schemes already in hand or proposed. A substantial proportion of the industrial effluent discharged in the area is process water. Almost all of the effluent discharged as solely cooling water is satisfactory, but only about 60 per cent of the remainder is discharged to a satisfactory standard. There are difficulties with discharge of minewater.

93 Expenditure is likely to be related mainly to sewage effluent discharges but a significant amount—more than £3 million—is likely to be required to deal with crude sewage. The cost per head is likely to be well above average. There is a forecast of a reduction of 75 miles in the lengths in Class 3 and Class 4 river but some 60 miles are likely to remain heavily polluted spread over several rivers.

94 Certain rivers, notably the Taff, the Taf Fawr and the Taf Fechan, receive discharges of effluent upstream of public water supply intakes.

95 It should be noted that the effluent from nearly two-thirds of the population of the area is not included in the survey as it is discharged direct to the Bristol Channel and that some very high expenditure, not included in the above figures, will be required to deal with those discharges.

Mersey and Weaver River Authority
96 With 340 river miles in Class 3 or Class 4 the area of the authority is one of the most heavily polluted in the survey in respect of both total river miles polluted and polluted miles per unit of area. Although there are signs of heavy pollution in the Weaver basin, it is in the Mersey basin with its rivers flowing through the highly populated and heavily industrialised conurbations of South East Lancashire and Merseyside that the worst and most extensive pollution is encountered. Many major rivers such as the Mersey, the Irwell, the Irk, the Roch and the Tame, as well as the Manchester Ship Canal, are heavily polluted throughout most or all of their lengths. There are nearly 190 discharges of sewage effluent, many of them from treatment works in the medium-sized and larger ranges. Just over half are considered to be satisfactory but only about one-third of all the effluent is discharged to a satisfactory standard. The Manchester Ship Canal receives about one-third of all the sewage effluent discharged in the whole area and it also receives nearly half of all the unsatisfactory effluent. Large quantities of crude sewage are discharged to the Mersey estuary from some 68 outfalls, the population served by these being in excess of one million. The returns also indicate very serious problems with storm overflows, with nearly 800 of them classed as unsatisfactory—more than one-third of the total recorded in the whole survey. It is clear that if all these overflows are to be brought to the standard the river authority consider to be acceptable, many major schemes of sewerage will be required at great cost. The rivers and canals in the area receive nearly 1,900 mgd of industrial effluent with a significant proportion being process water. About 93 per cent of the effluent discharged as solely cooling water is satisfactory but only just over a half of the remainder is discharged to a satisfactory standard. The area is conspicuous in that the discharges from several industries represent high proportions of the national totals for these industries. Discharges of mine water are also significant.

97 Heavy expenditure is envisaged on all types of discharge with more than £70 million likely to be required on crude sewage and sewage effluent discharges and more than £10 million on industrial effluent, but because of high population density, the cost per head will not be greatly above the average. The forecasts suggest that there should be a reduction of about 145 miles in the length of Class 4 rivers but more than 200 miles of river are likely to remain heavily polluted. A small improvement is forecast on the more polluted lengths of canals, with the elimination of Class 4.

98 A special Committee has been set up to study the problem of the Mersey estuary and to make recommendations.

Lancashire River Authority
99 Serious and widespread pollution occurs in the southern half of the authority's area. It is particularly bad on the lower parts of the

Ribble and on the Douglas and on the rivers flowing through those densely populated industrial areas of South East Lancashire which lie beyond the limits of the Mersey basin. Pollution on a more limited scale is also encountered further north. Altogether there are nearly 280 miles of river in Class 3 or Class 4. The survey records 157 discharges of sewage effluent, many from treatment works in the small ranges, several in the medium-sized ranges and a few from large works. Only about one-third of the discharges are considered to be satisfactory and a good number, it would seem, could be substantially improved by better maintenance. Only about 15 per cent of all the effluent is discharged to a satisfactory standard. The returns suggest that in the future there will be much grouping of discharges on a regional basis with a substantial reduction in the number of treatment works. Excluding a small number of discharges to the seaward end of Morecambe Bay, there are 30 discharges of crude sewage to rivers. Several schemes for dealing with these are already under construction and others are being prepared. Many storm overflows give trouble and 268 are reported as unsatisfactory. It is evident that many major schemes will be required to meet the situation as most of the overflows appear to be affected by inadequate sewer capacity, and costs are likely to be high. By comparison with some other river authorities whose areas are partly industrial in character, the amount of industrial effluent discharged is not excessive, but a significant proportion of it is process water. Effluent discharged as solely cooling water is mostly satisfactory but only some 22 per cent of the balance is discharged to a satisfactory standard.

100 Fairly large expenditure is envisaged, with some £20 million or so likely to be required to deal with sewage effluent, about £4 million to deal with crude sewage and over £1 million needed for industrial effluent and the cost per head is likely to be above average. Very big reductions are expected in the lengths of heavily polluted rivers, with a forecast that the present Class 3 and Class 4 length of 277 miles should be reduced to about 40 miles.

101 The Lune is a river which is significant because of discharges of effluent upstream of public water supply intakes.

Port of London Authority (including the London Excluded Area)

102 This area, where pollution control responsibilities are shared between the Port of London Authority and the Greater London Council, covers much of the densely populated area of Greater London. More than half of the river miles are in Class 3 or Class 4, including most of the length of the Thames which is Class 3 from Teddington to Canvey Island. In the London area, the most polluted tributaries are those flowing northwards from the Wimbledon and Croydon areas. None of the rivers are better than Class 2. Although there are only 25 discharges of sewage effluent, most of the treatment works are in the medium and large-sized ranges, with the total population served exceeding 8 million and a sewage flow of more than 500 mgd. Only about one-quarter of the sewage effluent is considered to be satisfactory. An indication of the problems which the authorities have to meet can be obtained from the fact that the population served by sewage works in the area is more than one-fifth of the total population served by all the treatment works in the survey and the sewage effluent discharged direct to rivers in the area represents nearly one-quarter of all the sewage effluent discharged to all the rivers and canals in the survey. If the discharges to rivers in the non-tidal Thames basin and in the Lee basin are added it can be seen that about 760 mgd of sewage effluent from some 13 million population ultimately passes down the lower reaches of the tidal Thames. These figures are about one-third of the corresponding totals in the survey for England and Wales. There are no discharges of crude sewage in the area, but there are 25 unsatisfactory storm overflows. The cost of dealing with these would be very heavy and the authority have some reservations about whether the benefits that would result would justify the expenditure. Nearly 3,400 mgd of industrial effluent (the largest volume recorded in the survey) is discharged to rivers and canals in the area, with cooling water accounting for about 98 per cent of the flow. Nearly all of the effluent discharged as solely cooling water is satisfactory, and about 86 per cent of the balance is discharged to a satisfactory standard. Discharges from several industries represent high proportions of the national totals for these industries.

103 A number of schemes are already under construction but, excluding these, it is likely that more than £30 million will have to be spent on discharges of sewage effluent and about £1.5 million on discharges of industrial effluent. Because of the very high population density, the cost per head is likely to be well below average. A moderate improvement by way of 11 miles of Class 4 upgraded to Class 2 is expected. Some 70 miles will remain in Class 3 or Class 4.

Appendix 1 Note on biological classification of rivers

1 A biological classification system was employed in the river pollution survey and the river authorities submitted biological classifications for many non-tidal rivers. Of the 22,313 miles of non-tidal rivers for which chemical classifications were submitted, 16,532 miles were also biologically classified. The four biological classes employed in the survey were described in Chapter 4 of Volume 1 and comparisons were made between the biological classifications and the chemical classifications of the miles of non-tidal rivers for each river authority.

2 Biological methods of pollution assessment are employed in addition to and not as a substitute for chemical measurements. A survey of the animal and plant communities of a river will show the recent history of the pollution of the river and may indicate mild or intermittent pollution which chemical techniques, even automatic monitoring, could fail to detect.

3 Macro-invertebrates are generally considered to be the most useful aquatic organisms for a biological assessment of pollution as they are relatively easily sampled and identified in comparison with, say, micro-organisms.

4 The four biological classes employed in the survey represent a simplified form of biological index of pollution. It was hoped that its simplicity would enable it to be used on the wide variety of rivers being surveyed from upland, flashy streams to lowlands drains. A biological classification was not employed in the previous and first river pollution survey of 1958.

5 The absence of biological classifications for some river mileage is likely to be due to human factors, eg, lack of biological facilities, whereas disagreements are more likely to relate to variations due to regional, environmental and other reasons.

6 The biological data in Volume 1 showed that the percentage of miles classified biologically varied between chemical classes and river authorities. The percentage of miles on which biological and chemical classifications agree also varied between chemical classes and river authorities. The percentages overall were higher in both cases for chemical classes 1 and 4 than 2 or 3 (see Volume 1, Table 69). This is not unexpected as both biological classes A and D represent more extreme conditions and the biotic criteria are more clearly defined than the intermediate classes B and C. Overall, about 75 per cent of river miles that were chemically classified were also biologically classified and of that 75 per cent a similar percentage showed agreement between corresponding biological and chemical classes. Both classification systems showed that in general the mileage in the four classes decreased from Class 1/A to 4/D. River mileage of the four biological classes was represented in each of the four chemical classes except that no biological class A miles were placed in chemical class 4.

7 The biological class assigned to a stretch of river will be determined by the invertebrate and fish fauna present. The biological community will reflect many environmental factors among which are pollution load and velocity of flow which interact in terms of dissolved oxygen and deposited solids. It appears that velocity of flow, which is a function of gradient, strongly influences the effect of pollution and hence the biological class assigned. The gradient of a particular river can be fairly constant over a considerable distance and as such can be measured.

8 The chemical classification has a built-in tendency to place a proportion of stretches in a higher chemical class than the corresponding biological class (see Volume 1, Table 39). In other words in chemical classes 2 and 3 those stretches that were not placed in the corresponding biological class tended to be placed in a lower rather than a higher biological class. The chemical classification therefore appears to favour slower velocities more than does the biological classification.

9 It is proposed to investigate further these differences between the biological and chemical classification systems in conjunction with river authority biologists to determine whether the two classification systems should be treated as criteria complementing each other but having a distinct role or whether the biological classification system should have a correction factor applied to obtain better correspondence with the chemical classification system for which, of course, there is a precedent in that the latter system only was used in the previous and first survey.

10 It could be said, in support of using the two classification systems for different purposes, that the chemical classification system measures the suitability of river water for water supply and the effect on river water of effluent disposal whereas the biological system monitors the ecological and amenity value of the river. The chemical system is perhaps best used to reflect short term changes whereas the biological system is perhaps best used as a continuous monitor of long term trends.

Appendix 2

Questionnaire circulated by The Confederation of British Industries

---

PART 2
ORIGINAL

SERIAL NO.
015442

Please return this form intact to the Confederation of British Industry **OR** a trade association, whichever is appropriate. Name and address to be detached by either the C.B.I. or trade association and retained.

FROM (insert name and address of company)

Tear on this fold

Fold and tear on this line

Questionnaire

Please provide answers to the following questions in respect of the factory referred to above. A separate questionnaire should be completed for each premises within Group or Company organisation.

Please enter figures in appropriate boxes using right correction — (Note 2(a))

JOB NO. **4 5 5** (1– 3)

SERIAL NO. **015442** (4– 8)

INDUSTRIAL EFFLUENT CLASSIFICATION CODE NO. (Note 2(b)) ☐☐ (9–10)

1. Please indicate Pollution Authority in whose area trade effluent is discharged (Note 3) by Code No.

   ☐☐ (11–12)

2. In respect of trade effluent discharged to public sewers:

   Card No. **1** (13)

   a. i. What average daily volume of total effluent **in thousands of gallons**, was discharged during the year 1968 (Note 2(c), 2(d), 2(e))

   ☐☐☐☐☐☐ (14–19)

   ii. What average daily volume of cooling water **in thousands of gallons**, was discharged during the year 1968 (Note 2(c), 2(e), 2(f))

   ☐☐☐☐☐ (20–24)

   b. Pre-treatment — capital costs

   i. What was total capital expenditure, **in thousands of pounds**, on plant installed during the period January 1 1960 to December 31 1968.

   ☐☐☐☐☐ (25–29)

   ii. What is forecast of capital expenditure, **in thousands of pounds**, during next 2 year period.

   ☐☐☐☐☐ (30–34)

c. Pre-treatment plant operating costs.
[Note 2(e), 2(g), 2(h)]

    i. What were operating costs, **in hundreds of pounds,** for the year 1968, excluding contractors' services. ☐☐☐☐ (35–38)

    ii. What were costs, **in hundreds of pounds,** of contractors services for the year 1968? ☐☐☐☐ (39–42)

d   What was total charge, **in hundreds of pounds,** for trade effluent disposal imposed by your local authority, including any contribution to provision of additional sewage treatment capacity for the year 1968? ☐☐☐☐☐ (43–47)

3. In respect of trade effluent discharged direct to:-
    (a) non-tidal rivers
    (b) controlled tidal waters
    (c) canals

    a. i. What average daily volume of total effluent, **in thousands of gallons,** was discharged during the year 1968? [Note 2(c), 2(d), 2(e)]
      (a) to (48–52)
      (b) (53–57)
      (c) (58–62)

    ii. What average daily volume of cooling water, **in thousands of gallons,** was discharged during the year 1968? [Note 2(c), 2(e), 2(f)]
      (a) to (63–67)
      (b) (68–72)
      (c) (73–77)

b. Treatment Plant – capital costs

New Card. Repeat Cols. 1–12 of Card 1.

☐2☐ (13)

   i. What was total capital expenditure, **in thousands of pounds**, on plant installed during the period January 1 1960 to December 31 1968.
     on (a)   (14–18)
        (b)   (19–23)
        (c)   (24–28)

   ii. What is forecast of capital expenditure, **in thousands of pounds,** during next 2 year period
     on (a)   (29–33)
        (b)   (34–38)
        (c)   (39–43)

c. Treatment Plant – operating costs
[Note 2(e), 2(g), 2(h)]

   i. What were operating costs, **in hundreds of pounds,** for year 1968 excluding disposal contractors' services.
     on (a)   (44–47)
        (b)   (48–51)
        (c)   (52–55)

   ii. What were costs, **in hundreds of pounds,** of disposal contractors' services for the year 1968.   (56–59)

4. In respect of trade effluent not disposed of according to 2. or 3. above.

   a. Incineration Plant – (Note 4) (Notes 2(c), 2(e), 2(g))

     i. What average daily volume **in thousands of gallons** was disposed of during the year 1968?   (60–64)

     ii. What has been capital expenditure **in thousands of pounds** during the period January 1 1960 to December 31 1968?   (65–69)

     iii. What is forecast of total capital expenditure **in thousands of pounds** for the next two years?   (70–74)

     vi. What were operating costs **in hundreds of pounds** for the year 1968?   (75–78)

New Card. Repeat Cols. 1–12 of Card 1.

☐3☐ (13)

   b. Pipelines to uncontrolled waters (Note 5)

     i. What average daily volume **in thousands of gallons** was discharged during the year 1968? – (Note 2(c), 2(e)).   (14–18)

     ii. What has been capital expenditure **in thousands of pounds** on installations during the period January 1 1960 to December 31 1968?   (19–23)

     iii. What is forecast of total capital expenditure **in thousands of pounds** for next 2 years?   (24–28)

     vi. What were operating costs **in hundreds of pounds** for the year 1968?   (29–32)

P.T.O.

c. Contractors' Services (including sea disposal) (Note 2(h)).

   i. What average daily volume **in thousands of gallons** was disposed of during the year 1968? (Note 2(c), 2(e)). ☐☐☐☐☐ (33–37)

   ii. What was expenditure **in hundreds of pounds** on disposal contractors' services for the year 1968? (Note 2(e), 2(h)). ☐☐☐☐ (38–41)

Printed by the Confederation of British Industry, London

Additional tables

The following tables give a further breakdown of details given in Chapters 2—6 of the Report.

Chapter 2. Discharges of sewage effluent. Tables 170 to 184.

Chapter 3. Discharges of crude sewage. Tables 185 and 186.

Chapter 4. Discharges of storm sewage from unsatisfactory storm overflows. Table 187.

Chapter 5. Discharges of industrial effluent. Tables 188 to 217.

Chapter 6. Expenditure. Tables 218 to 220.

Table 170 Sewage treatment works. Total number of discharges and percentage considered satisfactory

River Authority		Non-tidal rivers					Tidal rivers					All rivers
		Class 1	Class 2	Class 3	Class 4	Total non-tidal	Class 1	Class 2	Class 3	Class 4	Total tidal	Class 1
1 Northumbrian	No.	147	42	10	4	203	4	3	1	1	9	151
	%	73	74	40	50	71	75	67	100	0	67	73
2 Yorkshire	No.	283	39	37	54	413	3	2	10	2	17	286
	%	69	51	51	46	62	100	100	80	100	88	69
3 Trent	No.	188	136	59	75	458	0	2	0	10	12	188
	%	56	64	56	36	55		50		50	50	56
4 Lincolnshire	No.	107	6	3	0	116	1	0	1	0	2	108
	%	85	67	67		84	0		100		50	84
5 Welland and Nene	No.	131	21	1	0	153	0	2	0	0	2	131
	%	47	29	0		44		50			50	47
6 Great Ouse	No.	165	81	8	0	254	9	0	0	0	9	174
	%	73	67	50		70	100				100	74
7 East Suffolk and Norfolk	No.	66	17	0	0	83	13	19	0	0	32	79
	%	83	76			82	100	74			84	86
8 Essex	No.	99	23	8	1	131	16	3	7	1	27	115
	%	76	78	63	0	75	75	67	43	100	67	76
9 Kent	No.	134	0	0	0	134	10	3	7	2	22	144
	%	71				71	70	33	57	100	64	71
10 Sussex	No.	82	20	1	1	104	14	0	0	0	14	96
	%	68	60	100	100	67	50				50	67
11 Hampshire	No.	22	8	3	0	33	2	8	2	2	14	24
	%	77	88	0		73	100	50	0	0	43	79
12 Isle of Wight	No.	10	1	0	0	11	1	0	1	0	2	11
	%	50	100			55	100		0		50	55
13 Avon and Dorset	No.	68	10	1	0	79	6	0	0	0	6	74
	%	85	80	100		85	100				100	86
14 Devon	No.	116	4	3	0	123	4	3	6	0	13	120
	%	63	75	67		63	50	67	0		30	63
15 Cornwall	No.	72	5	7	8	92	7	2	0	0	9	79
	%	68	60	43	63	65	71	0			56	68
16 Somerset	No.	91	12	1	0	104	0	0	0	0	0	91
	%	66	33	0		62						66
17 Bristol Avon	No.	55	28	3	0	86	0	0	0	1	1	55
	%	65	68	33		65				0	0	65
18 Severn	No.	301	27	1	13	342	4	0	0	0	4	305
	%	66	56	100	54	65	50				50	65
19 Wye	No.	47	3	2	0	52	2	0	0	0	2	49
	%	87	67	50		85	100				100	88
20 Usk	No.	26	0	1	0	27	1	2	1	0	4	27
	%	85		0		81	0	100	0		50	81
21 Glamorgan	No.	19	7	9	5	40	0	0	0	0	0	19
	%	47	14	11	40	33						47
22 South West Wales	No.	85	6	3	0	94	14	6	3	5	28	99
	%	68	50	0		65	71	67	100	100	79	69
23 Gwynedd	No.	33	2	0	0	35	8	0	0	0	8	41
	%	64	50			63	63				63	63
24 Dee and Clwyd	No.	72	21	3	4	100	5	4	0	0	9	77
	%	75	48	67	50	68	20	50			33	71
25 Mersey and Weaver	No.	57	64	21	36	178	0	0	6	4	10	57
	%	56	53	43	44	51			67	75	70	56
26 Lancashire	No.	54	32	25	19	130	14	4	6	2	26	68
	%	43	41	28	5	34	43	0	50	50	38	43
27 Cumberland	No.	77	9	4	0	90	3	1	0	0	4	80
	%	19	22	25		20	0	0			0	19
28 Thames Conservancy	No.	304	29	12	0	345	0	0	0	0	0	304
	%	75	76	50		74						75
29 Lee Conservancy	No.	27	2	1	2	32	0	0	0	0	0	27
	%	89	100	100	0	84						89
30 Port of London Authority (incl. London Excluded Area)	No.	0	0	0	5	5	0	1	19	0	20	0
	%				40	40		100	37		40	
Total England and Wales	No.	2938	655	227	227	4047	141	65	70	30	306	3079
	%	68	60	46	40	64	68	58	49	63	61	68

130

	s 2	Class 3	Class 4	Total all rivers	Canals					All rivers and canals					
					Class 1	Class 2	Class 3	Class 4	Total canals	Class 1	Class 2	Class 3	Class 4	Total all rivers and canals	
		11	5	212	0	0	0	0	0	151	45	11	5	212	1
		45	40	71						73	73	45	40	71	
		47	56	430	0	0	0	2	2	286	41	47	58	432	2
		57	48	63				100	100	69	54	57	50	64	
		59	85	470	1	5	7	0	13	189	143	66	85	483	3
		56	38	55	0	60	43		46	56	64	55	38	55	
		4	0	118	1	0	0	0	1	109	6	4	0	119	4
		75		83	100				100	84	67	75		83	
		1	0	155	2	0	0	0	2	133	23	1	0	157	5
		0		44	50				50	47	30	0		45	
		8	0	263	0	0	0	0	0	174	81	8	0	263	6
		50		71						74	67	50		71	
		0	0	115	0	0	0	0	0	79	36	0	0	115	7
				83						86	75			83	
		15	2	158	0	0	0	0	0	115	26	15	2	158	8
		53	50	73						76	77	53	50	73	
		7	2	156	0	0	0	0	0	144	3	7	2	156	9
		57	100	70						71	33	57	100	70	
		1	1	118	0	0	0	0	0	96	20	1	1	118	10
		100	100	65						67	60	100	100	65	
		5	2	47	0	0	0	0	0	24	16	5	2	47	11
		0	0	64						79	69	0	0	64	
		1	0	13	0	0	0	0	0	11	1	1	0	13	12
		0		54						55	100	0		54	
		1	0	85	0	0	0	0	0	74	10	1	0	85	13
		100		86						86	80	100		86	
		9	0	136	0	0	0	0	0	120	7	9	0	136	14
		22		60						63	71	22		60	
		7	8	101	0	0	0	0	0	79	7	7	8	101	15
		43	63	64						68	43	43	63	64	
		1	0	104	0	0	0	0	0	91	12	1	0	104	16
		0		62						66	33	0		62	
		3	1	87	0	0	0	0	0	55	28	3	1	87	17
		33	0	64						65	68	33	0	64	
		1	13	346	0	1	0	0	1	305	28	1	13	347	18
		100	54	64		100			100	65	57	100	54	65	
		2	0	54	0	0	0	0	0	49	3	2	0	54	19
		50		85						88	67	50		85	
		2	0	31	0	0	0	0	0	27	2	2	0	31	20
		0		77						81	100	0		77	
		9	5	40	0	0	0	0	0	19	7	9	5	40	21
		11	40	33						47	14	11	40	33	
		6	5	122	0	0	0	0	0	99	12	6	5	122	22
		50	100	68						69	58	50	100	68	
		0	0	43	0	0	0	0	0	41	2	0	0	43	23
				63						63	50			63	
		3	4	109	0	0	0	0	0	77	25	3	4	109	24
		67	50	65						71	48	67	50	65	
		27	40	188	0	0	1	0	1	57	64	28	40	189	25
		48	48	52			0		0	56	53	46	48	52	
		31	21	156	0	1	0	0	1	68	37	31	21	157	26
		32	10	35		0			0	43	35	32	10	34	
		4	0	94	0	0	0	0	0	80	10	5	0	94	27
		25		19						19	20	20		19	
		12	0	345	5	0	0	0	5	309	29	12	0	350	28
		50		74	80				80	75	76	50		75	
		1	2	32	0	0	0	0	0	27	2	1	2	32	29
		100	0	84						89	100	100	0	84	
		19	5	25	0	0	0	0	0	0	1	19	5	25	30
		37	40	40							100	37	40	40	
		297	257	4353	9	7	8	2	26	3088	726	306	259	4379	
		46	42	63	67	57	38	100	58	68	60	46	43	63	

131

Table 171 Sewage treatment works. Total population served and percentage served by sewage treatment works considered satisfactory

River authority		Non-tidal rivers					Tidal rivers					All rivers
		Class 1	Class 2	Class 3	Class 4	Total non-tidal	Class 1	Class 2	Class 3	Class 4	Total tidal	Class 1
1 Northumbrian	Popn. %	319,071 63	231,868 63	182,729 22	26,469 11	760,137 51	12,838 95	9,532 35	25,200 100	21,546 0	69,116 59	331,9
2 Yorkshire	Popn. %	1,356,273 36	419,241 52	767,805 16	1,332,123 36	3,875,442 34	2,615 100	7,260 100	164,831 46	2,776 100	177,482 50	1,358,8
3 Trent	Popn. %	407,896 59	1,709,008 77	698,111 49	2,611,424 58	5,426,439 63	0	2,090 70	0	12,305 41	14,395 45	407,8
4 Lincolnshire	Popn. %	308,398 86	91,202 93	6,036 13	0	405,636 86	888 0	0	200 100	0	1,088 18	309,2
5 Welland and Nene	Popn. %	420,100 62	63,134 69	1,408 0	0	484,642 63	0	91,450 82	0	0	91,450 82	420,1
6 Great Ouse	Popn. %	495,044 54	383,013 48	71,175 60	0	949,232 52	16,497 100	0	0	0	16,497 100	511,5
7 East Suffolk and Norfolk	Popn. %	85,245 75	41,124 70	0	0	126,369 74	47,117 100	342,462 18	0	0	389,579 28	132,3
8 Essex	Popn. %	217,639 63	169,069 91	107,024 48	6,006 0	499,738 69	71,041 47	105,621 13	113,681 71	1,998 100	292,341 44	288,6
9 Kent	Popn. %	414,264 71	0	0	0	414,264 71	272,112 92	24,307 41	119,395 93	35,880 100	451,694 90	686,3
10 Sussex	Popn. %	229,111 56	77,210 68	1,200 100	300 100	307,821 59	54,229 26	0	0	0	54,229 26	283,3
11 Hampshire	Popn. %	106,657 42	17,772 98	2,419 0	0	126,848 49	91,017 100	259,901 7	10,520 0	73,500 0	434,938 25	197,6
12 Isle of Wight	Popn. %	8,169 47	150 33	0	0	8,319 48	743 100	0	21,340 0	0	22,083 3	8,9
13 Avon and Dorset	Popn. %	292,132 79	6,247 33	1,200 100	0	299,579 78	126,545 100	0	0	0	126,545 100	418,6
14 Devon	Popn. %	120,883 57	1,286 48	6,224 60	0	128,393 57	8,151 90	14,469 57	134,750 0	0	157,370 10	129,03
15 Cornwall	Popn. %	144,106 78	8,613 84	4,285 53	25,124 82	182,128 79	14,040 90	17,364 0	0	0	31,404 40	158,14
16 Somerset	Popn. %	165,305 39	56,584 6	900 0	0	222,789 31	0	0	0	0	0	165,30
17 Bristol Avon	Popn. %	147,835 51	193,916 40	9,330 11	0	351,081 44	0	0	0	1,199 0	1,199 0	147,83
18 Severn	Popn. %	1,298,312 56	226,389 46	175,000 100	556,528 53	2,256,229 57	74,218 3	0	0	0	74,218 3	1,372,53
19 Wye	Popn. %	61,755 79	1,433 50	47,800 1	0	110,988 45	2,905 100	0	0	0	2,905 100	64,66
20 Usk	Popn. %	47,418 81	0	220 0	0	47,638 81	93,280 0	5,675 100	16,218 0	0	115,173 5	140,69
21 Glamorgan	Popn. %	33,659 44	65,776 32	156,184 5	23,354 13	278,963 17	0	0	0	0	0	33,659
22 South West Wales	Popn. %	93,749 75	30,102 90	7,065 0	0	130,916 74	16,750 60	32,949 25	2,331 100	50,243 100	102,273 69	110,499
23 Gwynedd	Popn. %	38,778 60	2,254 16	0	0	41,032 58	7,496 41	0	0	0	7,496 41	46,274
24 Dee and Clwyd	Popn. %	110,143 88	101,274 69	8,825 20	32,452 96	252,694 79	16,611 19	99,112 89	0	0	115,723 79	126,754
25 Mersey and Weaver	Popn. %	359,973 63	903,241 45	673,676 46	1,883,022 21	3,819,912 35	0	0	49,551 54	135,535 85	185,086 76	359,973
26 Lancashire	Popn. %	95,406 26	201,180 18	154,158 44	390,541 6	841,285 18	22,795 34	3,549 0	247,148 11	2,508 54	276,000 13	118,201
27 Cumberland	Popn. %	75,060 20	99,869 6	15,050 89	0	189,979 18	1,297 0	500 0	0	0	1,797 0	76,357
28 Thames Conservancy	Popn. %	2,283,953 49	632,697 89	330,938 21	0	3,247,588 54	0	0	0	0	0	2,283,953
29 Lee Conservancy	Popn. %	446,392 92	40,880 100	4,000 100	717,000 0	1,208,272 38	0	0	0	0	0	446,392
30 PLA (incl. London Excluded Area)	Popn. %	0	0	0	697,910 55	697,910 55	0	174,590 100	7,549,923 21	0	7,724,513 22	0
Total England and Wales	Popn. %	10,182,726 57	5,774,522 62	3,432,762 36	8,302,253 38	27,692,263 50	953,185 68	1,190,832 40	8,455,088 23	337,490 62	10,936,594 30	11,135,911 57

s 2	Class 3	Class 4	Total all rivers	Canals					All rivers and canals					
				Class 1	Class 2	Class 3	Class 4	Total canals	Class 1	Class 2	Class 3	Class 4	Total all rivers and canals	
1,400	207,929	48,015	829,253	0	0	0	0	0	331,909	241,400	207,929	48,015	829,253	1
62	31	6	52						64	62	31	6	52	
6,501	932,636	1,334,899	4,052,924	0	0	0	26,874	26,874	1,358,888	426,501	932,636	1,361,773	4,079,798	2
53	21	36	35				100	100	36	53	21	37	35	
1,098	698,111	2,623,729	5,440,834	3,002	75,380	169,256	0	247,638	410,898	1,786,478	867,367	2,623,729	5,688,472	3
77	49	58	63	0	57	93		81	58	76	57	58	63	
1,202	6,236	0	406,724	770	0	0	0	770	310,056	91,202	6,236	0	407,494	4
93	16		86	100				100	85	93	16		86	
4,584	1,408	0	576,092	1,071	0	0	0	1,071	421,171	154,584	1,408	0	577,163	5
77	0		66	28				28	62	77	0		66	
3,013	71,175	0	965,729	0	0	0	0	0	511,541	383,013	71,175	0	965,729	6
48	60		53						55	48	60		53	
3,586	0	0	515,948	0	0	0	0	0	132,362	383,586	0	0	515,948	7
23			39						84	23			39	
74,690	220,705	8,004	792,079	0	0	0	0	0	288,680	274,690	220,705	8,004	792,079	8
61	60	25	60						59	61	60	25	60	
24,307	119,395	35,880	865,958	0	0	0	0	0	686,376	24,307	119,395	35,880	865,958	9
41	93	100	81						79	41	93	100	81	
77,210	1,200	300	362,050	0	0	0	0	0	283,340	77,210	1,200	300	362,050	10
68	100	100	54						50	68	100	100	54	
77,673	12,939	73,500	561,786	0	0	0	0	0	197,674	277,673	12,939	73,500	561,786	11
13	0	0	31						69	13	0	0	31	
150	21,340	0	30,402	0	0	0	0	0	8,912	150	21,340	0	30,402	12
33	0		16						51	33	0		16	
6,247	1,200	0	426,124	0	0	0	0	0	418,677	6,247	1,200	0	426,124	13
33	100		85						85	48	100		85	
15,755	140,974	0	285,763	0	0	0	0	0	129,034	15,755	140,974	0	285,763	14
56	3		31						59	56	3		31	
25,977	4,285	25,124	213,532	0	0	0	0	0	158,146	25,977	4,285	25,124	213,532	15
28	53	82	73						80	28	53	82	73	
56,584	900	0	222,789	0	0	0	0	0	165,305	56,584	900	0	222,789	16
13	0		31						39	16	0		31	
93,916	9,330	1,199	352,280	0	0	0	0	0	147,835	193,916	9,330	1,199	352,280	17
40	11	0	43						51	40	11	0	43	
226,389	175,000	556,528	2,330,447	0	2,100	0	0	2,100	1,372,530	228,489	175,000	556,528	2,332,547	18
46	100	53	56		100			100	53	46	100	53	56	
1,433	47,800	0	113,893	0	0	0	0	0	64,660	1,433	47,800	0	113,893	19
50	1		47						80	50	1		47	
5,675	16,438	0	162,811	0	0	0	0	0	140,698	5,675	16,438	0	162,811	20
100	0		27						100	100	0		27	
65,766	156,184	23,354	278,963	0	0	0	0	0	33,659	65,766	156,184	23,354	278,963	21
32	5	13	17						44	32	5	13	17	
63,051	9,396	50,243	233,189	0	0	0	0	0	110,499	63,051	9,396	50,243	233,189	22
56	25	100	72						73	56	25	100	72	
2,254	0	0	48,528	0	0	0	0	0	46,274	2,254	0	0	48,528	23
16			55						57	16			55	
200,386	8,825	32,452	368,417	0	0	0	0	0	126,754	200,386	8,825	32,452	368,417	24
79	20	96	79						79	79	20	96	79	
903,241	723,227	2,018,557	4,004,998	0	0	730	0	730	359,973	903,241	723,957	2,018,557	4,005,728	25
45	46	25	37			0		0	63	45	46	25	37	
204,729	401,306	393,049	1,117,285	0	351	0	0	351	118,201	205,080	401,306	393,049	1,117,636	26
18	24	6	17		0			0	28	18	24	6	17	
100,369	15,050	0	191,776	0	0	0	0	0	76,357	100,369	15,050	0	191,776	27
6	89		18						20	6	89		18	
632,697	330,938	0	3,247,688	22,708	0	0	0	22,708	2,306,661	632,697	330,938	0	3,270,296	28
89	21		54	65				65	49	89	21		54	
40,880	4,000	717,000	1,208,272	0	0	0	0	0	446,392	40,880	4,000	717,000	1,208,272	29
100	100	0	38						92	100	100	0	38	
174,590	7,549,923	697,910	8,422,423	0	0	0	0	0	0	174,590	7,549,923	697,910	8,422,423	30
100	21	55	25						0	100	21	55	25	
6,965,353	11,887,850	8,639,743	38,628,857	27,551	77,831	169,986	26,874	302,242	11,163,462	7,043,184	12,057,836	8,666,617	38,931,099	
58	26	39	44	57	58	93	100	81	57	58	27	39	44	

133

Table 172 Sewage treatment works. Total dry weather flow of sewage effluent and percentage discharged from sewage treatment works considered satisfactory (Volumes x 1,000 gpd)

River Authority		Non-tidal rivers				Total non-tidal	Tidal rivers				Total tidal	All rivers
		Class 1	Class 2	Class 3	Class 4		Class 1	Class 2	Class 3	Class 4		Class 1
1 Northumbrian	DWF	13,238	9,872	8,239	1,051	32,400	590	297	658	1,266	2,811	13,828
	%	57	67	14	9	48	97	39	100	0	48	59
2 Yorkshire	DWF	77,257	22,007	44,550	83,446	227,260	77	1,879	10,277	82	12,315	77,334
	%	31	43	8	26	26	100	100	47	100	56	31
3 Trent	DWF	18,871	107,469	33,334	167,825	327,499	0	94	0	497	591	18,871
	%	55	73	50	62	64		77		44	49	55
4 Lincolnshire	DWF	10,660	4,684	375	0	15,719	16	0	15	0	31	10,676
	%	88	92	7		87	6		100		50	88
5 Welland and Nene	DWF	22,263	2,755	43	0	25,061	0	5,046	0	0	5,046	22,263
	%	65	62	0		65		89			89	65
6 Great Ouse	DWF	24,308	20,279	3,257	0	47,844	822	0	0	0	822	25,130
	%	47	42	67		46	100				100	49
7 East Suffolk and Norfolk	DWF	2,703	1,652	0	0	4,355	1,689	16,165	0	0	17,854	4,392
	%	77	66			73	100	10			18	86
8 Essex	DWF	8,145	6,716	4,866	209	19,936	3,547	6,261	4,799	59	14,666	11,692
	%	68	93	45	0	70	51	13	74	100	43	62
9 Kent	DWF	19,504	0	0	0	19,504	11,029	1,031	5,690	1,456	19,206	30,533
	%	65				65	84	43	95	100	86	72
10 Sussex	DWF	9,265	3,014	55	3	12,337	3,556	0	0	0	3,556	12,821
	%	53	72	100	100	57	14				14	42
11 Hampshire	DWF	6,404	615	297	0	7,316	6,026	12,894	351	3,799	23,070	12,430
	%	38	98	1		42	100	5	0	0	29	68
12 Isle of Wight	DWF	483	4	0	0	487	110	0	781	0	891	593
	%	42	100			42	100		0		12	52
13 Avon and Dorset	DWF	12,495	140	9	0	12,644	5,766	0	0	0	5,766	18,261
	%	79	39	100		79	100				100	86
14 Devon	DWF	6,540	48	311	0	6,899	314	1,334	5,693	0	7,341	6,854
	%	54	39	68		55	91	25	0		9	55
15 Cornwall	DWF	8,198	512	134	1,122	9,966	621	948	0	0	1,569	8,819
	%	79	93	55	85	80	92	0			36	80
16 Somerset	DWF	10,979	4,343	30	0	15,352	0	0	0	0	0	10,979
	%	31	3	0		23						31
17 Bristol Avon	DWF	6,921	10,454	545	0	17,920	0	0	0	59	59	6,921
	%	42	45	9		42				0	0	42
18 Severn	DWF	65,644	12,538	12,240	31,766	122,188	5,209	0	0	0	5,209	70,853
	%	49	50	100	59	57	3				3	46
19 Wye	DWF	3,324	72	5,582	0	8,978	107	0	0	0	107	3,431
	%	62	35	1		24	100				100	63
20 Usk	DWF	2,523	0	8	0	2,531	3,924	171	597	0	4,692	6,447
	%	79		0		79	0	100	0		4	31
21 Glamorgan	DWF	1,855	3,821	19,868	891	26,435	0	0	0	0	0	1,855
	%	42	19	2	12	8						42
22 South West Wales	DWF	4,755	1,726	288	0	6,769	655	2,448	92	1,954	5,149	5,410
	%	75	98	0		78	58	29	100	100	61	73
23 Gwynedd	DWF	1,460	87	0	0	1,547	388	0	0	0	388	1,848
	%	59	12			56	21				21	50
24 Dee and Clwyd	DWF	4,018	4,117	429	1,191	9,755	675	4,889	0	0	5,564	4,693
	%	88	76	6	97	80	13	91			81	77
25 Mersey and Weaver	DWF	20,825	49,239	40,329	138,386	248,779	0	0	2,611	6,381	8,992	20,825
	%	53	50	42	18	31			71	89	84	53
26 Lancashire	DWF	7,531	13,378	8,904	23,108	52,921	963	145	15,968	115	17,191	8,494
	%	29	13	43	3	16	30	1	14	47	15	29
27 Cumberland	DWF	4,640	6,222	1,377	0	12,239	42	17	0	0	59	4,682
	%	42	3	95		28	7	0			5	41
28 Thames Conservancy	DWF	124,801	32,310	17,361	0	174,472	0	0	0	0	0	124,801
	%	47	90	16		52						47
29 Lee Conservancy	DWF	28,029	1,678	100	40,198	70,005	0	0	0	0	0	28,029
	%	91	100	100	0	39						91
30 PLA (incl. London Excluded Area)	DWF	0	0	0	36,975	36,975	0	7,965	466,819	0	474,784	0
	%				49	49		100	20		21	
Total England and Wales	DWF	527,639	319,752	202,531	526,171	1,576,093	46,126	61,584	514,351	15,668	637,729	573,765
	%	52	61	32	36	46	62	38	22	61	27	53

	Class 3	Class 4	Total all rivers	Canals				Total canals	All rivers and canals				Total all rivers and canals	
2				Class 1	Class 2	Class 3	Class 4		Class 1	Class 2	Class 3	Class 4		
69	8,897	2,317	35,211	0	0	0	0	0	13,828	10,169	8,897	2,317	35,211	1
66	20	4	47						59	66	20	4	47	
86	54,827	83,528	239,575	0	0	0	841	841	77,334	23,886	54,827	84,369	240,416	2
48	15	26	28				100	100	31	48	15	27	28	
63	33,334	168,322	328,090	84	5,232	10,956	0	16,272	18,955	112,795	44,290	168,322	344,362	3
73	50	62	64	0	42	96		78	55	71	61	62	65	
84	390	0	15,750	23	0	0	0	23	10,699	4,684	390	0	15,773	4
92	10		87	100				100	88	92	10		87	
01	43	0	30,107	96	0	0	0	96	22,359	7,801	43	0	30,203	5
79	0		69	5				5	65	79	0		69	
79	3,257	0	48,666	0	0	0	0	0	25,130	20,279	3,257	0	48,666	6
42	67		47						49	42	67		47	
17	0	0	22,209	0	0	0	0	0	4,392	17,817	0	0	22,209	7
15			29						86	15			29	
77	9,665	268	34,602	0	0	0	0	0	11,692	12,977	9,665	268	34,602	8
55	60	22	58						62	55	60	22	58	
31	5,690	1,456	38,710	0	0	0	0	0	30,533	1,031	5,690	1,456	38,710	9
43	95	100	76						72	43	95	100	76	
14	55	3	15,893	0	0	0	0	0	12,821	3,014	55	3	15,893	10
72	100	100	48						42	72	100	100	48	
09	648	3,799	30,386	0	0	0	0	0	12,430	13,509	648	3,799	30,386	11
9	0	0	32						68	9	0	0	32	
4	781	0	1,378	0	0	0	0	0	593	4	781	0	1,378	12
00	0		23						52	100	0		23	
40	9	0	18,410	0	0	0	0	0	18,261	140	9	0	18,410	13
39	100		85						86	39	100		85	
82	6,004	0	14,240	0	0	0	0	0	6,854	1,382	6,004	0	14,240	14
25	2		31						55	25	2		31	
60	134	1,122	11,535	0	0	0	0	0	8,819	1,460	134	1,122	11,535	15
32	55	85	74						80	32	55	85	74	
43	30	0	15,352	0	0	0	0	0	10,979	4,343	30	0	15,352	16
3	0		23						31	3	0		23	
54	545	59	17,979	0	0	0	0	0	6,921	10,454	545	59	17,979	17
45	9	0	42						42	45	9	0	42	
38	12,240	31,766	127,397	0	240	0	0	240	70,853	12,778	12,240	31,766	127,637	18
50	100	59	55		100			100	46	51	100	59	55	
72	5,582	0	9,085	0	0	0	0	0	3,431	72	5,582	0	9,085	19
35	1		25						63	35	1		25	
71	605	0	7,223	0	0	0	0	0	6,447	171	605	0	7,223	20
00	98		30						31	100	98		30	
21	19,868	891	26,435	0	0	0	0	0	1,855	3,821	19,868	891	26,435	21
19	2	12	8						42	19	2	12	8	
74	380	1,954	11,918	0	0	0	0	0	5,410	4,174	380	1,954	11,918	22
58	24	100	70						73	58	24	100	70	
37	0	0	1,935	0	0	0	0	0	1,848	87	0	0	1,935	23
12			48						50	12			48	
06	429	1,191	15,319	0	0	0	0	0	4,693	9,006	429	1,191	15,319	24
34	6	97	81						77	84	6	97	81	
39	42,940	144,767	257,771	0	0	25	0	25	20,825	49,239	42,965	144,767	257,796	25
50	44	21	33			0		0	53	50	44	21	33	
23	24,872	23,223	70,112	0	15	0	0	15	8,494	13,538	24,872	23,223	70,127	26
13	24	3	14		0			0	29	13	24	3	14	
39	1,377	0	12,298	0	0	0	0	0	4,682	6,239	1,377	0	12,298	27
3	95		28						41	3	95		28	
0	17,361	0	174,472	853	0	0	0	853	125,654	32,310	17,361	0	175,325	28
90	16		52	60				60	47	90	16		52	
78	100	40,198	70,005	0	0	0	0	0	28,029	1,678	100	40,198	70,005	29
00	100	0	39						91	100	100	0	39	
55	466,819	36,975	511,759	0	0	0	0	0	0	7,965	466,819	36,975	511,759	30
00	20	49	23						0	100	20	49	23	
6	716,882	541,839	2,213,822	1,056	5,487	10,981	841	18,365	574,821	386,823	727,863	542,680	2,232,187	
7	25	37	40	51	44	95	100	78	53	57	26	37	40	

135

Table 173 Sewage treatment works. Total number of discharges required to comply with Royal Commission Standard and percentage considered satisfactory

River Authority		Non-tidal rivers					Tidal rivers					All rivers
		Class 1	Class 2	Class 3	Class 4	Total non-tidal	Class 1	Class 2	Class 3	Class 4	Total tidal	Class 1
1 Northumbrian	No.	143	40	10	3	196	3	2	0	0	5	146
	%	73	78	40	67	72	67	50			60	73
2 Yorkshire	No.	273	39	37	54	403	1	1	1	1	4	274
	%	69	51	51	46	63	100	100	0	100	75	69
3 Trent	No.	188	133	59	64	444	0	2	0	10	12	188
	%	56	64	56	39	56		50		50	50	56
4 Lincolnshire	No.	90	5	3	0	98	0	0	1	0	1	90
	%	86	60	67		84			100		100	86
5 Welland and Nene	No.	115	21	1	0	137	0	1	0	0	1	115
	%	46	29	0		43		100			100	46
6 Great Ouse	No.	150	72	8	0	230	9	0	0	0	9	159
	%	71	64	50		68	100				100	73
7 East Suffolk and Norfolk	No.	59	12	0	0	71	4	12	0	0	16	63
	%	81	83			82	100	83			88	83
8 Essex	No.	72	15	4	1	92	13	3	7	0	23	85
	%	71	73	75	0	71	77	67	43		65	72
9 Kent	No.	107	0	0	0	107	6	2	7	1	16	113
	%	67				67	83	0	57	100	63	68
10 Sussex	No.	74	19	1	1	95	11	0	0	0	11	85
	%	65	58	100	100	64	45				45	62
11 Hampshire	No.	20	8	2	0	30	2	4	1	0	7	22
	%	80	88	0		77	100	25	0		43	82
12 Isle of Wight	No.	10	0	0	0	10	1	0	1	0	2	11
	%	50				50	100		0		50	55
13 Avon and Dorset	No.	62	10	0	0	72	6	0	0	0	6	68
	%	87	80			86	100				100	88
14 Devon	No.	106	4	2	0	112	2	0	4	0	6	108
	%	66	75	100		67	50		0		17	66
15 Cornwall	No.	71	5	7	7	90	7	2	0	0	9	78
	%	68	60	43	57	64	71	0			56	68
16 Somerset	No.	52	8	0	0	60	0	0	0	0	0	52
	%	71	38			67						71
17 Bristol Avon	No.	52	25	3	0	80	0	0	0	0	0	52
	%	65	68	33		65						65
18 Severn	No.	273	27	1	6	307	2	0	0	0	2	275
	%	66	56	100	50	64	50				50	65
19 Wye	No.	46	2	2	0	50	1	0	0	0	1	47
	%	89	50	50		86	100				100	89
20 Usk	No.	24	0	1	0	25	0	1	0	0	1	24
	%	88		0		84		100			100	88
21 Glamorgan	No.	18	7	8	5	38	0	0	0	0	0	18
	%	44	14	13	40	32						44
22 South West Wales	No.	80	6	3	0	89	10	2	3	0	15	90
	%	69	50	0		65	90	100	100		93	71
23 Gwynedd	No.	32	2	0	0	34	4	0	0	0	4	36
	%	63	50			62	75				75	64
24 Dee and Clwydd	No.	70	19	2	3	94	1	4	0	0	5	71
	%	74	47	100	67	69	0	50			40	73
25 Mersey and Weaver	No.	56	63	21	35	175	0	0	5	3	8	56
	%	55	52	43	46	51			60	67	63	55
26 Lancashire	No.	49	28	23	19	119	3	1	3	1	8	52
	%	39	36	30	5	31	100	0	100	100	88	42
27 Cumberland	No.	77	8	4	0	89	3	1	0	0	4	80
	%	19	13	25		19	0	0			0	19
28 Thames Conservancy	No.	289	22	9	0	320	0	0	0	0	0	289
	%	75	73	56		74						75
29 Lee Conservancy	No.	11	0	0	2	13	0	0	0	0	0	11
	%	91			0	77						91
30 PLA (incl. London Excluded Area)	No.	0	0	0	4	4	0	0	10	0	10	0
	%				25	25			20		20	
Total England and Wales	No.	2,669	600	211	204	3,684	89	38	43	16	186	2,758
	%	67	59	47	40	63	76	58	44	63	64	67

	Class 3	Class 4	Total all rivers	Canals Class 1	Class 2	Class 3	Class 4	Total canals	All rivers and canals Class 1	Class 2	Class 3	Class 4	Total all rivers and canals	
	10	3	201	0	0	0	0	0	146	42	10	3	201	1
	40	*67*	*72*						*73*	*76*	*40*	*67*	*72*	
	38	55	407	0	0	0	2	2	274	40	38	57	409	2
	50	*47*	*63*				*100*	*100*	*69*	*53*	*50*	*49*	*63*	
	59	74	456	1	5	7	0	13	189	140	66	74	469	3
	56	*41*	*56*	*0*	*60*	*43*		*46*	*56*	*64*	*55*	*41*	*56*	
	4	0	99	1	0	0	0	1	91	5	4	0	100	4
	75		*84*	*100*				*100*	*86*	*60*	*75*		*84*	
	1	0	138	2	0	0	0	2	117	22	1	0	140	5
	0		*43*	*50*				*50*	*46*	*32*	*0*		*44*	
	8	0	239	0	0	0	0	0	159	72	8	0	239	6
	50		*69*						*73*	*64*	*50*		*69*	
	0	0	87	0	0	0	0	0	63	24	0	0	87	7
			83						*83*	*83*			*83*	
	11	1	115	0	0	0	0	0	85	18	11	1	115	8
	55	*0*	*70*						*72*	*72*	*55*	*0*	*70*	
	7	1	123	0	0	0	0	0	113	2	7	1	123	9
	57	*100*	*67*						*68*	*0*	*57*	*100*	*67*	
	1	1	106	0	0	0	0	0	85	19	1	1	106	10
	100	*100*	*62*						*62*	*58*	*100*	*100*	*62*	
	3	0	37	0	0	0	0	0	22	12	3	0	37	11
	0		*70*						*82*	*67*	*0*		*70*	
	1	0	12	0	0	0	0	0	11	0	1	0	12	12
	0		*50*						*55*		*0*		*50*	
	0	0	78	0	0	0	0	0	68	10	0	0	78	13
			87						*88*	*80*			*87*	
	6	0	118	0	0	0	0	0	108	4	6	0	118	14
	33		*64*						*66*	*75*	*33*		*64*	
	7	7	99	0	0	0	0	0	78	7	7	7	99	15
	43	*57*	*64*						*68*	*43*	*43*	*57*	*64*	
	0	0	60	0	0	0	0	0	52	8	0	0	60	16
			67						*71*	*38*			*67*	
	3	0	80	0	0	0	0	0	52	25	3	0	80	17
	33		*65*						*65*	*68*	*33*		*65*	
	1	6	309	0	1	0	0	1	275	28	1	6	310	18
	100	*50*	*64*		*100*			*100*	*65*	*57*	*100*	*50*	*64*	
	2	0	51	0	0	0	0	0	47	2	2	0	51	19
	50		*86*						*89*	*50*	*50*		*86*	
	1	0	26	0	0	0	0	0	24	1	1	0	26	20
	0		*85*						*88*	*100*	*0*		*85*	
	8	5	38	0	0	0	0	0	18	7	8	5	38	21
	13	*40*	*32*						*44*	*14*	*13*	*40*	*32*	
	6	0	104	0	0	0	0	0	90	8	6	0	104	22
	50		*69*						*71*	*62*	*50*		*69*	
	0	0	38	0	0	0	0	0	36	2	0	0	38	23
			63						*64*	*50*			*63*	
	2	3	99	0	0	0	0	0	71	23	2	3	99	24
	100	*67*	*68*						*73*	*48*	*100*	*67*	*68*	
	26	38	183	0	0	1	0	1	56	63	27	38	184	25
	46	*47*	*51*			*0*		*0*	*55*	*52*	*44*	*47*	*51*	
	26	20	127	0	1	0	0	1	52	30	26	20	128	26
	38	*10*	*35*		*0*			*0*	*42*	*33*	*38*	*10*	*34*	
	4	0	93	0	0	0	0	0	80	9	4	0	93	27
	25		*18*						*19*	*11*	*25*		*18*	
	9	0	320	5	0	0	0	5	294	22	9	0	325	28
	56		*74*	*80*				*80*	*75*	*73*	*56*		*74*	
	0	2	13	0	0	0	0	0	11	0	0	2	13	29
		0	*77*						*91*			*0*	*77*	
	10	4	14	0	0	0	0	0	0	0	10	4	14	30
	20	*25*	*21*								*20*	*25*	*21*	
	254	220	3,870	9	7	8	2	26	2,767	645	262	222	3,896	
	46	*42*	*63*	*67*	*57*	*38*	*100*	*58*	*69*	*59*	*46*	*42*	*63*	

137

Table 174 Sewage treatment works. Total number of discharges required to comply with more stringent than Royal Commission Standard and percent considered satisfactory

River Authority			Non-tidal rivers					Tidal rivers					All rivers
			Class 1	Class 2	Class 3	Class 4	Total non-tidal	Class 1	Class 2	Class 3	Class 4	Total tidal	Class 1
1	Northumbrian	No.	1	0	0	0	1	0	0	0	0	0	1
		%	100				100						100
2	Yorkshire	No.	9	0	0	0	9	0	0	0	0	0	9
		%	67				67						67
3	Trent	No.	0	2	0	0	2	0	0	0	0	0	0
		%		50			50						
4	Lincolnshire	No.	16	1	0	0	17	0	0	0	0	0	16
		%	81	100			82						81
5	Welland and Nene	No.	6	0	0	0	6	0	0	0	0	0	6
		%	83				83						83
6	Great Ouse	No.	15	9	0	0	24	0	0	0	0	0	15
		%	87	89			88						87
7	East Suffolk and Norfolk	No.	7	5	0	0	12	0	3	0	0	3	7
		%	100	60			83		67			67	100
8	Essex	No.	27	8	4	0	39	0	0	0	0	0	27
		%	89	88	50		85						89
9	Kent	No.	27	0	0	0	27	0	0	0	0	0	27
		%	85				85						85
10	Sussex	No.	8	1	0	0	9	0	0	0	0	0	8
		%	100	100			100						100
11	Hampshire	No.	2	0	1	0	3	0	0	0	0	0	2
		%	50		0		33						50
12	Isle of Wight	No.	0	1	0	0	1	0	0	0	0	0	0
		%		100			100						
13	Avon and Dorset	No.	5	0	1	0	6	0	0	0	0	0	5
		%	80		100		83						80
14	Devon	No.	10	0	1	0	11	0	0	1	0	1	10
		%	30		0		27			0		0	30
15	Cornwall	No.	1	0	0	1	2	0	0	0	0	0	1
		%	100			100	100						100
16	Somerset	No.	39	4	1	0	44	0	0	0	0	0	39
		%	59	25	0		55						59
17	Bristol Avon	No.	3	3	0	0	6	0	0	0	0	0	3
		%	67	67			67						67
18	Severn	No.	27	0	0	7	34	0	0	0	0	0	27
		%	67			57	65						67
19	Wye	No.	1	1	0	0	2	0	0	0	0	0	1
		%	0	100			50						0
20	Usk	No.	1	0	0	0	1	0	0	0	0	0	1
		%	0				0						0
21	Glamorgan	No.	0	0	1	0	1	0	0	0	0	0	0
		%			0		0						
22	South West Wales	No.	4	0	0	0	4	0	0	0	0	0	4
		%	75				75						75
23	Gwynedd	No.	0	0	0	0	0	0	0	0	0	0	0
		%											
24	Dee and Clwyd	No.	2	2	1	1	6	0	0	0	0	0	2
		%	100	50	0	0	50						100
25	Mersey and Weaver	No.	1	1	0	0	2	0	0	0	0	0	1
		%	100	100			100						100
26	Lancashire	No.	0	3	0	0	3	0	0	0	0	0	0
		%		67			67						
27	Cumberland	No.	0	0	0	0	0	0	0	0	0	0	0
		%											
28	Thames Conservancy	No.	15	7	3	0	25	0	0	0	0	0	15
		%	87	86	33		80						87
29	Lee Conservancy	No.	16	2	1	0	19	0	0	0	0	0	16
		%	88	100	100		89						88
30	PLA (incl. London Excluded Area)	No.	0	0	0	1	1	0	0	3	0	3	0
		%				100	100			67		67	
	Total England and Wales	No.	243	50	14	10	317	0	3	4	0	7	243
		%	76	76	36	60	74		67	50		57	76

138

	Class 3	Class 4	Total all rivers	Canals Class 1	Class 2	Class 3	Class 4	Total canals	All rivers and canals Class 1	Class 2	Class 3	Class 4	Total all rivers and canals	
	0	0	1 *100*	0	0	0	0	0	1 *100*	0	0	0	1 *100*	1
	0	0	9 *67*	0	0	0	0	0	9 *67*	0	0	0	9 *67*	2
	0	0	2 *50*	0	0	0	0	0	0	2 *50*	0	0	2 *50*	3
	0	0	17 *82*	0	0	0	0	0	16 *81*	1 *100*	0	0	17 *82*	4
	0	0	6 *83*	0	0	0	0	0	6 *83*	0	0	0	6 *83*	5
	0	0	24 *88*	0	0	0	0	0	15 *87*	9 *89*	0	0	24 *88*	6
	0	0	15 *80*	0	0	0	0	0	7 *100*	8 *63*	0	0	15 *80*	7
	4 *50*	0	39 *85*	0	0	0	0	0	27 *89*	8 *88*	4 *50*	0	39 *84*	8
	0	0	27 *85*	0	0	0	0	0	27 *85*	0	0	0	27 *85*	9
	0	0	9 *100*	0	0	0	0	0	8 *100*	1 *100*	0	0	9 *100*	10
	1 *0*	0	3 *33*	0	0	0	0	0	2 *50*	0	1 *0*	0	3 *33*	11
	0	0	1 *100*	0	0	0	0	0	0	1 *100*	0	0	1 *100*	12
	1 *100*	0	6 *83*	0	0	0	0	0	5 *80*	0	1 *100*	0	6 *83*	13
	2 *0*	0	12 *25*	0	0	0	0	0	10 *30*	0	2 *0*	0	12 *25*	14
	0	1 *100*	2 *100*	0	0	0	0	0	1 *100*	0	0	1 *100*	2 *100*	15
	1 *0*	0	44 *55*	0	0	0	0	0	39 *59*	4 *25*	1 *0*	0	44 *55*	16
	0	0	6 *67*	0	0	0	0	0	3 *67*	3 *67*	0	0	6 *67*	17
	0	7 *57*	34 *65*	0	0	0	0	0	27 *67*	0	0	7 *57*	34 *65*	18
	0	0	2 *50*	0	0	0	0	0	1 *0*	1 *100*	0	0	2 *50*	19
	0	0	1 *0*	0	0	0	0	0	1 *0*	0	0	0	1 *0*	20
	1 *0*	0	1 *0*	0	0	0	0	0	0	0	1 *0*	0	1 *0*	21
	0	0	4 *75*	0	0	0	0	0	4 *75*	0	0	0	4 *75*	22
	0	0	0	0	0	0	0	0	0	0	0	0	0	23
	1 *0*	1 *0*	6 *50*	0	0	0	0	0	2 *100*	2 *50*	1 *0*	1 *0*	6 *50*	24
	0	0	2 *100*	0	0	0	0	0	1 *100*	1 *100*	0	0	2 *100*	25
	0	0	3 *67*	0	0	0	0	0	0	3 *67*	0	0	3 *67*	26
	0	0	0	0	0	0	0	0	0	0	0	0	0	27
	3 *33*	0	25 *80*	0	0	0	0	0	15 *87*	7 *86*	3 *33*	0	25 *80*	28
	1 *100*	0	19 *89*	0	0	0	0	0	16 *88*	2 *100*	1 *100*	0	19 *89*	29
	3 *67*	1 *100*	4 *75*	0	0	0	0	0	0	0	3 *67*	1 *100*	4 *75*	30
	18 *39*	10 *60*	324 *73*	0	0	0	0	0	243 *76*	53 *75*	18 *39*	10 *60*	324 *73*	

Table 175 Sewage treatment works. Total number of discharges required to comply with less stringent than Royal Commission Standard and perce[ntage] considered satisfactory

River Authority			Non-tidal rivers					Tidal rivers					All riv[ers]
			Class 1	Class 2	Class 3	Class 4	Total non-tidal	Class 1	Class 2	Class 3	Class 4	Total tidal	Class
1	Northumbrian	No.	3	2	0	1	6	1	1	1	1	4	4
		%	33	0		0	17	100	100	100	0	75	50
2	Yorkshire	No.	1	0	0	0	1	2	1	9	1	13	3
		%	0				0	100	100	89	100	92	67
3	Trent	No.	0	1	0	11	12	0	0	0	0	0	0
		%		100		18	25						
4	Lincolnshire	No.	0	0	0	0	0	1	0	0	0	1	1
		%						0				0	0
5	Welland and Nene	No.	10	0	0	0	10	0	1	0	0	1	10
		%	40				40		0			0	40
6	Great Ouse	No.	0	0	0	0	0	0	0	0	0	0	0
		%											
7	East Suffolk and Norfolk	No.	0	0	0	0	0	9	4	0	0	13	9
		%						100	50			85	100
8	Essex	No.	0	0	0	0	0	3	0	0	1	4	3
		%						67			100	75	67
9	Kent	No.	0	0	0	0	0	4	1	0	1	6	4
		%						50	100		100	67	50
10	Sussex	No.	0	0	0	0	0	3	0	0	0	3	3
		%						67				67	67
11	Hampshire	No.	0	0	0	0	0	0	2	0	0	2	0
		%							50			50	
12	Isle of Wight	No.	0	0	0	0	0	0	0	0	0	0	0
		%											
13	Avon and Dorset	No.	1	0	0	0	1	0	0	0	0	0	1
		%	0				0						0
14	Devon	No.	0	0	0	0	0	2	3	1	0	6	2
		%						50	67	0		50	50
15	Cornwall	No.	0	0	0	0	0	0	0	0	0	0	0
		%											
16	Somerset	No.	0	0	0	0	0	0	0	0	0	0	0
		%											
17	Bristol Avon	No.	0	0	0	0	0	0	0	0	1	1	0
		%									0	0	
18	Severn	No.	1	0	0	0	1	2	0	0	0	2	3
		%	100				100	50				50	67
19	Wye	No.	0	0	0	0	0	1	0	0	0	1	1
		%						100				100	100
20	Usk	No.	1	0	0	0	1	1	1	1	0	3	2
		%	100				100	0	100	0		33	50
21	Glamorgan	No.	1	0	0	0	1	0	0	0	0	0	1
		%	100				100						100
22	South West Wales	No.	1	0	0	0	1	4	4	0	5	13	5
		%	0				0	25	50		100	62	20
23	Gwynedd	No.	1	0	0	0	1	4	0	0	0	4	5
		%	100				100	50				50	60
24	Dee and Clwyd	No.	0	0	0	0	0	4	0	0	0	4	4
		%						25				25	25
25	Mersey and Weaver	No.	0	0	0	1	1	0	0	1	1	2	0
		%				0	0			100	100	100	
26	Lancashire	No.	5	1	2	0	8	11	3	3	1	18	16
		%	80	100	0		63	27	0	0	0	17	44
27	Cumberland	No.	0	1	0	0	1	0	0	0	0	0	0
		%		100			100						
28	Thames Conservancy	No.	0	0	0	0	0	0	0	0	0	0	0
		%											
29	Lee Conservancy	No.	0	0	0	0	0	0	0	0	0	0	0
		%											
30	PLA (incl. London Excluded Area)	No.	0	0	0	0	0	0	1	6	0	7	0
		%							100	50		57	
	Total England and Wales	No.	25	5	2	13	45	52	22	22	12	108	77
		%	52	60	0	15	40	54	55	59	75	57	53

	Class 3	Class 4	Total all rivers	Canals					All rivers and canals					
				Class 1	Class 2	Class 3	Class 4	Total canals	Class 1	Class 2	Class 3	Class 4	Total all rivers and canals	
	1 100	2 0	10 40	0	0	0	0	0	4 50	3 33	1 100	2 0	10 40	1
	9 89	1 100	14 86	0	0	0	0	0	3 67	1 100	9 89	1 100	14 86	2
	0	11 18	12 25	0	0	0	0	0	0	1 100	0	11 18	12 25	3
	0	0	1 0	0	0	0	0	0	1 0	0	0	0	1 0	4
	0	0	11 36	0	0	0	0	0	10 40	1 0	0	0	11 36	5
	0	0	0	0	0	0	0	0	0	0	0	0	0	6
	0	0	13 85	0	0	0	0	0	9 100	4 50	0	0	13 85	7
	0	1 100	4 75	0	0	0	0	0	3 67	0	0	1 100	4 75	8
	0	1 100	6 67	0	0	0	0	0	4 50	1 100	0	1 100	6 67	9
	0	0	3 67	0	0	0	0	0	3 67	0	0	0	3 67	10
	0	0	2 50	0	0	0	0	0	0	2 50	0	0	2 50	11
	0	0	0	0	0	0	0	0	0	0	0	0	0	12
	0	0	1 0	0	0	0	0	0	1 0	0	0	0	1 0	13
	1 0	0	6 50	0	0	0	0	0	2 50	3 67	1 0	0	6 50	14
	0	0	0	0	0	0	0	0	0	0	0	0	0	15
	0	0	0	0	0	0	0	0	0	0	0	0	0	16
	0	1 0	1 0	0	0	0	0	0	0	0	0	1 0	1 0	17
	0	0	3 67	0	0	0	0	0	3 67	0	0	0	3 67	18
	0	0	1 100	0	0	0	0	0	1 100	0	0	0	1 100	19
	1 0	0	4 50	0	0	0	0	0	2 50	1 100	1 0	0	4 50	20
	0	0	1 100	0	0	0	0	0	1 100	0	0	0	1 100	21
	0	5 100	14 57	0	0	0	0	0	5 20	4 50	0	5 100	14 57	22
	0	0	5 60	0	0	0	0	0	5 60	0	0	0	5 60	23
	0	0	4 25	0	0	0	0	0	4 25	0	0	0	4 25	24
	1 100	2 50	3 67	0	0	0	0	0	0	0	1 100	2 50	3 67	25
	5 0	1 0	26 31	0	0	0	0	0	16 44	4 25	5 0	1 0	26 31	26
	0	0	1 100	0	0	0	0	0	0	1 100	0	0	1 100	27
	0	0	0	0	0	0	0	0	0	0	0	0	0	28
	0	0	0	0	0	0	0	0	0	0	0	0	0	29
	6 50	0	7 57	0	0	0	0	0	0	1 100	6 50	0	7 57	30
	24 54	25 44	153 52	0	0	0	0	0	77 53	27 55	24 54	25 44	153 52	

Table 176 Sewage treatment works. Total population served by treatment works required to comply with Royal Commission Standard and percentage s[cut] by works considered satisfactory

River Authority			Non-tidal rivers					Tidal rivers					All rive[cut]
			Class 1	Class 2	Class 3	Class 4	Total non-tidal	Class 1	Class 2	Class 3	Class 4	Total tidal	Class 1
1	Northumbrian	Popn. %	298,893 62	226,578 65	182,729 22	23,969 12	732,169 51	3,688 83	7,012 12	0	0	10,700 36	302,[cut]
2	Yorkshire	Popn. %	1,276,610 34	419,241 52	767,805 16	1,332,123 36	3,795,779 33	1,194 100	7,260 100	85 000 0	1,001 100	94,455 10	1,277,[cut]
3	Trent	Popn. %	407,896 59	1,052,018 81	698,111 49	2,183,762 67	4,341,787 66	0	2,090 70	0	12,305 41	14,395 46	407,[cut]
4	Lincolnshire	Popn. %	233,056 83	11,102 44	6,036 13	0	250,194 80	0	0	200 100	0	200 100	233,0[cut]
5	Welland and Nene	Popn. %	402,212 61	63,134 69	1,408 0	0	466,754 62	0	75,250 100	0	0	75,250 100	402,2[cut]
6	Great Ouse	Popn. %	406,076 46	293,364 34	71,175 60	0	770,615 43	16,497 100	0	0	0	16,497 100	422,5[cut]
7	East Suffolk and Norfolk	Popn. %	76,881 72	27,157 72	0	0	104,038 72	13,763 100	35,510 83	0	0	49,273 88	90,6[cut]
8	Essex	Popn. %	121,694 53	38,265 66	31,977 86	6,006 0	197,942 59	47,961 36	105,621 13	113,681 71	0	267,263 42	169,6[cut]
9	Kent	Popn. %	273,649 67	0	0	0	273,649 67	35,725 61	14,297 0	119,395 93	1,200 100	170,617 78	309,3[cut]
10	Sussex	Popn. %	184,276 45	61,410 60	1,200 100	300 100	247,186 49	31,413 15	0	0	0	31,413 15	215,6[cut]
11	Hampshire	Popn. %	100,657 43	17,772 98	2,118 0	0	120,547 51	91,017 100	187,031 0	4,760 0	0	282,808 32	191,6[cut]
12	Isle of Wight	Popn. %	8,169 47	0	0	0	8,169 47	743 100	0	21,340 0	0	22,083 3	8,9[cut]
13	Avon and Dorset	Popn. %	284,903 80	6,247 33	0	0	291,150 79	126,545 100	0	0	0	126,545 100	411,4[cut]
14	Devon	Popn. %	111,335 61	1,286 48	3,731 100	0	116,352 62	997 40	0	122,790 0	0	123,787 0	112,3[cut]
15	Cornwall	Popn. %	143,470 78	8,613 84	4,285 53	21,695 79	178,063 78	14,040 90	17,364 0	0	0	31,404 40	157,5[cut]
16	Somerset	Popn. %	51,659 64	11,535 14	0	0	63,194 55	0	0	0	0	0	51,6[cut]
17	Bristol Avon	Popn. %	142,151 49	191,822 40	9,330 11	0	343,303 43	0	0	0	0	0	142,1[cut]
18	Severn	Popn. %	869,459 42	226,389 46	175,000 100	199,704 17	1,470,552 46	2,740 32	0	0	0	2,740 32	872,19[cut]
19	Wye	Popn. %	60,615 81	929 23	47,800 1	0	109,344 46	2,380 100	0	0	0	2,380 100	62,99[cut]
20	Usk	Popn. %	41,138 93	0	220 0	0	41,358 93	0	200 100	0	0	200 100	41,13[cut]
21	Glamorgan	Popn. %	30,139 38	65,766 32	142,884 5	23,354 13	262,143 16	0	0	0	0	0	30,13[cut]
22	South West Wales	Popn. %	91,048 75	30,102 90	7,065 0	0	128,215 74	8,904 87	1,389 100	2,331 100	0	12,624 91	99,95[cut]
23	Gwynedd	Popn. %	36,228 57	2,254 16	0	0	38,482 55	3,306 78	0	0	0	3,306 78	39,53[cut]
24	Dee and Clwyd	Popn. %	108,053 87	76,239 74	1,785 100	31,390 99	217,467 85	2,841 0	99,112 89	0	0	101,953 87	110,89[cut]
25	Mersey and Weaver	Popn. %	345,553 62	850,541 41	673,676 46	1,861,022 21	3,730,792 34	0	0	32,611 31	27,535 24	60,146 28	345,55[cut]
26	Lancashire	Popn. %	93,358 25	82,141 34	152,529 44	390,541 6	718,569 20	4,100 100	249 0	27,662 100	1,350 100	33,361 99	97,458 28[cut]
27	Cumberland	Popn. %	75,060 20	95,252 1	15,050 89	0	185,362 16	1,297 0	500 0	0	0	1,797 0	76,357 2[cut]
28	Thames Conservancy	Popn. %	1,459,702 59	326,743 81	233,914 18	0	2,020,359 58	0	0	0	0	0	1,459,702 59
29	Lee Conservancy	Popn. %	28,597 55	0	0	717,000 0	745,597 2	0	0	0	0	0	28,597 55
30	PLA (incl. London Excluded Area)	Popn. %	0	0	0	358,460 13	358,460 13	0	0	4,249,206 1	0	4,249,206 1	0
	Total England and Wales	Popn. %	7,762,537 54	4,185,900 57	3,229,828 37	7,149,326 35	22,327,591 46	409,151 80	552,885 39	4,778,976 5	43,391 35	5,784,403 14	8,171,688 55

| | Class 3 | Class 4 | Total all rivers | Canals | | | | | All rivers and canals | | | | | |
				Class 1	Class 2	Class 3	Class 4	Total canals	Class 1	Class 2	Class 3	Class 4	Total all rivers and canals	
,590 *63*	182,729 *22*	23,969 *12*	742,869 *51*	0	0	0	0	0	302,581 *62*	233,590 *63*	182,729 *22*	23,969 *12*	742,869 *45*	1
,501 *53*	852,805 *14*	1,333,124 *37*	3,890,234 *33*	0	0	0	26,874 *100*	26,874 *100*	1,277,804 *34*	426,501 *53*	852,805 *14*	1,359,998 *38*	3,917,108 *33*	2
,108 *81*	698,111 *49*	2,196,067 *66*	4,356,182 *66*	3,002 *0*	75,380 *57*	169,256 *93*	0	247,638 *81*	410,898 *58*	1,129,488 *79*	867,367 *57*	2,196,067 *66*	4,603,820 *67*	3
,102 *44*	6,236 *16*	0	250,394 *80*	770 *100*	0	0	0	770 *100*	233,826 *83*	11,102 *44*	6,236 *16*	0	251,164 *80*	4
,384 *86*	1,408 *0*	0	542,004 *68*	1,071 *28*	0	0	0	1,071 *28*	403,283 *61*	138,384 *86*	1,408 *0*	0	543,075 *67*	5
,364 *34*	71,175 *60*	0	787,112 *44*	0	0	0	0	0	422,573 *48*	293,364 *34*	71,175 *60*	0	787,112 *44*	6
,667 *79*	0	0	153,311 *77*	0	0	0	0	0	90,644 *77*	62,667 *79*	0	0	153,311 *77*	7
,886 *27*	145,658 *74*	6,006 *0*	465,205 *49*	0	0	0	0	0	169,655 *48*	143,886 *27*	145,658 *74*	6,006 *0*	465,205 *49*	8
,297 *0*	119,395 *93*	1,200 *100*	444,266 *71*	0	0	0	0	0	309,374 *66*	14,297 *0*	119,395 *93*	1,200 *100*	444,266 *71*	9
,410 *60*	1,200 *100*	300 *100*	278,599 *45*	0	0	0	0	0	215,689 *40*	61,140 *60*	1,200 *100*	300 *100*	278,599 *45*	10
,803 *9*	6,878 *0*	0	403,355 *38*	0	0	0	0	0	191,674 *70*	204,803 *9*	6,878 *0*	0	403,355 *38*	11
0	21,340 *0*	0	30,252 *15*	0	0	0	0	0	8,912 *51*	0	21,340 *0*	0	30,252 *15*	12
,247 *33*	0	0	417,695 *85*	0	0	0	0	0	411,448 *86*	6,247 *33*	0	0	417,695 *85*	13
,286 *48*	126,521 *3*	0	240,139 *30*	0	0	0	0	0	112,332 *61*	1,286 *48*	126,521 *3*	0	240,139 *30*	14
,977 *28*	4,285 *53*	21,695 *79*	209,467 *72*	0	0	0	0	0	157,510 *79*	25,977 *28*	4,285 *53*	21,695 *79*	209,467 *72*	15
,535 *14*	0	0	63,194 *55*	0	0	0	0	0	51,659 *64*	11,535 *14*	0	0	63,194 *55*	16
,822 *40*	9,330 *11*	0	343,303 *43*	0	0	0	0	0	142,151 *49*	191,822 *40*	9,330 *11*	0	343,303 *43*	17
,389 *46*	175,000 *100*	199,704 *17*	1,473,292 *46*	0	2,100 *100*	0	0	2,100 *100*	872,199 *42*	228,489 *46*	175,000 *100*	199,704 *17*	1,475,392 *46*	18
929 *23*	47,800 *1*	0	111,724 *47*	0	0	0	0	0	62,995 *82*	929 *23*	47,800 *1*	0	111,724 *47*	19
200 *100*	220 *0*	0	41,558 *93*	0	0	0	0	0	41,138 *93*	200 *100*	220 *0*	0	41,558 *93*	20
,766 *32*	142,884 *5*	23,354 *13*	262,143 *16*	0	0	0	0	0	30,139 *38*	65,766 *32*	142,884 *5*	23,354 *13*	262,143 *16*	21
,491 *90*	9,396 *25*	0	140,839 *76*	0	0	0	0	0	99,952 *76*	31,491 *90*	9,396 *25*	0	140,839 *76*	22
2,254 *16*	0	0	41,788 *57*	0	0	0	0	0	39,534 *59*	2,254 *16*	0	0	41,788 *57*	23
,351 *83*	1,785 *100*	31,390 *99*	319,420 *85*	0	0	0	0	0	110,894 *85*	175,351 *83*	1,785 *100*	31,390 *99*	319,420 *85*	24
,541 *41*	706,287 *45*	1,888,557 *21*	3,790,938 *34*	0	0	730 *0*	0	730 *0*	345,553 *62*	850,541 *41*	707,017 *45*	1,888,557 *21*	3,791,668 *34*	25
,390 *34*	180,191 *53*	391,891 *6*	751,930 *23*	0	351 *0*	0	0	351 *0*	97,458 *28*	82,741 *34*	180,191 *53*	391,891 *6*	752,281 *23*	26
,752 *1*	15,050 *89*	0	187,159 *16*	0	0	0	0	0	76,357 *20*	95,752 *1*	15,050 *89*	0	187,159 *16*	27
,743 *81*	233,914 *18*	0	2,020,359 *58*	22,708 *65*	0	0	0	22,708 *65*	1,482,410 *59*	326,743 *81*	233,914 *18*	0	2,043,067 *58*	28
0	0	717,000 *0*	745,597 *2*	0	0	0	0	0	28,597 *55*	0	0	717,000 *0*	745,597 *2*	29
0	4,249,206 *1*	358,460 *13*	4,607,666 *1*	0	0	0	0	0	0	0	4,249,206 *1*	358,460 *13*	4,607,666 *1*	30
,785 *55*	8,008,804 *18*	7,192,717 *35*	28,111,994 *40*	27,551 *57*	77,831 *58*	169,986 *93*	26,874 *100*	302,242 *81*	8,199,239 *55*	4,816,616 *55*	8,178,790 *19*	7,219,591 *35*	28,414,236 *40*	

Table 177 Sewage treatment works. Total population served by treatment works required to comply with more stringent than Royal Commission Standard percentage served by works considered satisfactory

River Authority		Non-tidal rivers					Tidal rivers					All rivers
		Class 1	Class 2	Class 3	Class 4	Total non-tidal	Class 1	Class 2	Class 3	Class 4	Total tidal	Class 1
1 Northumbrian	Popn.	14,080	0	0	0	14,080	0	0	0	0	0	14,080
	%	100				100						100
2 Yorkshire	Popn.	79,438	0	0	0	79,438	0	0	0	0	0	79,438
	%	69				69						69
3 Trent	Popn.	0	196,990	0	0	196,990	0	0	0	0	0	
	%		3			3						
4 Lincolnshire	Popn.	73,342	80,100	0	0	153,442	0	0	0	0	0	73,342
	%	94	100			97						94
5 Welland and Nene	Popn.	12,991	0	0	0	12,991	0	0	0	0	0	12,991
	%	93				93						93
6 Great Ouse	Popn.	88,968	89,649	0	0	178,617	0	0	0	0	0	88,968
	%	89	95			92						89
7 East Suffolk and Norfolk	Popn.	8,364	13,967	0	0	22,331	0	4,162	0	0	4,162	8,364
	%	100	66			78		81			81	100
8 Essex	Popn.	95,945	130,804	75,047	0	301,796	0	0	0	0	0	95,945
	%	76	98	32		75						76
9 Kent	Popn.	140,615	0	0	0	140,615	0	0	0	0	0	140,615
	%	78				78						78
10 Sussex	Popn.	44,835	15,800	0	0	60,635	0	0	0	0	0	44,835
	%	100	100			100						100
11 Hampshire	Popn.	6,000	0	301	0	6,301	0	0	0	0	0	6,000
	%	30		0		29						30
12 Isle of Wight	Popn.	0	150	0	0	150	0	0	0	0	0	0
	%		100			100						
13 Avon and Dorset	Popn.	5,229	0	1,200	0	6,429	0	0	0	0	0	5,229
	%	72		100		77						72
14 Devon	Popn.	9,548	0	2,493	0	12,041	0	0	4,920	0	4,920	9,548
	%	16		0		13			0		0	16
15 Cornwall	Popn.	636	0	0	3,429	4,065	0	0	0	0	0	636
	%	100			100	100						100
16 Somerset	Popn.	113,646	45,049	900	0	159,595	0	0	0	0	0	113,646
	%	28	4	0		21						28
17 Bristol Avon	Popn.	5,684	2,094	0	0	7,778	0	0	0	0	0	5,684
	%	96	43			81						96
18 Severn	Popn.	152,353	0	0	356,824	509,177	0	0	0	0	0	152,353
	%	57			72	68						57
19 Wye	Popn.	1,140	504	0	0	1,644	0	0	0	0	0	1,140
	%	0	100			31						0
20 Usk	Popn.	6,000	0	0	0	6,000	0	0	0	0	0	6,000
	%	0				0						0
21 Glamorgan	Popn.	0	0	13,300	0	13,300	0	0	0	0	0	0
	%			0		0						
22 South West Wales	Popn.	2,373	0	0	0	2,373	0	0	0	0	0	2,373
	%	69				69						69
23 Gwynedd	Popn.	0	0	0	0	0	0	0	0	0	0	0
	%											
24 Dee and Clwyd	Popn.	2,090	25,035	7,040	1,062	35,227	0	0	0	0	0	2,090
	%	100	52	0	0	43						100
25 Mersey and Weaver	Popn.	14,420	52,700	0	0	67,120	0	0	0	0	0	14,420
	%	100	100			100						100
26 Lancashire	Popn.	0	118,299	0	0	118,299	0	0	0	0	0	0
	%		7			7						
27 Cumberland	Popn.	0	0	0	0	0	0	0	0	0	0	0
	%											
28 Thames Conservancy	Popn.	824,251	305,954	97,024	0	1,227,229	0	0	0	0	0	824,251
	%	30	98	27		47						30
29 Lee Conservancy	Popn.	417,795	40,880	4,000	0	462,675	0	0	0	0	0	417,795
	%	95	100	100		95						95
30 PLA (incl. London Excluded Area)	Popn.	0	0	0	339,450	339,450	0	0	3,107,800	0	3,107,800	0
	%				100	100			49		49	
Total England and Wales	Popn.	2,119,743	1,117,975	201,305	700,765	4,139,788	0	4,162	3,112,720	0	3,116,882	2,119,743
	%	59	66	27	86	64		81	48		48	59

	2	Class 3	Class 4	Total all rivers	Canals Class 1	Class 2	Class 3	Class 4	Total canals	All rivers and canals Class 1	Class 2	Class 3	Class 4	Total all rivers and canals	
	0	0	0	14,080	0	0	0	0	0	14,080	0	0	0	14,080	1
				100						100				100	
	0	0	0	79,438	0	0	0	0	0	79,438	0	0	0	79,438	2
				69						69				69	
	,990	0	0	196,990	0	0	0	0	0	0	196,990	0	0	196,990	3
	3			3							3			3	
	,100	0	0	153,442	0	0	0	0	0	73,342	80,100	0	0	153,442	4
	100			100						94	100			97	
	0	0	0	12,991	0	0	0	0	0	12,991	0	0	0	12,991	5
				93						93				93	
	,649	0	0	178,617	0	0	0	0	0	88,968	89,649	0	0	178,617	6
	95			92						89	95			92	
	8,129	0	0	26,493	0	0	0	0	0	8,364	18,129	0	0	26,493	7
	69			79						100	69			79	
	,804	75,047	0	301,796	0	0	0	0	0	95,945	130,804	75,047	0	301,796	8
	98	32		75						76	98	32		75	
	0	0	0	140,615	0	0	0	0	0	140,615	0	0	0	140,615	9
				78						78				78	
	,800	0	0	60,635	0	0	0	0	0	44,835	15,800	0	0	60,635	10
	100			100						100	100			100	
	0	301	0	6,301	0	0	0	0	0	6,000	0	301	0	6,301	11
		0		29						30		0		29	
	150	0	0	150	0	0	0	0	0	0	150	0	0	150	12
	100			100							100			100	
	0	1,200	0	6,429	0	0	0	0	0	5,229	0	1,200	0	6,429	13
		100		77						72		100		77	
	0	7,413	0	16,961	0	0	0	0	0	9,548	0	7,413	0	16,961	14
		0		9						16		0		9	
	0	0	3,429	4,065	0	0	0	0	0	636	0	0	3,429	4,065	15
			100	100						100			100	100	
	45,049	900	0	159,595	0	0	0	0	0	113,646	45,049	900	0	159,595	16
	4	0		21						28	4	0		21	
	2,094	0	0	7,778	0	0	0	0	0	5,684	2,094	0	0	7,778	17
	43			81						96	43			81	
	0	0	356,824	509,177	0	0	0	0	0	152,353	0	0	356,824	509,177	18
			72	68						57			72	68	
	504	0	0	1,644	0	0	0	0	0	1,140	504	0	0	1,644	19
	100			31						0	100			31	
	0	0	0	6,000	0	0	0	0	0	6,000	0	0	0	6,000	20
				0						0				0	
	0	13,300	0	13,300	0	0	0	0	0	0	0	13,300	0	13,300	21
		0		0								0		0	
	0	0	0	2,373	0	0	0	0	0	2,373	0	0	0	2,373	22
				69						69				69	
	0	0	0	0	0	0	0	0	0	0	0	0	0	0	23
	25,035	7,040	1,062	35,227	0	0	0	0	0	2,090	25,035	7,040	1,062	35,227	24
	52	0	0	43						100	52	0	0	43	
	52,700	0	0	67,120	0	0	0	0	0	14,420	52,700	0	0	67,120	25
	100			100						100	100			100	
	18,299	0	0	118,299	0	0	0	0	0	0	118,299	0	0	118,299	26
	7			7							7			7	
	0	0	0	0	0	0	0	0	0	0	0	0	0	0	27
	305,954	97,024	0	1,227,229	0	0	0	0	0	824,251	305,954	97,024	0	1,277,229	28
	98	27		47						30	98	27		47	
	40,880	4,000	0	462,675	0	0	0	0	0	417,795	40,880	4,000	0	462,675	29
	100	100		95						95	100	100		95	
	0	3,107,800	339,450	3,447,200	0	0	0	0	0	0	0	3,107,800	339,450	3,447,200	30
		49	100	54								49	100	54	
	122,137	3,314,025	700,765	7,256,670	0	0	0	0	0	2,119,743	1,122,137	3,314,025	700,765	7,256,670	
	66	47	86	57						59	66	47	86	57	

145

Table 178 Sewage treatment works. Total population served by treatment works required to comply with less stringent than Royal Commission Standard percentage considered satisfactory

River Authority		Non-tidal rivers					Tidal rivers					All rivers
		Class 1	Class 2	Class 3	Class 4	Total non-tidal	Class 1	Class 2	Class 3	Class 4	Total tidal	Class 1
1 Northumbrian	Popn. %	6,098 *3*	5,290 *0*	0	2,500 *0*	13,888 *1*	9,150 *100*	2,520 *100*	25,200 *100*	21,546 *0*	58,416 *63*	15,248 *6*
2 Yorkshire	Popn. %	225 *0*	0	0	0	225 *0*	1,421 *100*	0	79,831 *95*	1,775 *100*	83,027 *95*	1,646 *8*
3 Trent	Popn. %	0	460,000 *100*	0	427,662 *12*	887,662 *58*	0	0	0	0	0	
4 Lincolnshire	Popn. %	0	0	0	0	0	888 *0*	0	0	0	888 *0*	888
5 Welland and Nene	Popn. %	4,897 *43*	0	0	0	4,897 *43*	0	16,200 *0*	0	0	16,200 *0*	4,897 *43*
6 Great Ouse	Popn. %	0	0	0	0	0	0	0	0	0	0	0
7 East Suffolk and Norfolk	Popn. %	0	0	0	0	0	33,354 *100*	302,790 *9*	0	0	336,144 *18*	33,354 *100*
8 Essex	Popn. %	0	0	0	0	0	23,080 *70*	0	0	1,998 *100*	25,078 *72*	23,080 *70*
9 Kent	Popn. %	0	0	0	0	0	236,387 *97*	10,010 *100*	0	34,680 *100*	281,077 *97*	236,387 *97*
10 Sussex	Popn. %	0	0	0	0	0	22,816 *43*	0	0	0	22,816 *43*	22,816 *43*
11 Hampshire	Popn. %	0	0	0	0	0	0	70,800 *23*	0	0	70,800 *23*	0
12 Isle of Wight	Popn. %	0	0	0	0	0	0	0	0	0	0	0
13 Avon and Dorset	Popn. %	2,000 *0*	0	0	0	2,000 *0*	0	0	0	0	0	2,000 *0*
14 Devon	Popn. %	0	0	0	0	0	7,154 *97*	14,469 *57*	7,040 *0*	0	28,663 *53*	7,154 *97*
15 Cornwall	Popn. %	0	0	0	0	0	0	0	0	0	0	0
16 Somerset	Popn. %	0	0	0	0	0	0	0	0	0	0	0
17 Bristol Avon	Popn. %	0	0	0	0	0	0	0	0	1,199 *0*	1,199 *0*	0
18 Severn	Popn. %	276,500 *100*	0	0	0	276,500 *100*	71,478 *2*	0	0	0	71,478 *2*	347,978 *80*
19 Wye	Popn. %	0	0	0	0	0	525 *100*	0	0	0	525 *100*	525 *100*
20 Usk	Popn. %	280 *100*	0	0	0	280 *100*	93,280 *0*	5,475 *100*	16,218 *0*	0	114,973 *5*	93,560
21 Glamorgan	Popn. %	3,520 *100*	0	0	0	3,520 *100*	0	0	0	0	0	3,520 *100*
22 South West Wales	Popn. %	328 *0*	0	0	0	328 *0*	7,846 *29*	31,560 *21*	0	50,243 *100*	89,649 *66*	8,174 *28*
23 Gwynedd	Popn. %	2,550 *100*	0	0	0	2,550 *100*	4,190 *11*	0	0	0	4,190 *11*	6,740 *45*
24 Dee and Clwyd	Popn. %	0	0	0	0	0	13,770 *23*	0	0	0	13,770 *23*	13,770 *23*
25 Mersey and Weaver	Popn. %	0	0	0	22,000 *0*	22,000 *0*	0	0	16,940 *100*	108,000 *100*	124,940 *100*	0
26 Lancashire	Popn. %	2,048 *89*	740 *100*	1,629 *0*	0	4,417 *58*	18,695 *19*	3,300 *0*	219,486 *0*	1,158 *0*	242,639 *1*	20,743 *26*
27 Cumberland	Popn. %	0	4,617 *100*	0	0	4,617 *100*	0	0	0	0	0	0
28 Thames Conservancy	Popn. %	0	0	0	0	0	0	0	0	0	0	0
29 Lee Conservancy	Popn. %	0	0	0	0	0	0	0	0	0	0	0
30 PLA (incl. London Excluded Area)	Popn. %	0	0	0	0	0	0	174,590 *100*	192,917 *15*	0	367,507 *55*	0
Total England and Wales	Popn. %	298,446 *96*	470,647 *99*	1,629 *0*	452,162 *12*	1,222,884 *66*	544,034 *58*	631,714 *40*	557,632 *26*	220,599 *89*	1,953,979 *47*	842,480 *72*

	Class 3	Class 4	Total all rivers	Canals Class 1	Class 2	Class 3	Class 4	Total canals	All rivers and canals Class 1	Class 2	Class 3	Class 4	Total all rivers and canals	
,810 *32*	25,200 *100*	24,046 *0*	72,304 *51*	0	0	0	0	0	15,248 *61*	7,810 *32*	25,200 *100*	24,046 *0*	72,304 *51*	1
0	79,831 *95*	1,775 *100*	83,252 *95*	0	0	0	0	0	1,646 *86*	0	79,831 *95*	1,775 *100*	83,252 *95*	2
,000 *100*	0	427,662 *12*	887,662 *58*	0	0	0	0	0	0	460,000 *100*	0	427,667 *12*	887,662 *58*	3
0	0	0	888 *0*	0	0	0	0	0	888 *0*	0	0	0	888 *0*	4
,200 *0*	0	0	21,097 *10*	0	0	0	0	0	4,897 *43*	16,200 *0*	0	0	21,097 *10*	5
0	0	0	0	0	0	0	0	0	0	0	0	0	0	6
,790 *9*	0	0	336,144 *18*	0	0	0	0	0	33,354 *100*	302,790 *9*	0	0	336,144 *18*	7
0	0	1,998 *100*	25,078 *72*	0	0	0	0	0	23,080 *70*	0	0	1,998 *100*	24,078 *72*	8
,010 *100*	0	34,680 *100*	281,077 *97*	0	0	0	0	0	236,397 *97*	10,010 *100*	0	34,680 *100*	281,077 *97*	9
0	0	0	22,816 *43*	0	0	0	0	0	22,816 *43*	0	0	0	22,816 *43*	10
,800 *23*	0	0	70,800 *23*	0	0	0	0	0	0	70,800 *23*	0	0	70,800 *23*	11
0	0	0	0	0	0	0	0	0	0	0	0	0	0	12
0	0	0	2,000 *0*	0	0	0	0	0	2,000 *0*	0	0	0	2,000 *0*	13
4,469 *57*	7,040 *0*	0	28,663 *53*	0	0	0	0	0	7,154 *97*	14,469 *57*	7,040 *0*	0	28,663 *53*	14
0	0	0	0	0	0	0	0	0	0	0	0	0	0	15
0	0	0	0	0	0	0	0	0	0	0	0	0	0	16
0	0	1,199 *0*	1,199 *0*	0	0	0	0	0	0	0	0	1,199 *0*	1,199 *0*	17
0	0	0	347,978 *80*	0	0	0	0	0	347,978 *80*	0	0	0	347,978 *80*	18
0	0	0	525 *100*	0	0	0	0	0	525 *100*	0	0	0	525 *100*	19
5,475 *100*	16,218 *0*	0	115,253 *5*	0	0	0	0	0	93,560 *0*	5,475 *100*	16,218 *0*	0	115,253 *5*	20
0	0	0	3,520 *100*	0	0	0	0	0	3,520 *100*	0	0	0	3,520 *100*	21
31,560 *21*	0	50,243 *100*	89,977 *66*	0	0	0	0	0	8,174 *28*	31,560 *21*	0	50,243 *100*	89,977 *66*	22
0	0	0	6,740 *45*	0	0	0	0	0	6,740 *45*	0	0	0	6,740 *45*	23
0	0	0	13,770 *23*	0	0	0	0	0	13,770 *23*	0	0	0	13,770 *23*	24
0	16,940 *100*	130,000 *83*	146,940 *85*	0	0	0	0	0	0	0	16,940 *100*	130,000 *83*	146,940 *85*	25
4,040 *18*	221,115 *0*	1,158 *0*	247,056 *2*	0	0	0	0	0	20,743 *26*	4,040 *18*	221,115 *0*	1,158 *0*	247,056 *2*	26
4,617 *100*	0	0	4,617 *100*	0	0	0	0	0	0	4,617 *100*	0	0	4,617 *100*	27
0	0	0	0	0	0	0	0	0	0	0	0	0	0	28
0	0	0	0	0	0	0	0	0	0	0	0	0	0	29
74,590 *100*	192,917 *15*	0	367,507 *55*	0	0	0	0	0	0	174,590 *100*	192,917 *15*	0	367,507 *55*	30
02,361 *65*	559,261 *26*	672,761 *37*	3,176,863 *54*	0	0	0	0	0	842,980 *72*	1,102,361 *65*	559,261 *26*	672,761 *37*	3,176,863 *54*	

Table 179 Sewage treatment works. Total dry weather flow of sewage effluent required to comply with Royal Commission Standard and perce discharged from treatment works considered satisfactory (Volumes x 1,000 gpd)

	River Authority		Non-tidal rivers					Tidal rivers					All riv
			Class 1	Class 2	Class 3	Class 4	Total non-tidal	Class 1	Class 2	Class 3	Class 4	Total tidal	Class
1	Northumbrian	DWF	12,401	9,753	8,239	811	31,204	133	222	0	0	355	12,53
		%	55	68	14	11	47	89	18			44	5
2	Yorkshire	DWF	71,582	22,007	44,550	83,446	221,585	35	279	5,300	26	5,640	71,61
		%	29	43	8	26	25	100	100	0	100	6	2
3	Trent	DWF	18,871	55,100	33,334	141,381	248,686	0	94	0	497	591	18,87
		%	55	85	50	71	70		77		44	49	5
4	Lincolnshire	DWF	6,852	684	375	0	7,911	0	0	15	0	15	6,85
		%	84	45	7		77			100		100	8
5	Welland and Nene	DWF	21,601	2,755	43	0	24,399	0	4,473	0	0	4,473	21,60
		%	65	62	0		64		100			100	6
6	Great Ouse	DWF	20,871	15,873	3,257	0	40,001	822	0	0	0	822	21,69
		%	40	27	67		37	100				100	4
7	East Suffolk and Norfolk	DWF	2,435	1,021	0	0	3,456	377	1,207	0	0	1,584	2,81
		%	74	70			73	100	86			89	7
8	Essex	DWF	4,028	1,464	1,300	209	7,001	2,311	6,261	4,799	0	13,371	6,33
		%	58	75	86	0	65	36	13	74		39	5
9	Kent	DWF	12,629	0	0	0	12,629	2,365	592	5,690	24	8,671	14,99
		%	60				60	36	0	95	100	73	5
10	Sussex	DWF	8,186	2,397	55	3	10,641	2,240	0	0	0	2,240	10,42
		%	46	64	100	100	51	6				6	3
11	Hampshire	DWF	6,058	615	274	0	6,947	6,026	10,421	153	0	16,600	12,084
		%	37	98	0		41	100	0	0		36	6
12	Isle of Wight	DWF	483	0	0	0	483	110	0	781	0	891	59
		%	42				42	100		0		12	5
13	Avon and Dorset	DWF	12,197	140	0	0	12,337	5,766	0	0	0	5,766	17,963
		%	80	37			80	100				100	8
14	Devon	DWF	6,136	48	211	0	6,395	32	0	5,276	0	5,308	6,168
		%	57	38	100		58	38		0		0	56
15	Cornwall	DWF	8,179	512	134	1,002	9,827	621	948	0	0	1,569	8,800
		%	79	93	54	83	79	92	0			36	80
16	Somerset	DWF	2,234	848	0	0	3,082	0	0	0	0	0	2,234
		%	63	7			48						63
17	Bristol Avon	DWF	6,677	10,391	545	0	17,613	0	0	0	0	0	6,677
		%	40	45	9		42						40
18	Severn	DWF	45,896	12,538	12,240	8,715	79,389	128	0	0	0	128	46,024
		%	35	50	100	20	46	27				27	35
19	Wye	DWF	3,285	57	5,582	0	8,924	83	0	0	0	83	3,368
		%	62	18	1		24	100				100	63
20	Usk	DWF	2,176	0	8	0	2,184	0	6	0	0	6	2,176
		%	91		0		91		100			100	91
21	Glamorgan	DWF	1,693	3,821	19,468	891	25,873	0	0	0	0	0	1,693
		%	36	19	2	12	7						36
22	South West Wales	DWF	4,630	1,726	288	0	6,644	336	46	92	0	474	4,966
		%	76	98	0		78	85	100	100		89	76
23	Gwynedd	DWF	1,342	87	0	0	1,429	87	0	0	0	87	1,429
		%	54	11			52	75				75	56
24	Dee and Clwyd	DWF	3,956	3,454	22	1,166	8,598	110	4,889	0	0	4,999	4,066
		%	87	79	100	99	86	0	91			89	85
25	Mersey and Weaver	DWF	19,723	45,736	40,329	137,094	242,882	0	0	1,182	973	2,155	19,723
		%	50	46	42	18	30			36	28	32	50
26	Lancashire	DWF	7,447	4,430	8,839	23,108	43,824	150	8	2,199	54	2,411	7,597
		%	29	30	43	3	18	100	0	100	100	100	30
27	Cumberland	DWF	4,640	6,040	1,377	0	12,057	42	17	0	0	59	4,682
		%	42	0	95		27	0	0			0	41
28	Thames Conservancy	DWF	79,438	15,160	13,907	0	108,505	0	0	0	0	0	79,438
		%	59	81	13		56						59
29	Lee Conservancy	DWF	1,799	0	0	40,198	41,997	0	0	0	0	0	1,799
		%	31			0	1						31
30	PLA (incl. London Excluded Area)	DWF	0	0	0	21,003	21,003	0	0	273,539	0	273,539	0
		%				10	10			0		0	
	Total England and Wales	DWF	397,445	216,657	194,377	459,027	1,267,506	21,774	29,463	299,026	1,574	351,837	419,219
		%	50	58	32	33	42	75	38	4	38	11	51

	Class 3	Class 4	Total all rivers	Canals				Total canals	All rivers and canals				Total all rivers and canals	
2				Class 1	Class 2	Class 3	Class 4		Class 1	Class 2	Class 3	Class 4		
975	8,239	811	31,559	0	0	0	0	0	12,534	9,975	8,239	811	31,559	1
67	14	11	47						56	67	14	11	47	
286	49,850	83,472	227,225	0	0	0	841	841	71,617	22,286	49,850	84,313	228,066	2
44	7	26	25				100	100	29	44	7	27	25	
194	33,334	141,878	249,277	84	5,232	10,956	0	16,272	18,955	60,426	44,290	141,878	265,549	3
85	50	71	70	0	42	96		78	55	81	61	71	70	
684	390	0	7,926	23	0	0	0	23	6,875	684	390	0	7,949	4
45	11		77	100				100	84	45	11		77	
228	43	0	28,872	96	0	0	0	96	21,697	7,228	43	0	28,968	5
86	0		70	4				4	64	86	0		69	
873	3,257	0	40,823	0	0	0	0	0	21,693	15,873	3,257	0	40,823	6
27	67		39						43	27	67		39	
228	0	0	5,040	0	0	0	0	0	2,812	2,228	0	0	5,040	7
79			78						77	79			78	
725	6,099	209	20,372	0	0	0	0	0	6,339	7,725	6,099	209	20,372	8
25	77	0	48						50	25	77	0	48	
592	5,690	24	21,300	0	0	0	0	0	14,994	592	5,690	24	21,300	9
0	95	100	65						56	0	95	100	65	
397	55	3	12,881	0	0	0	0	0	10,426	2,397	55	3	12,881	10
64	100	100	43						37	64	100	100	43	
036	427	0	23,547	0	0	0	0	0	12,084	11,036	427	0	23,547	11
5	0		38						68	5	0		38	
0	781	0	1,374	0	0	0	0	0	593	0	781	0	1,374	12
	0		23						52		0		23	
140	0	0	18,103	0	0	0	0	0	17,963	140	0	0	18,103	13
37			86						87	37			86	
48	5,487	0	11,703	0	0	0	0	0	6,168	48	5,487	0	11,703	14
38	4		32						56	38	4		32	
460	134	1,002	11,396	0	0	0	0	0	8,800	1,460	134	1,002	11,396	15
32	54	83	74						80	32	54	83	74	
848	0	0	3,082	0	0	0	0	0	2,234	848	0	0	3,082	16
7			48						63	7			48	
0,391	545	0	17,613	0	0	0	0	0	6,677	10,391	545	0	17,613	17
45	9		42						40	45	9		42	
2,538	12,240	8,715	79,517	0	240	0	0	240	46,024	12,778	12,240	8,715	79,757	18
50	100	20	46		100			100	35	51	100	20	46	
57	5,582	0	9,007	0	0	0	0	0	3,368	57	5582	0	9,007	19
18	1		25						63	18	1		25	
6	8	0	2,190	0	0	0	0	0	2,176	6	8	0	2,190	20
100	0		91						91	100	0		91	
3,821	19,468	891	25,873	0	0	0	0	0	1,693	3,821	19,468	891	25,873	21
19	2	12	7						36	19	2	12	7	
1,772	380	0	7,118	0	0	0	0	0	4,966	1,772	380	0	7,118	22
98	24		79						76	98	24		79	
87	0	0	1,516	0	0	0	0	0	1,429	87	0	0	1,516	23
11			53						56	11			53	
8,343	22	1,166	13,597	0	0	0	0	0	4,066	8,343	22	1,166	13,597	24
86	100	99	87						85	86	100	99	87	
45,736	41,511	138,067	245,037	0	0	25	0	25	19,723	45,736	41,536	138,067	245,062	25
46	42	18	30			0		0	50	46	42	18	30	
4,438	11,038	23,162	46,235	0	15	0	0	15	7,597	4,453	11,038	23,162	46,250	26
30	54	3	22		0			0	30	30	54	3	22	
6,057	1,377	0	12,116	0	0	0	0	0	4,682	6,057	1,377	0	12,116	27
0	95		27						41	0	95		27	
15,160	13,907	0	108,505	853	0	0	0	853	80,291	15,160	13,907	0	109,358	28
81	13		56	60				60	59	81	13		56	
0	0	40,198	41,997	0	0	0	0	0	1,799	0	0	40,198	41,997	29
		0	1						31			0	1	
0	273,539	21,003	294,542	0	0	0	0	0	0	0	273,539	21,003	294,542	30
	0	10	1								0	10	1	
46,120	493,403	460,601	1,619,343	1,056	5,487	10,981	841	18,365	420,275	251,607	504,384	461,442	1,637,708	
55	15	34	36	51	44	95	100	78	51	54	17	34	36	

149

Table 180 Sewage treatment works. Total dry weather flow of sewage effluent required to comply with more stringent than Royal Commission Standard: percentage discharged from treatment works considered satisfactory (Volumes x 1,000 gpd)

River Authority			Non-tidal rivers					Tidal rivers					All rivers
			Class 1	Class 2	Class 3	Class 4	Total non-tidal	Class 1	Class 2	Class 3	Class 4	Total tidal	Class 1
1	Northumbrian	DWF	659	0	0	0	659	0	0	0	0	0	659
		%	100				100						100
2	Yorkshire	DWF	5,666	0	0	0	5,666	0	0	0	0	0	5,666
		%	56				56						56
3	Trent	DWF	0	21,219	0	0	21,219	0	0	0	0	0	0
		%		1			1						
4	Lincolnshire	DWF	3,723	4,000	0	0	7,723	0	0	0	0	0	3,723
		%	95	100			98						95
5	Welland and Nene	DWF	481	0	0	0	481	0	0	0	0	0	481
		%	91				91						91
6	Great Ouse	DWF	3,437	4,406	0	0	7,843	0	0	0	0	0	3,437
		%	87	95			91						87
7	East Suffolk and Norfolk	DWF	268	631	0	0	899	0	124	0	0	124	268
		%	100	59			72		81			81	100
8	Essex	DWF	4,117	5,252	3,566	0	12,935	0	0	0	0	0	4,117
		%	76	98	30		73						76
9	Kent	DWF	6,875	0	0	0	6,875	0	0	0	0	0	6,875
		%	73				73						73
10	Sussex	DWF	1,079	617	0	0	1,696	0	0	0	0	0	1,079
		%	100	100			100						100
11	Hampshire	DWF	346	0	23	0	369	0	0	0	0	0	346
		%	57		0		53						57
12	Isle of Wight	DWF	0	4	0	0	4	0	0	0	0	0	0
		%		100			100						
13	Avon and Dorset	DWF	232	0	9	0	241	0	0	0	0	0	232
		%	53		100		55						53
14	Devon	DWF	404	0	100	0	504	0	0	196	0	196	404
		%	12		0		10			0		0	12
15	Cornwall	DWF	19	0	0	120	139	0	0	0	0	0	19
		%	100			100	100						100
16	Somerset	DWF	8,745	3,495	30	0	12,270	0	0	0	0	0	8,745
		%	23	1	0		17						23
17	Bristol Avon	DWF	244	63	0	0	307	0	0	0	0	0	244
		%	98	40			86						98
18	Severn	DWF	7,568	0	0	23,051	30,619	0	0	0	0	0	7,568
		%	53			74	69						53
19	Wye	DWF	39	15	0	0	54	0	0	0	0	0	39
		%	0	100			28						0
20	Usk	DWF	340	0	0	0	340	0	0	0	0	0	340
		%	0				0						0
21	Glamorgan	DWF	0	0	400	0	400	0	0	0	0	0	0
		%			0		0						
22	South West Wales	DWF	104	0	0	0	104	0	0	0	0	0	104
		%	66				66						66
23	Gwynedd	DWF	0	0	0	0	0	0	0	0	0	0	0
		%											
24	Dee and Clwyd	DWF	62	663	407	25	1,157	0	0	0	0	0	62
		%	100	60	0	0	40						100
25	Mersey and Weaver	DWF	1,102	3,503	0	0	4,605	0	0	0	0	0	1,102
		%	100	100			100						100
26	Lancashire	DWF	0	8,924	0	0	8,924	0	0	0	0	0	0
		%		5			5						
27	Cumberland	DWF	0	0	0	0	0	0	0	0	0	0	0
		%											
28	Thames Conservancy	DWF	45,363	17,150	3,454	0	65,967	0	0	0	0	0	45,363
		%	26	98	26		45						26
29	Lee Conservancy	DWF	26,230	1,678	100	0	28,008	0	0	0	0	0	26,230
		%	95	100	100		96						95
30	PLA (incl. London Excluded Area)	DWF	0	0	0	15,972	15,972	0	0	186,347	0	186,347	0
		%				100	100			49		49	
	Total England and Wales	DWF	117,103	71,620	8,089	39,168	235,980	0	124	186,543	0	186,667	117,103
		%	56	52	26	84	58		81	49		49	56

150

2	Class 3	Class 4	Total all rivers	Canals Class 1	Class 2	Class 3	Class 4	Total canals	All rivers and canals Class 1	Class 2	Class 3	Class 4	Total all rivers and canals	
0	0	0	659 *100*	0	0	0	0	0	659 *100*	0	0	0	659 *100*	1
0	0	0	5,666 *56*	0	0	0	0	0	5,666 *56*	0	0	0	5,666 *56*	2
19 *1*	0	0	21,219 *1*	0	0	0	0	0	0	21,219 *1*	0	0	21,219 *1*	3
00 *00*	0	0	7,723 *98*	0	0	0	0	0	3,723 *95*	4,000 *100*	0	0	7,723 *98*	4
0	0	0	481 *91*	0	0	0	0	0	481 *91*	0	0	0	481 *91*	5
06 *95*	0	0	7,843 *91*	0	0	0	0	0	3,437 *87*	4,406 *95*	0	0	7,843 *91*	6
55 *63*	0	0	1,023 *73*	0	0	0	0	0	268 *100*	755 *63*	0	0	1,023 *73*	7
52 *98*	3,566 *30*	0	12,935 *73*	0	0	0	0	0	4,117 *76*	5,252 *98*	3,566 *30*	0	12,935 *73*	8
0	0	0	6,875 *73*	0	0	0	0	0	6,875 *73*	0	0	0	6,875 *73*	9
17 *00*	0	0	1,696 *100*	0	0	0	0	0	1,079 *100*	617 *100*	0	0	1,696 *100*	10
0	23 *0*	0	369 *53*	0	0	0	0	0	346 *57*	0	23 *0*	0	369 *53*	11
4 *00*	0	0	4 *100*	0	0	0	0	0	0	4 *100*	0	0	4 *100*	12
0	9 *100*	0	241 *55*	0	0	0	0	0	232 *53*	0	9 *100*	0	241 *55*	13
0	296 *0*	0	700 *7*	0	0	0	0	0	404 *12*	0	296 *0*	0	700 *7*	14
0	0	120 *100*	139 *100*	0	0	0	0	0	19 *100*	0	0	120 *100*	139 *100*	15
495 *1*	30 *0*	0	12,270 *17*	0	0	0	0	0	8,745 *23*	3,495 *1*	30 *0*	0	12,270 *17*	16
63 *40*	0	0	307 *86*	0	0	0	0	0	244 *98*	63 *40*	0	0	307 *86*	17
0	0	23,051 *74*	30,619 *69*	0	0	0	0	0	7,568 *53*	0	0	23,051 *74*	30,619 *69*	18
15 *100*	0	0	54 *28*	0	0	0	0	0	39 *0*	15 *100*	0	0	54 *28*	19
0	0	0	340 *0*	0	0	0	0	0	340 *0*	0	0	0	340 *0*	20
0	400 *0*	0	400 *0*	0	0	0	0	0	0	0	400 *0*	0	400 *0*	21
0	0	0	104 *66*	0	0	0	0	0	104 *66*	0	0	0	104 *66*	22
0	0	0	0	0	0	0	0	0	0	0	0	0	0	23
663 *60*	407 *0*	25 *0*	1,157 *40*	0	0	0	0	0	62 *100*	663 *60*	407 *0*	25 *0*	1,157 *40*	24
503 *100*	0	0	4,605 *100*	0	0	0	0	0	1,102 *100*	3,503 *100*	0	0	4,605 *100*	25
924 *5*	0	0	8,924 *5*	0	0	0	0	0	0	8,924 *5*	0	0	8,924 *5*	26
0	0	0	0	0	0	0	0	0	0	0	0	0	0	27
,150 *98*	3,454 *26*	0	65,967 *45*	0	0	0	0	0	45,363 *26*	17,150 *98*	3,454 *26*	0	65,967 *45*	28
,678 *100*	100 *100*	0	28,008 *96*	0	0	0	0	0	26,230 *95*	1,678 *100*	100 *100*	0	28,008 *96*	29
0	186,347 *49*	15,972 *100*	202,319 *53*	0	0	0	0	0	0	0	186,347 *49*	15,972 *100*	202,319 *53*	30
,744 *52*	194,632 *48*	39,168 *84*	422,647 *54*	0	0	0	0	0	117,103 *56*	71,744 *52*	194,632 *48*	39,168 *84*	422,647 *54*	

Table 181 Sewage treatment works. Total dry weather flow of sewage effluent required to comply with less stringent than Royal Commission and percent discharged from treatment works considered satisfactory (Volume x 1,000 gpd)

River Authority		Non-tidal rivers					Tidal rivers					All rivers
		Class 1	Class 2	Class 3	Class 4	Total non-tidal	Class 1	Class 2	Class 3	Class 4	Total tidal	Class 1
1 Northumbrian	DWF	178	119	0	240	537	457	75	658	1,266	2,456	635
	%	3	0		0	1	100	100	100	0	48	73
2 Yorkshire	DWF	9	0	0	0	9	42	1,600	4,977	56	6,675	51
	%	0				0	100	100	97	100	98	82
3 Trent	DWF	0	31,150	0	26,444	57,594	0	0	0	0	0	0
	%		100		15	61						
4 Lincolnshire	DWF	0	0	0	0	0	16	0	0	0	16	16
	%						0				0	0
5 Welland and Nene	DWF	181	0	0	0	181	0	573	0	0	573	181
	%	45				45		0			0	45
6 Great Ouse	DWF	0	0	0	0	0	0	0	0	0	0	0
	%											
7 East Suffolk and Norfolk	DWF	0	0	0	0	0	1,312	14,834	0	0	16,146	1,312
	%						100	3			11	100
8 Essex	DWF	0	0	0	0	0	1,236	0	0	59	1,295	1,236
	%						79			100	80	79
9 Kent	DWF	0	0	0	0	0	8,664	439	0	1,432	10,535	8,664
	%						97	100		100	98	97
10 Sussex	DWF	0	0	0	0	0	1,316	0	0	0	1,316	1,316
	%						28				28	28
11 Hampshire	DWF	0	0	0	0	0	0	2,394	0	0	2,394	0
	%							21			21	
12 Isle of Wight	DWF	0	0	0	0	0	0	0	0	0	0	0
	%											
13 Avon and Dorset	DWF	66	0	0	0	66	0	0	0	0	0	66
	%	0				0						0
14 Devon	DWF	0	0	0	0	0	282	1,334	221	0	1,837	282
	%						98	34	0		33	98
15 Cornwall	DWF	0	0	0	0	0	0	0	0	0	0	0
	%											
16 Somerset	DWF	0	0	0	0	0	0	0	0	0	0	0
	%											
17 Bristol Avon	DWF	0	0	0	0	0	0	0	0	59	59	0
	%									0	0	
18 Severn	DWF	12,180	0	0	0	12,180	5,081	0	0	0	5,081	17,261
	%	100				100	2				2	71
19 Wye	DWF	0	0	0	0	0	24	0	0	0	24	24
	%						100				190	100
20 Usk	DWF	7	0	0	0	7	3,924	165	597	0	4,686	3,931
	%	100				100	0	100	0		4	*
21 Glamorgan	DWF	162	0	0	0	162	0	0	0	0	0	162
	%	100				100						100
22 South West Wales	DWF	21	0	0	0	21	319	2,402	0	1,954	4,675	340
	%	0				0	29	28		100	58	27
23 Gwynedd	DWF	118	0	0	0	118	301	0	0	0	301	419
	%	100				100	4				4	31
24 Dee and Clwyd	DWF	0	0	0	0	0	565	0	0	0	565	565
	%						16				16	16
25 Mersey and Weaver	DWF	0	0	0	1,292	1,292	0	0	1,429	5,408	6,837	0
	%				0	0			100	100	100	
26 Lancashire	DWF	84	24	65	0	173	813	137	13,769	61	14,780	897
	%	83	100	0		54	17	0	0	0	1	23
27 Cumberland	DWF	0	182	0	0	182	0	0	0	0	0	0
	%		100			100						
28 Thames Conservancy	DWF	0	0	0	0	0	0	0	0	0	0	0
	%											
29 Lee Conservancy	DWF	0	0	0	0	0	0	0	0	0	0	0
	%											
30 PLA (incl. London Excluded Area)	DWF	0	0	0	0	0	0	7,965	6,933	0	14,898	0
	%							100	14		60	
Total England and Wales	DWF	13,006	31,475	65	27,976	72,522	24,352	31,918	28,584	10,295	95,149	37,358
	%	97	100	0	14	66	50	38	27	87	43	67

* 0.2%

	Class 2	Class 3	Class 4	Total all rivers	Canals					All rivers and canals					
					Class 1	Class 2	Class 3	Class 4	Total canals	Class 1	Class 2	Class 3	Class 4	Total all rivers and canals	
	194	658	1,506	2,993	0	0	0	0	0	635	194	658	1,506	2,993	1
	39	100	0	40						73	39	100	0	40	
	1,600	4,977	56	6,684	0	0	0	0	0	51	1,600	4,977	56	6,684	2
	100	97	100	97						82	100	97	100	97	
	31,150	0	26,444	57,594	0	0	0	0	0	0	31,150	0	26,444	57,594	3
	100		15	61							100		15	61	
	0	0	0	16	0	0	0	0	0	16	0	0	0	16	4
				0						0				0	
	573	0	0	754	0	0	0	0	0	181	573	0	0	754	5
	0			11						45	0			11	
	0	0	0	0	0	0	0	0	0	0	0	0	0	0	6
	14,834	0	0	16,146	0	0	0	0	0	1,312	14,834	0	0	16,146	7
	3			11						100	3			11	
	0	0	59	1,295	0	0	0	0	0	1,236	0	0	59	1,295	8
			100	80						79			100	80	
	439	0	1,432	10,535	0	0	0	0	0	8,664	439	0	1,432	10,535	9
	100		100	98						97	100		100	98	
	0	0	0	1,316	0	0	0	0	0	1,316	0	0	0	1,316	10
				28						28				28	
	2,394	0	0	2,394	0	0	0	0	0	0	2,394	0	0	2,394	11
	21			21							21			21	
	0	0	0	0	0	0	0	0	0	0	0	0	0	0	12
	0	0	0	66	0	0	0	0	0	66	0	0	0	66	13
				0						0				0	
	1,334	221	0	1,837	0	0	0	0	0	282	1,334	221	0	1,837	14
	34	0		33						98	34	0		33	
	0	0	0	0	0	0	0	0	0	0	0	0	0	0	15
	0	0	0	0	0	0	0	0	0	0	0	0	0	0	16
	0	0	59	59	0	0	0	0	0	0	0	0	59	59	17
			0	0									0	0	
	0	0	0	17,261	0	0	0	0	0	17,261	0	0	0	17,261	18
				71						71				71	
	0	0	0	24	0	0	0	0	0	24	0	0	0	24	19
				100						100				100	
	165	597	0	4,693	0	0	0	0	0	3,931	165	597	0	4,693	20
	100	0		4						*	100	0		4	
	0	0	0	162	0	0	0	0	0	162	0	0	0	162	21
				100						100				100	
	2,402	0	1,954	4,696	0	0	0	0	0	340	2,402	0	1,954	4,696	22
	28		100	58						27	28		100	58	
	0	0	0	419	0	0	0	0	0	419	0	0	0	419	23
				31						31				31	
	0	0	0	565	0	0	0	0	0	565	0	0	0	565	24
				16						16				16	
	0	1,429	6,700	8,129	0	0	0	0	0	0	0	1,429	6,700	8,129	25
		100	81	84								100	81	84	
	161	13,834	61	14,953	0	0	0	0	0	897	161	13,834	61	14,953	26
	15	0	0	2						23	15	0	0	2	
	182	0	0	182	0	0	0	0	0	0	182	0	0	182	27
	100			100							100			100	
	0	0	0	0	0	0	0	0	0	0	0	0	0	0	28
	0	0	0	0	0	0	0	0	0	0	0	0	0	0	29
	7,965	6,933	0	14,898	0	0	0	0	0	0	7,965	6,933	0	14,898	30
	100	14		60							100	14		60	
	63,393	28,649	38,271	167,671	0	0	0	0	0	37,358	63,393	28,649	38,271	167,671	
	69	27	34	53						67	69	27	34	53	

RP—F*

Table 182 Sewage treatment works by population ranges and percentage satisfactory (population up to 50,000)

River Authority			Non-tidal rivers					Tidal rivers					All rivers
			1,000 and Under	1,001 to 5,000	5,001 to 10,000	10,001 to 50,000	Total to 50,000	1,000 and Under	1,001 to 5,000	5,001 to 10,000	10,001 to 50,000	Total to 50,000	1,000 and Under
1	Northumbrian	Popn.	109	59	13	21	202	2	3	2	2	9	111
		%	76	73	54	52	71	50	100	50	50	67	76
2	Yorkshire	Popn.	180	122	44	54	400	5	6	1	4	16	185
		%	72	63	48	46	63	100	83	100	100	94	72
3	Trent	Popn.	161	169	46	62	438	8	4	0	0	12	169
		%	58	50	61	56	55	50	50			50	58
4	Lincolnshire	Popn.	53	50	5	7	115	2	0	0	0	2	55
		%	91	78	60	86	83	50				50	89
5	Welland and Nene	Popn.	93	47	6	6	152	0	0	0	1	1	93
		%	42	49	33	50	44				0	0	42
6	Great Ouse	Popn.	125	97	15	14	251	4	4	1	0	9	129
		%	74	71	53	50	70	100	100	100		100	75
7	East Suffolk and Norfolk	Popn.	45	33	5	0	83	13	10	4	3	30	58
		%	82	85	60		82	85	100	75	100	90	83
8	Essex	Popn.	60	44	13	13	130	9	8	3	5	25	69
		%	82	70	69	62	75	78	75	67	40	68	81
9	Kent	Popn.	63	51	13	7	134	3	7	5	5	20	66
		%	68	73	85	57	71	67	43	60	80	60	68
10	Sussex	Popn.	59	31	4	10	104	8	3	1	2	14	67
		%	71	65	50	60	67	50	67	100	0	50	69
11	Hampshire	Popn.	17	11	3	2	33	2	3	1	3	9	19
		%	76	73	67	50	73	100	67	0	33	56	79
12	Isle of Wight	Popn.	6	5	0	0	11	1	0	0	1	2	7
		%	67	40			55	100			0	50	71
13	Avon and Dorset	Popn.	36	32	3	8	79	2	2	0	1	5	38
		%	89	78	100	88	85	100	100		100	100	89
14	Devon	Pop.	85	35	2	1	123	4	3	4	1	12	89
		%	69	51	0	100	63	50	0	50	0	33	68
15	Cornwall	Popn.	61	25	3	2	91	5	2	0	2	9	66
		%	66	60	67	100	65	60	50		50	56	65
16	Somerset	Popn.	64	30	8	2	104	0	0	0	0	0	64
		%	72	53	25	0	62						72
17	Bristol Avon	Popn.	40	29	11	5	85	0	1	0	0	1	40
		%	70	62	64	60	66		0			0	70
18	Severn	Popn.	188	94	21	28	331	1	2	0	0	3	189
		%	75	56	33	54	65	100	50			67	75
19	Wye	Popn.	36	11	4	1	52	1	1	0	0	2	37
		%	92	73	75	0	85	100	100			100	92
20	Usk	Popn.	17	7	2	1	27	1	0	1	1	3	18
		%	82	86	50	100	81	100		100	0	67	83
21	Glamorgan	Popn.	13	14	4	8	39	0	0	0	0	0	13
		%	46	29	50	13	33						46
22	South West Wales	Popn.	70	20	1	3	94	14	8	3	3	28	84
		%	70	40	100	100	65	86	63	100	67	79	73
23	Gwynedd	Popn.	23	12	0	0	35	6	2	0	0	8	29
		%	65	58			63	67	50			63	65
24	Dee and Clwyd	Popn.	60	26	7	7	100	1	4	2	1	8	61
		%	63	73	71	86	68	0	25	0	100	25	62
25	Mersey and Weaver	Popn.	45	46	21	49	161	2	3	1	3	9	47
		%	67	52	38	45	52	50	100	100	33	67	66
26	Lancashire	Popn.	63	34	11	19	127	10	12	1	1	24	73
		%	43	29	18	26	35	40	42	0	100	42	42
27	Cumberland	Popn.	68	18	1	2	89	4	0	0	0	4	72
		%	18	22	100	50	20	0				0	17
28	Thames Conservancy	Popn.	159	101	23	47	330	0	0	0	0	0	159
		%	84	71	61	64	76						84
29	Lee Conservancy	Popn.	13	11	0	4	28	0	0	0	0	0	13
		%	100	91		50	89						100
30	PLA (incl. London Excluded Area)	Popn.	0	0	0	1	1	2	2	1	5	10	2
		%				100	100	50	100	0	40	50	50
	Total England and Wales	Popn.	2,012	1,264	289	384	3,949	110	90	31	44	275	2,122
		%	69	62	53	54	64	67	66	61	55	64	69

				Canals					All rivers and canals					
01 ... 00	5,001 to 10,000	10,001 to 50,000	Total to 50,000	1,000 and Under	1,001 to 5,000	5,001 to 10,000	10,001 to 50,000	Total to 50,000	1,000 and Under	1,001 to 5,000	5,001 to 10,000	10,001 to 50,000	Total to 50,000	
62	15	23	211	0	0	0	0	0	111	62	15	23	211	1
74	53	52	71						76	74	53	52	71	
28	45	58	416	0	0	1	1	2	185	128	46	59	418	2
64	49	50	64			100	100	100	72	64	50	51	64	
73	46	62	450	2	2	2	5	11	171	175	48	67	461	3
50	61	56	55	0	0	50	60	36	57	50	60	57	55	
50	5	7	117	1	0	0	0	1	56	50	5	7	118	4
78	60	86	83	100				100	89	78	60	86	83	
47	6	7	153	2	0	0	0	2	95	47	6	7	155	5
49	33	43	44	50				50	42	49	33	43	44	
101	16	14	260	0	0	0	0	0	129	101	16	14	260	6
72	56	50	72						75	72	56	50	72	
43	9	3	113	0	0	0	0	0	58	43	9	3	113	7
88	67	100	84						83	88	67	100	84	
52	16	18	155	0	0	0	0	0	69	52	16	18	155	8
71	69	56	74						81	71	69	56	74	
58	18	12	154	0	0	0	0	0	66	58	18	12	154	9
69	78	67	69						68	69	78	67	69	
34	5	12	118	0	0	0	0	0	67	34	5	12	118	10
65	60	50	65						69	65	60	50	65	
14	4	5	42	0	0	0	0	0	19	14	4	5	42	11
71	50	40	69						79	71	50	40	69	
5	0	1	13	0	0	0	0	0	7	5	0	1	13	12
40		0	54						71	40		0	54	
34	3	9	84	0	0	0	0	0	38	34	3	9	84	13
79	100	89	86						89	79	100	89	86	
38	6	2	135	0	0	0	0	0	89	38	6	2	135	14
47	33	50	61						68	47	33	50	61	
27	3	4	100	0	0	0	0	0	66	27	3	4	100	15
59	67	75	64						65	59	67	75	64	
30	8	2	104	0	0	0	0	0	64	30	8	2	104	16
53	25	0	62						72	53	25	0	62	
30	11	5	86	0	0	0	0	0	40	30	11	5	86	17
60	64	60	65						70	60	64	60	65	
96	21	28	334	0	1	0	0	1	189	97	21	28	335	18
56	33	54	65		100			100	75	57	33	54	65	
12	4	1	54	0	0	0	0	0	37	12	4	1	54	19
75	75	0	85						92	75	75	0	85	
7	3	2	30	0	0	0	0	0	18	7	3	2	30	20
86	67	50	80						83	86	67	50	80	
14	4	8	39	0	0	0	0	0	13	14	4	8	39	21
29	50	13	33						46	29	50	13	33	
28	4	6	122	0	0	0	0	0	84	28	4	56	122	22
46	100	83	68						73	46	100	83	68	
14	0	0	43	0	0	0	0	0	29	14	0	0	43	23
57			63						65	57			63	
30	9	8	108	0	0	0	0	0	61	30	9	8	108	24
67	56	88	65						62	67	56	88	65	
49	22	52	170	1	0	0	0	1	48	49	22	52	171	25
55	41	44	43	0				0	65	55	41	44	53	
46	12	20	151	1	0	0	0	1	74	46	12	20	152	26
33	17	30	36	0				0	42	33	17	30	36	
18	1	2	93	0	0	0	0	0	72	18	1	2	93	27
22	100	50	19						17	22	100	50	19	
101	23	47	330	2	1	1	1	5	161	102	24	48	335	28
71	61	64	76	100	100	0	100	80	84	72	58	65	76	
11	0	4	28	0	0	0	0	0	13	11	0	4	28	29
91		50	89						100	91		50	89	
2	1	6	11	0	0	0	0	0	2	2	1	6	11	30
100	0	50	54						50	100	0	50	54	
1,354	320	428	4,224	9	4	4	7	24	2,131	1,358	324	435	4,248	
62	54	54	64	44	50	50	71	54	69	62	54	54	64	

Table 183 Sewage treatment works by population ranges and percentage satisfactory (population over 50,000)

River Authority			Non-tidal rivers				Tidal rivers			
			50,001 to 100,000	100,001 to 500,000	Over 500,000	Total Over 50,000	50,001 to 100,000	100,001 to 500,000	Over 500,000	Total Over 50,000
1	Northumbrian	No.	1	0	0	1	0	0	0	0
		%	0			0				
2	Yorkshire	No.	7	5	1	13	1	0	0	1
		%	71	20	0	46	0			0
3	Trent	No.	12	7	1	20	0	0	0	0
		%	50	43	100	50				
4	Lincolnshire	No.	1	0	0	1	0	0	0	0
		%	100			100				
5	Welland and Nene	No.	0	1	0	1	1	0	0	1
		%		100		100	100			100
6	Great Ouse	No.	2	1	0	3	0	0	0	0
		%	50	0		33				0
7	East Suffolk and Norfolk	No.	0	0	0	0	0	2	0	2
		%						0		0
8	Essex	No.	1	0	0	1	2	0	0	2
		%	100			100	50			50
9	Kent	No.	0	0	0	0	1	1	0	2
		%					100	100		100
10	Sussex	No.	0	0	0	0	0	0	0	0
		%								
11	Hampshire	No.	0	0	0	0	4	1	0	5
		%					25	0		20
12	Isle of Wight	No.	0	0	0	0	0	0	0	0
		%								
13	Avon and Dorset	No.	0	0	0	0	1	0	0	1
		%					100			100
14	Devon	No.	0	0	0	0	1	0	0	1
		%					0			0
15	Cornwall	No.	1	0	0	1	0	0	0	0
		%	100			100				
16	Somerset	No.	0	0	0	0	0	0	0	0
		%								
17	Bristol Avon	No.	1	0	0	1	0	0	0	0
		%	0			0				
18	Severn	No.	7	4	0	11	1	0	0	1
		%	29	75		45	0			0
19	Wye	No.	0	0	0	0	0	0	0	0
		%								
20	Usk	No.	0	0	0	0	1	0	0	1
		%					0			0
21	Glamorgan	No.	1	0	0	1	0	0	0	0
		%	0			0				
22	South West Wales	No.	0	0	0	0	0	0	0	0
		%								
23	Gwynedd	No.	0	0	0	0	0	0	0	0
		%								
24	Dee and Clwyd	No.	0	0	0	0	1	0	0	1
		%					100			100
25	Mersey and Weaver	No.	10	6	1	17	0	1	0	1
		%	50	33	0	41		100		100
26	Lancashire	No.	1	2	0	3	1	1	0	2
		%	0	0		0	0	0		0
27	Cumberland	No.	1	0	0	1	0	0	0	0
		%	0			0				
28	Thames Conservancy	No.	9	5	1	15	0	0	0	0
		%	33	80	0	47				
29	Lee Conservancy	No.	1	2	1	4	0	0	0	0
		%	0	100	0	50				
30	PLA (incl. London Excluded Area)	No.	2	2	0	4	4	2	4	10
		%	0	50		25	25	50	25	30
	Total England and Wales	No.	58	35	5	98	19	8	4	31
		%	43	49	20	44	37	38	25	35

rivers				Canals				All rivers and canals				
...001 to ...,000	100,001 to 500,000	Over 500,000	Total Over 50,000	50,001 to 100,000	100,001 to 500,000	Over 500,000	Total Over 50,000	50,001 to 100,000	100,001 to 500,000	Over 500,000	Total Over 50,000	
1 / 0	0	0	1 / 0	0	0	0	0	1 / 0	0	0	1 / 0	1
8 / 62	5 / 20	1 / 0	14 / 43	0	0	0	0	8 / 62	5 / 20	1 / 0	14 / 43	2
12 / 0	7 / 43	1 / 100	20 / 50	2 / 100	0	0	2 / 100	14 / 57	7 / 43	1 / 100	22 / 55	3
1 / 0	0	0	1 / 100	0	0	0	0	1 / 100	0	0	1 / 100	4
1 / 0	1 / 100	0	2 / 100	0	0	0	0	1 / 100	1 / 100	0	2 / 100	5
2 / 0	1 / 0	0	3 / 33	0	0	0	0	2 / 50	1 / 0	0	3 / 33	6
0	2 / 0	0	2 / 0	0	0	0	0	0	2 / 0	0	2 / 0	7
3 / 67	0	0	3 / 67	0	0	0	0	3 / 67	0	0	3 / 67	8
1 / 0	1 / 100	0	2 / 100	0	0	0	0	1 / 100	1 / 100	0	2 / 100	9
0	0	0	0	0	0	0	0	0	0	0	0	10
4 / 25	1 / 0	0	5 / 20	0	0	0	0	4 / 25	1 / 0	0	5 / 20	11
0	0	0	0	0	0	0	0	0	0	0	0	12
1 / 100	0	0	1 / 100	0	0	0	0	1 / 100	0	0	1 / 100	13
1 / 0	0	0	1 / 0	0	0	0	0	1 / 0	0	0	1 / 0	14
1 / 100	0	0	1 / 100	0	0	0	0	1 / 100	0	0	1 / 100	15
0	0	0	0	0	0	0	0	0	0	0	0	16
1 / 0	0	0	1 / 0	0	0	0	0	1 / 0	0	0	1 / 0	17
8 / 25	4 / 75	0	12 / 42	0	0	0	0	8 / 25	4 / 75	0	12 / 42	18
0	0	0	0	0	0	0	0	0	0	0	0	19
1 / 0	0	0	1 / 0	0	0	0	0	1 / 0	0	0	1 / 0	20
1 / 0	0	0	1 / 0	0	0	0	0	1 / 0	0	0	1 / 0	21
0	0	0	0	0	0	0	0	0	0	0	0	22
0	0	0	0	0	0	0	0	0	0	0	0	23
1 / 100	0	0	1 / 100	0	0	0	0	1 / 100	0	0	1 / 100	24
10 / 50	7 / 43	1 / 0	18 / 44	0	0	0	0	10 / 50	7 / 43	1 / 0	18 / 44	25
2 / 0	3 / 0	0	5 / 0	0	0	0	0	2 / 0	3 / 0	0	5 / 0	26
1 / 0	0	0	1 / 0	0	0	0	0	1 / 0	0	0	1 / 0	27
9 / 33	5 / 80	1 / 0	15 / 47	0	0	0	0	9 / 33	5 / 80	1 / 0	15 / 47	28
1 / 0	2 / 100	1 / 0	4 / 50	0	0	0	0	1 / 0	2 / 100	1 / 0	4 / 50	29
6 / 17	4 / 50	4 / 25	14 / 29	0	0	0	0	6 / 17	4 / 50	4 / 25	14 / 29	30
77 / 42	43 / 47	9 / 22	129 / 42	2 / 100	0	0	2 / 100	79 / 43	43 / 47	9 / 22	131 / 43	

Table 184 Sewage treatment works. 1980 situation and changes, 1970-1980

	River Authority	Estimated number of discharges 1980	Estimated population served 1980	Estimated dry weather flow 1980 (x 1,000 gpd)	Number of discharges likely to cease 1970-1980	Number of new discharges 1970-1980
1	Northumbrian	198	1,140,854	69,997	15	1
2	Yorkshire	393	4,869,221	327,413	40	1
3	Trent	393	6,768,421	456,803	99	9
4	Lincolnshire	116	489,681	21,745	3	0
5	Welland and Nene	151	1,009,751	65,379	14	8
6	Great Ouse	245	1,541,438	99,378	24	6
7	East Suffolk and Norfolk	115	719,429	36,727	0	0
8	Essex	138	1,027,638	52,488	24	4
9	Kent	141	1,226,762	64,725	23	8
10	Sussex	108	503,351	24,593	13	3
11	Hampshire	39	838,010	56,468	10	2
12	Isle of Wight	15	39,685	2,474	0	2
13	Avon and Dorset	84	719,499	32,071	1	0
14	Devon	129	368,401	22,481	10	3
15	Cornwall	95	262,671	14,210	8	2
16	Somerset	90	366,681	24,799	16	2
17	Bristol Avon	81	475,418	26,876	8	2
18	Severn	311	3,268,784	209,541	62	26
19	Wye	53	164,279	12,404	1	0
20	Usk	29	285,607	16,111	3	1
21	Glamorgan	22	463,627	39,714	22	4
22	South West Wales	122	365,801	19,323	5	5
23	Gwynedd	39	73,556	3,372	7	3
24	Dee and Clwyd	88	503,030	25,887	23	2
25	Mersey and Weaver	155	4,630,791	360,290	41	7
26	Lancashire	114	1,493,294	117,150	50	7
27	Cumberland	93	210,672	13,276	1	0
28	Thames Conservancy	314	4,212,847	249,851	43	7
29	Lee Conservancy	28	1,414,750	95,152	4	0
30	Port of London Authority (incl. London Excluded Area)	22	8,594,050	600,005	3	0
	Total England and Wales	3,921	48,047,999	3,160,703	573	115

Note: Number of new discharges refers to new works on new sites and does not include reconstruction of existing works on existing sites. New works to deal with existing discharges of crude sewage are not included.

...wage treatment works where improved effluent standards will be required

mber	Estimated population served 1980	Estimated dry weather flow 1980 (x 1,000 gpd)	
3	255,004	16,931	1
2	4,004,558	274,198	2
0	4,407,900	334,200	3
4	39,205	572	4
7	482,860	31,324	5
4	358,760	20,560	6
3	217,800	14,292	7
2	87,450	3,966	8
9	149,800	6,889	9
4	58,200	2,488	10
5	134,570	9,220	11
1	2,200	132	12
2	30,350	1,494	13
1	350	18	14
0	0	0	15
27	138,222	10,213	16
3	203,172	12,738	17
8	839,876	55,396	18
2	16,000	750	19
1	126,500	6,720	20
3	5,768	248	21
5	30,880	1,451	22
3	8,900	753	23
3	106,000	8,171	24
5	466,788	37,920	25
26	918,673	75,697	26
1	5,700	220	27
36	1,509,235	94,055	28
6	813,840	52,535	29
12	5,547,200	408,642	30
33	20,965,761	1,481,793	

Table 185 Discharges of crude sewage. Estimated number, population served, dry weather flow (x 1,000 gpd) and percentage satisfactory

River Authority	Subject	Non-tidal rivers					
		Class 1	Class 2	Class 3	Class 4	Total non-tidal
1 Northumbrian	Number and % satisfactory	2 (0)	0	0	0	2 (0)
	Population and % satisfactory	3,269 (0)	0	0	0	3,269 (0)
	DWF and % satisfactory	98 (0)	0	0	0	98 (0)
2 Yorkshire	Number and % satisfactory	5 (0)	0	0	0	5 (0)
	Population and % satisfactory	2,610 (0)	0	0	0	2,610 (0)
	DWF and % satisfactory	79 (0)	0	0	0	79 (0)
3 Trent	Number and % satisfactory	0	0	0	0	0
	Population and % satisfactory	0	0	0	0	0
	DWF and % satisfactory	0	0	0	0	0
4 Lincolnshire	Number and % satisfactory	0	0	0	0	0
	Population and % satisfactory	0	0	0	0	0
	DWF and % satisfactory	0	0	0	0	0
6 Great Ouse	Number and % satisfactory	0	1 (0)	0	0	1 (0)
	Population and % satisfactory	0	5,400 (0)	0	0	5,400 (0)
	DWF and % satisfactory	0	324 (0)	0	0	324 (0)
7 East Suffolk and Norfolk	Number and % satisfactory	0	0	0	0	0
	Population and % satisfactory	0	0	0	0	0
	DWF and % satisfactory	0	0	0	0	0
8 Essex	Number and % satisfactory	0	0	0	0	0
	Population and % satisfactory	0	0	0	0	0
	DWF and % satisfactory	0	0	0	0	0
9 Kent	Number and % satisfactory	0	0	0	0	0
	Population and % satisfactory	0	0	0	0	0
	DWF and % satisfactory	0	0	0	0	0
10 Sussex	Number and % satisfactory	0	0	0	0	0
	Population and % satisfactory	0	0	0	0	0
	DWF and % satisfactory	0	0	0	0	0
11 Hampshire	Number and % satisfactory	0	0	0	0	0
	Population and % satisfactory	0	0	0	0	0
	DWF and % satisfactory	0	0	0	0	0
12 Isle of Wight	Number and % satisfactory	0	0	0	0	0
	Population and % satisfactory	0	0	0	0	0
	DWF and % satisfactory	0	0	0	0	0
14 Devon	Number and % satisfactory	2 (0)	3 (0)	0	0	5 (0)
	Population and % satisfactory	530 (0)	1,160 (0)	0	0	1,690 (0)
	DWF and % satisfactory	20 (0)	53 (0)	0	0	73 (0)
15 Cornwall	Number and % satisfactory	1 (0)	0	0	0	1 (0)
	Population and % satisfactory	1,467 (0)	0	0	0	1,467 (0)
	DWF and % satisfactory	89 (0)	0	0	0	89 (0)
16 Somerset	Number and % satisfactory	8 (0)	0	0	1 (0)	9 (0)
	Population and % satisfactory	3,264 (0)	0	0	620 (0)	3,884 (0)
	DWF and % satisfactory	102 (0)	0	0	19 (0)	121 (0)
17 Bristol Avon	Number and % satisfactory	3 (0)	8 (0)	0	0	11 (0)
	Population and % satisfactory	950 (0)	6,110 (0)	0	0	7,060 (0)
	DWF and % satisfactory	27 (0)	178 (0)	0	0	205 (0)
18 Severn	Number and % satisfactory	0	0	0	0	0
	Population and % satisfactory	0	0	0	0	0
	DWF and % satisfactory	0	0	0	0	0
19 Wye	Number and % satisfactory	0	0	0	0	0
	Population and % satisfactory	0	0	0	0	0
	DWF and % satisfactory	0	0	0	0	0
20 Usk	Number and % satisfactory	0	0	0	0	0
	Population and % satisfactory	0	0	0	0	0
	DWF and % satisfactory	0	0	0	0	0
21 Glamorgan	Number and % satisfactory	0	0	0	0	0
	Population and % satisfactory	0	0	0	0	0
	DWF and % satisfactory	0	0	0	0	0
22 South West Wales	Number and % satisfactory	3 (0)	0	0	0	3 (0)
	Population and % satisfactory	2,025 (0)	0	0	0	2,025 (0)
	DWF and % satisfactory	53 (0)	0	0	0	53 (0)
23 Gwynedd	Number and % satisfactory	10 (0)	0	0	0	10 (0)
	Population and % satisfactory	11,584 (0)	0	0	0	11,584 (0)
	DWF and % satisfactory	184 (0)	0	0	0	184 (0)
24 Dee and Clwyd	Number and % satisfactory	0	0	0	0	0
	Population and % satisfactory	0	0	0	0	0
	DWF and % satisfactory	0	0	0	0	0
25 Mersey and Weaver	Number and % satisfactory	0	0	0	0	0
	Population and % satisfactory	0	0	0	0	0
	DWF and % satisfactory	0	0	0	0	0
26 Lancashire	Number and % satisfactory	0	0	0	0	0
	Population and % satisfactory	0	0	0	0	0
	DWF and % satisfactory	0	0	0	0	0
27 Cumberland	Number and % satisfactory	0	0	0	0	0
	Population and % satisfactory	0	0	0	0	0
	DWF and % satisfactory	0	0	0	0	0
Total England and Wales	Number and % satisfactory	34 (0)	12 (0)	0	1 (0)	47 (0)
	Population and % satisfactory	25,699 (0)	12,670 (0)	0	620 (0)	38,989 (0)
	DWF and % satisfactory	652 (0)	555 (0)	0	19 (0)	1,226 (0)

* 0.3%   ** 0.2%   *** 0.1%   Notes: There are no crude sewage discharges in the areas of Welland and Nene River Authority, Avon and Dorset River Authority, Thames

al rivers						All rivers				
ss 1	Class 2	Class 3	Class 4	Total tidal		Class 1	Class 2	Class 3	Class 4	Total all rivers
14 (0)	8 (0)	26 (0)	6 (0)	54 (0)		16 (0)	8 (0)	26 (0)	6 (0)	56 (0) 1
,665 (0)	34,406 (0)	294,273 (0)	1,075,985 (0)	1,416,329 (0)		14,934 (0)	34,406 (0)	294,273 (0)	1,075,985 (0)	1,419,598 (0)
712 (0)	1,354 (0)	14,010 (0)	48,900 (0)	64,976 (0)		810 (0)	1,354 (0)	14,010 (0)	48,900 (0)	65,074 (0)
3 (0)	0	4 (0)	0	7 (0)		8 (0)	0	4 (0)	0	12 (0) 2
5,290 (0)	0	20,890 (0)	0	26,180 (0)		7,900 (0)	0	20,890 (0)	0	28,790 (0)
293 (0)	0	1,127 (0)	0	1,420 (0)		372 (0)	0	1,127 (0)	0	1,499 (0)
0	0	0	10 (0)	10 (0)		0	0	0	10 (0)	10 (0) 3
0	0	0	25,000 (0)	25,000 (0)		0	0	0	25,000 (0)	25,000 (0)
0	0	0	1,420 (0)	1,420 (0)		0	0	0	1,420 (0)	1,420 (0)
0	0	3 (0)	0	3 (0)		0	0	3 (0)	0	3 (0) 4
0	0	25,090 (0)	0	25,090 (0)		0	0	25,090 (0)	0	25,090 (0)
0	0	1,500 (0)	0	1,500 (0)		0	0	1,500 (0)	0	1,500 (0)
7 (0)	1 (0)	0	0	8 (0)		7 (0)	2 (0)	0	0	9 (0) 6
7,600 (0)	5,400 (0)	0	0	23,000 (0)		17,600 (0)	10,800 (0)	0	0	28,400 (0)
1,054 (0)	324 (0)	0	0	1,378 (0)		1,054 (0)	648 (0)	0	0	1,702 (0)
1 (0)	30 (0)	0	0	31 (0)		1 (0)	30 (0)	0	0	31 (0) 7
1,000 (0)	51,734 (0)	0	0	62,734 (0)		11,000 (0)	51,734 (0)	0	0	62,734 (0)
575 (0)	5,931 (0)	0	0	6,506 (0)		575 (0)	5,931 (0)	0	0	6,506 (0)
1 (0)	0	1 (0)	3 (0)	5 (0)		1 (0)	0	1 (0)	3 (0)	5 (0) 8
10,000 (0)	0	1,600 (0)	4,200 (0)	15,800 (0)		10,000 (0)	0	1,600 (0)	4,200 (0)	15,800 (0)
400 (0)	0	51 (0)	1,383 (0)	1,834 (0)		400 (0)	0	51 (0)	1,383 (0)	1,834 (0)
5 (20)	1 (0)	2 (0)	0	8 (12)		5 (20)	1 (0)	2 (0)	0	8 (12) 9
20,875 (5)	450 (0)	4,600 (0)	0	25,925 (3)		20,875 (5)	450 (0)	4,600 (0)	0	25,925 (3)
1,928 (2)	35 (0)	320 (0)	0	2,283 (1)		1,928 (2)	35 (0)	320 (0)	0	2,283 (1)
2 (50)	0	0	0	2 (50)		2 (50)	0	0	0	2 (50) 10
32,878 (9)	0	0	0	32,878 (9)		32,878 (9)	0	0	0	32,878 (9)
1,884 (7)	0	0	0	1,884 (7)		1,884 (7)	0	0	0	1,884 (7)
5 (60)	0	0	0	5 (60)		5 (60)	0	0	0	5 (60) 11
246,290 (98)	0	0	0	246,290 (98)		246,290 (98)	0	0	0	246,290 (98)
14,576 (98)	0	0	0	14,576 (98)		14,576 (98)	0	0	0	14,576 (98)
0	13 (0)	0	0	13 (0)		0	13 (0)	0	0	13 (0) 12
0	13,000 (0)	0	0	13,000 (0)		0	13,000 (0)	0	0	13,000 (0)
0	495 (0)	0	0	495 (0)		0	495 (0)	0	0	495 (0)
3 (0)	9 (0)	12 (0)	0	24 (0)		5 (0)	12 (0)	12 (0)	0	29 (0) 14
16,950 (0)	14,105 (0)	74,483 (0)	0	105,538 (0)		17,480 (0)	15,265 (0)	74,483 (0)	0	107,228 (0)
552 (0)	672 (0)	3,991 (0)	0	5,215 (0)		572 (0)	725 (0)	3,991 (0)	0	5,288 (0)
22 (18)	14 (7)	0	0	36 (14)		23 (17)	14 (7)	0	0	37 (13) 15
31,758 (26)	79,668 (8)	0	0	211,426 (19)		133,225 (25)	79,668 (8)	0	0	212,893 (19)
6,335 (26)	4,397 (7)	0	0	10,732 (18)		6,424 (26)	4,397 (7)	0	0	10,821 (18)
0	3 (0)	6 (0)	9 (0)	18 (0)		8 (0)	3 (0)	6 (0)	10 (0)	27 (0) 16
0	8,560 (0)	17,210 (0)	23,410 (0)	49,180 (0)		3,264 (0)	8,560 (0)	17,210 (0)	24,030 (0)	53,064 (0)
0	447 (0)	1,051 (0)	3,879 (0)	5,377 (0)		102 (0)	447 (0)	1,051 (0)	3,898 (0)	5,498 (0)
0	0	0	6 (0)	6 (0)		3 (0)	8 (0)	0	6 (0)	17 (0) 17
0	0	0	301,300 (0)	301,300 (0)		950 (0)	6,110 (0)	0	301,300 (0)	308,360 (0)
0	0	0	19,460 (0)	19,460 (0)		27 (0)	178 (0)	0	19,460 (0)	19,665 (0)
8 (0)	0	0	0	8 (0)		8 (0)	0	0	0	8 (0) 18
15,547 (0)	0	0	0	15,547 (0)		15,547 (0)	0	0	0	15,547 (0)
8,744 (0)	0	0	0	8,744 (0)		8,744 (0)	0	0	0	8,744 (0)
4 (25)	0	0	0	4 (25)		4 (25)	0	0	0	4 (25) 19
9,150 (14)	0	0	0	9,150 (14)		9,150 (14)	0	0	0	9,150 (14)
442 (15)	0	0	0	442 (15)		442 (15)	0	0	0	442 (15)
0	4 (0)	13 (0)	1 (0)	18 (0)		0	4 (0)	13 (0)	1 (0)	18 (0) 20
0	25,650 (0)	86,390 (0)	9,000 (0)	121,040 (0)		0	25,050 (0)	86,390 (0)	9,000 (0)	121,040 (0)
0	1,324 (0)	5,052 (0)	450 (0)	6,826 (0)		0	1,324 (0)	5,052 (0)	450 (0)	6,826 (0)
0	5 (0)	22 (4)	2 (0)	29 (3)		0	5 (0)	22 (4)	2 (0)	29 (3) 21
0	5,900 (0)	79,018 (1)	6,964 (0)	91,882 (1)		0	5,900 (0)	79,018 (1)	6,964 (0)	91,882 (1)
0	368 (0)	11,561 (2)	404 (0)	12,333 (2)		0	368 (0)	11,561 (2)	404 (0)	12,333 (2)
8 (12)	13 (0)	1 (0)	9 (0)	31 (3)		11 (9)	13 (0)	1 (0)	9 (0)	34 (3) 22
38,300 (1)	40,716 (0)	230 (0)	75,219 (0)	154,465 (*)		40,325 (1)	40,716 (0)	230 (0)	75,219 (0)	156,490 (*)
1,948 (1)	1,606 (0)	20 (0)	5,563 (0)	9,137 (**)		2,001 (1)	1,606 (0)	20 (0)	5,563 (0)	9,190 (**)
10 (0)	0	0	0	10 (0)		20 (0)	0	0	0	20 (0) 23
24,960 (0)	0	0	0	24,960 (0)		36,544 (0)	0	0	0	36,544 (0)
813 (0)	0	0	0	813 (0)		997 (0)	0	0	0	997 (0)
3 (0)	1 (0)	0	1 (0)	5 (0)		3 (0)	1 (0)	0	1 (0)	5 (0) 24
14,000 (0)	11,000 (0)	0	8,600 (0)	33,600 (0)		14,000 (0)	11,000 (0)	0	8,600 (0)	33,600 (0)
700 (0)	550 (0)	0	317 (0)	1,567 (0)		700 (0)	550 (0)	0	317 (0)	1,567 (0)
0	11 (0)	36 (0)	21 (0)	68 (0)		0	11 (0)	36 (0)	21 (0)	68 (0) 25
0	385,080 (0)	653,087 (0)	121,814 (0)	1,159,981 (0)		0	385,080 (0)	653,087 (0)	121,814 (0)	1,159,981 (0)
0	20,994 (0)	31,079 (0)	12,865 (0)	64,938 (0)		0	20,994 (0)	31,079 (0)	12,865 (0)	64,938 (0)
9 (0)	10 (0)	3 (0)	8 (12)	30 (3)		9 (0)	10 (0)	3 (0)	8 (12)	30 (3) 26
18,635 (0)	117,655 (0)	21,430 (0)	75,496 (2)	233,216 (1)		18,635 (0)	117,655 (0)	21,430 (0)	75,496 (2)	233,216 (1)
887 (0)	11,484 (0)	1,048 (0)	3,992 (1)	17,411 (*)		887 (0)	11,484 (0)	1,048 (0)	3,992 (1)	17,411 (*)
4 (0)	0	0	0	4 (0)		4 (0)	0	0	0	4 (0) 27
4,080 (0)	0	0	0	4,080 (0)		4,080 (0)	0	0	0	4,080 (0)
1,516 (0)	0	0	0	1,516 (0)		1,516 (0)	0	0	0	1,516 (0)
109 (10)	123 (1)	129 (1)	76 (1)	437 (3)		143 (8)	135 (1)	129 (1)	77 (1)	484 (3)
628,978 (45)	793,324 (1)	1,278,301 (***)	1,726,988 (***)	4,427,591 (7)		654,677 (43)	805,994 (1)	127,830 (***)	1,727,608 (***)	4,466,580 (6)
43,359 (37)	49,981 (1)	70,810 (*)	98,633 (***)	262,783 (6)		44,011 (37)	50,536 (1)	70,810 (*)	98,652 (***)	264,009 (6)

Conservancy, Lee Conservancy and Port of London Authority (incl. London Excluded Area). Attention is drawn to the footnote on Table 34 and to paragraph 3 of Chapter 3.

Table 186 Discharges of crude sewage: Estimated number of outlets still remaining in 1980, population served and dry weather flow (1,000 gpd)

River Authority	Subject	Non-tidal rivers				Total non-tidal
		Class 1	Class 2	Class 3	Class 4	
1 Northumbrian	Number	0	0	0	0	0
	Population	0	0	0	0	0
	DWF	0	0	0	0	0
2 Yorkshire	Number	4	0	0	0	4
	Population	6,800	0	0	0	6,800
	DWF	195	0	0	0	195
3 Trent	Number	0	0	0	0	0
	Population	0	0	0	0	0
	DWF	0	0	0	0	0
4 Lincolnshire	Number	0	0	0	0	0
	Population	0	0	0	0	0
	DWF	0	0	0	0	0
7 East Suffolk and Norfolk	Number	0	0	0	0	0
	Population	0	0	0	0	0
	DWF	0	0	0	0	0
8 Essex	Number	0	0	0	0	0
	Population	0	0	0	0	0
	DWF	0	0	0	0	0
9 Kent	Number	0	0	0	0	0
	Population	0	0	0	0	0
	DWF	0	0	0	0	0
10 Sussex	Number	0	0	0	0	0
	Population	0	0	0	0	0
	DWF	0	0	0	0	0
11 Hampshire	Number	0	0	0	0	0
	Population	0	0	0	0	0
	DWF	0	0	0	0	0
14 Devon	Number	1	1	0	0	2
	Population	430	806	0	0	1,236
	DWF	17	32	0	0	49
15 Cornwall	Number	1	0	0	0	1
	Population	7,000	0	0	0	7,000
	DWF	240	0	0	0	240
16 Somerset	Number	2	0	0	0	2
	Population	2,950	0	0	0	2,950
	DWF	89	0	0	0	89
17 Bristol Avon	Number	1	2	0	0	3
	Population	500	4,890	0	0	5,390
	DWF	20	205	0	0	225
18 Severn	Number	0	0	0	0	0
	Population	0	0	0	0	0
	DWF	0	0	0	0	0
19 Wye	Number	0	0	0	0	0
	Population	0	0	0	0	0
	DWF	0	0	0	0	0
20 Usk	Number	0	0	0	0	0
	Population	0	0	0	0	0
	DWF	0	0	0	0	0
21 Glamorgan	Number	0	0	0	0	0
	Population	0	0	0	0	0
	DWF	0	0	0	0	0
22 South West Wales	Number	3	0	0	0	3
	Population	2,386	0	0	0	2,386
	DWF	82	0	0	0	82
23 Gwynedd	Number	2	0	0	0	2
	Population	15,200	0	0	0	15,200
	DWF	173	0	0	0	173
24 Dee and Clwyd	Number	0	0	0	0	0
	Population	0	0	0	0	0
	DWF	0	0	0	0	0
25 Mersey and Weaver	Number	0	0	0	0	0
	Population	0	0	0	0	0
	DWF	0	0	0	0	0
26 Lancashire	Number	0	0	0	0	0
	Population	0	0	0	0	0
	DWF	0	0	0	0	0
27 Cumberland	Number	0	0	0	0	0
	Population	0	0	0	0	0
	DWF	0	0	0	0	0
Total England and Wales	Number	14	3	0	0	17
	Population	35,266	5,696	0	0	40,962
	DWF	816	237	0	0	1,053

Notes: There are no crude sewage discharges at present in the areas of Welland and Nene River Authority, Avon and Dorset River Authority, Thames Conservancy, Lee Conservancy and Port of London Authority (including London Excluded Area).
All existing outlets of crude sewage in the areas of Great Ouse River Authority and Isle of Wight River and Water Authority will be diverted elsewhere before 1980.
Most of the discharges remaining are expected to have additional treatment provided before 1980. There are no discharges to canals and there will be none in 1980.

	rivers					All rivers					
s 1	Class 2	Class 3	Class 4	Total tidal	Class 1	Class 2	Class 3	Class 4	Total all rivers	
2	0	2	4	8	2	0	2	4	8	1
300	0	26,183	1,362,000	1,410,483	22,300	0	26,183	1,362,000	1,410,483	
327	0	1,273	82,016	84,616	1,327	0	1,273	82,016	84,616	
2	0	4	0	6	6	0	4	0	10	2
350	0	26,180	0	32,530	13,150	0	26,180	0	39,330	
416	0	1,385	0	1,801	611	0	1,385	0	1,996	
0	0	0	1	1	0	0	0	1	1	3
0	0	0	26,000	26,000	0	0	0	26,000	26,000	
0	0	0	1,893	1,893	0	0	0	1,893	1,893	
0	0	1	0	1	0	0	1	0	1	4
0	0	29,620	0	29,620	0	0	29,620	0	29,620	
0	0	2,600	0	2,600	0	0	2,600	0	2,600	
1	30	0	0	31	1	30	0	0	31	7
,500	52,780	0	0	67,280	14,500	52,780	0	0	67,280	
750	9,845	0	0	10,595	750	9,845	0	0	10,595	
11	0	0	1	2	1	0	0	1	2	8
,000	0	0	0	12,000	12,000	0	0	0	12,000	
560	0	0	1,213	1,773	560	0	0	1,213	1,773	
2	0	0	0	2	2	0	0	0	2	9
,600	0	0	0	9,600	9,600	0	0	0	9,600	
,270	0	0	0	1,270	1,270	0	0	0	1,270	
1	0	0	0	1	1	0	0	0	1	10
3,500	0	0	0	3,500	3,500	0	0	0	3,500	
175	0	0	0	175	175	0	0	0	175	
5	0	0	0	5	5	0	0	0	5	11
,100	0	0	0	251,100	251,100	0	0	0	251,100	
7,259	0	0	0	17,259	17,259	0	0	0	17,259	
1	6	4	0	11	2	7	4	0	13	14
3,000	40,550	19,250	0	72,800	13,430	41,356	19,250	0	74,036	
320	2,993	807	0	4,120	337	3,025	807	0	4,169	
11	7	0	0	18	12	7	0	0	19	15
8,007	115,146	0	0	273,153	165,007	115,146	0	0	280,153	
8,839	6,678	0	0	15,517	9,079	6,678	0	0	15,757	
0	1	0	2	3	2	1	0	2	5	16
0	2,700	0	89,510	92,210	2,950	2,700	0	89,510	95,160	
0	120	0	7,905	8,025	89	120	0	7,905	8,114	
0	0	0	0	0	1	2	0	0	3	17
0	0	0	0	0	500	4,890	0	0	5,390	
0	0	0	0	0	20	205	0	0	225	
1	0	0	0	1	1	0	0	0	1	18
1,400	0	0	0	1,400	1,400	0	0	0	1,400	
71	0	0	0	71	71	0	0	0	71	
2	0	0	0	2	2	0	0	0	2	19
3,300	0	0	0	13,300	13,300	0	0	0	13,300	
606	0	0	0	606	606	0	0	0	606	
0	2	4	0	6	0	2	4	0	6	20
0	7,800	139,030	0	146,830	0	7,800	139,030	0	146,830	
0	624	11,014	0	11,638	0	624	11,014	0	11,638	
0	0	1	0	1	0	0	1	0	1	21
0	0	1,000	0	1,000	0	0	1,000	0	1,000	
0	0	209	0	209	0	0	209	0	209	
4	3	0	1	8	7	3	0	1	11	22
35,210	38,015	0	40,000	113,225	37,596	38,015	0	40,000	115,611	
1,451	1,008	0	1,600	4,059	1,533	1,008	0	1,600	4,141	
4	0	0	0	4	6	0	0	0	6	23
26,850	0	0	0	26,850	42,050	0	0	0	42,050	
892	0	0	0	892	1,065	0	0	0	1,065	
1	1	0	0	2	1	1	0	0	2	24
13,000	15,000	0	0	28,000	13,000	15,000	0	0	28,000	
650	750	0	0	1,400	650	750	0	0	1,400	
0	5	17	0	22	0	5	17	0	22	25
0	689,840	270,316	0	960,156	0	689,840	270,316	0	960,156	
0	41,764	13,531	0	55,295	0	41,764	13,531	0	55,295	
6	5	1	4	16	6	5	1	4	16	26
19,825	127,500	25,000	82,881	255,206	19,825	127,500	25,000	82,881	255,206	
977	15,386	1,250	4,232	21,845	977	15,386	1,250	4,232	21,845	
2	0	0	0	2	2	0	0	0	2	27
4,150	0	0	0	4,150	4,150	0	0	0	4,150	
1,518	0	0	0	1,518	1,518	0	0	0	1,518	
46	60	34	13	153	60	63	34	13	170	
604,092	1,089,331	536,579	1,600,391	3,830,393	639,358	1,095,027	536,579	1,600,391	3,871,355	
37,081	79,168	32,069	98,859	247,177	37,897	79,405	32,069	98,859	248,230	

Table 187 Number of unsatisfactory storm overflows

River Authority		Non-tidal rivers					Tidal rivers				
		Class 1	Class 2	Class 3	Class 4	Total non-tidal	Class 1	Class 2	Class 3	Class 4	Total tidal
1	Northumbrian	2	4	2	0	8	0	0	0	0	0
2	Yorkshire	51	51	69	126	297	2	0	1	0	3
3	Trent	1	16	18	43	78	0	0	0	0	0
4	Lincolnshire	9	0	2	0	11	0	0	1	0	1
7	East Suffolk and Norfolk	9	3	0	0	12	1	6	0	0	7
8	Essex	10	8	5	1	24	0	3	2	0	5
9	Kent	7	0	0	0	7	0	0	3	1	4
10	Sussex	3	1	0	0	4	1	0	0	0	1
11	Hampshire	3	3	0	0	6	0	0	0	1	1
12	Isle of Wight	7	7	4	0	18	9	7	1	0	17
13	Avon and Dorset	7	5	0	2	14	0	0	0	0	0
14	Devon	9	13	5	0	27	0	2	4	0	6
15	Cornwall	32	10	3	4	49	12	22	0	0	34
16	Somerset	9	0	0	1	10	0	0	0	0	0
17	Bristol Avon	12	24	0	0	36	0	0	0	0	0
18	Severn	51	55	2	13	121	0	0	0	0	0
19	Wye	1	0	3	0	4	0	0	0	0	0
20	Usk	41	84	2	49	176	0	0	0	0	0
21	Glamorgan	0	1	3	0	4	0	0	0	1	1
22	South West Wales	17	2	1	0	20	0	1	2	0	3
23	Gwynedd	3	0	0	0	3	0	0	0	0	0
24	Dee and Clwyd	12	10	7	0	29	1	0	0	0	1
25	Mersey and Weaver	103	156	92	425	776	0	2	2	6	10
26	Lancashire	5	64	18	176	263	0	5	0	0	5
27	Cumberland	17	3	0	0	20	0	0	0	0	0
28	Thames Conservancy	9	0	0	0	9	0	0	0	0	9
30	Port of London Authority (incl. London Excluded Area)	0	0	0	0	0	0	0	25	0	25
	Total England and Wales	430	520	236	840	2,026	26	48	41	9	124

Note: There are no known unsatisfactory storm overflows in the areas of Welland and Nene River Authority, Great Ouse River Authority and Lee Conservancy.

	Class 2	Class 3	Class 4	Total all rivers	Canals				Total canals	All rivers and canals				Total all rivers and canals	
1					Class 1	Class 2	Class 3	Class 4		Class 1	Class 2	Class 3	Class 4		
	4	2	0	8	0	0	0	0	0	2	4	2	0	8	1
	51	70	126	300	0	0	0	2	2	53	51	70	128	302	2
	16	18	43	78	0	0	0	2	2	1	16	18	45	80	3
	0	3	0	12	0	0	0	0	0	9	0	3	0	12	4
	9	0	0	19	0	0	0	0	0	10	9	0	0	19	7
	11	7	1	29	0	0	0	0	0	10	11	7	1	29	8
	0	3	1	11	0	0	0	0	0	7	0	3	1	11	9
	1	0	0	5	0	0	0	0	0	4	1	0	0	5	10
	3	0	1	7	0	0	0	0	0	3	3	0	1	7	11
	14	5	0	35	0	0	0	0	0	16	14	5	0	35	12
	5	0	2	14	0	0	0	0	0	7	5	0	2	14	13
	15	9	0	33	0	0	0	0	0	9	15	9	0	33	14
	32	3	4	83	0	0	0	0	0	44	32	3	4	83	15
	0	0	1	10	0	0	0	0	0	9	0	0	1	10	16
	24	0	0	36	0	0	0	0	0	12	24	0	0	36	17
	55	2	13	121	0	2	0	0	2	51	57	2	13	123	18
	0	3	0	4	0	0	0	0	0	1	0	3	0	4	19
	84	2	49	176	0	0	0	0	0	41	84	2	49	176	20
	1	3	1	5	0	0	0	0	0	0	1	3	1	5	21
	3	3	0	23	0	0	0	0	0	17	3	3	0	23	22
	0	0	0	3	0	0	0	0	0	3	0	0	0	3	23
	10	7	0	30	0	0	0	0	0	13	10	7	0	30	24
	158	94	431	786	5	0	0	0	5	108	158	94	431	791	25
	69	18	176	268	0	0	0	0	0	5	69	18	176	268	26
	3	0	0	20	0	0	0	0	0	17	3	0	0	20	27
	0	0	0	9	1	0	0	0	1	10	0	0	0	10	28
	0	25	0	25	0	0	0	0	0	0	0	25	0	25	30
	568	277	849	2,150	6	2	0	4	12	462	570	277	853	2,162	

NORTHUMBRIAN RIVER AUTHORITY

TABLE 188 Main survey Numbers and volumes of discharges of industrial effluent to rivers and canals in England and Wales and percentage considered satisfactory

Industry		Discharges of total effluent (which may contain some cooling water) to											
		Non-tidal rivers			Tidal rivers			All rivers			Canals*		
CBI Code	Description	Total Number (% satisfactory)	Total Volume 1000s galls (% satisfactory)	Volume of Cooling Water 1000s galls	Total Number (% satisfactory)	Total Volume 1000s galls (% satisfactory)	Volume of Cooling Water 1000s galls	Total Number (% satisfactory)	Total Volume 1000s galls (% satisfactory)	Volume of Cooling Water 1000s galls	Total Number (% satisfactory)	Total Volume 1000s galls (% satisfactory)	Volume of Cooling Water 1000s galls
01	Brewing												
02	Brickmaking												
03	Cement making	1(Nil)	6(Nil)	Nil				1(Nil)	6(Nil)	Nil			
04	Chemical and allied industries	5(80)	1920(78)	Nil	12(Nil)	214115(Nil)	186764	17(23)	216035(1)	186764			
05	Coal mining	5(20)	570(60)	Nil	1(100)	7(100)	Nil	6(33)	577(60)	Nil			
06	Distillation of ethanol												
07	Electricity generation				3(100)	300(100)	Nil	3(100)	300(100)	Nil			
08	Engineering	2(50)	610(98)	Nil	3(33)	565(64)	Nil	5(40)	1175(82)	Nil			
09	Food processing and manufacture	1(Nil)	60(Nil)	Nil	3(33)	251(68)	Nil	4(25)	311(55)	Nil			
10	Gas and coke	2(Nil)	1826(Nil)	Nil	1(Nil)	35(Nil)	Nil	3(Nil)	1861(Nil)	Nil			
11	Glass making												
12	Glue and gelatine				1(Nil)	30(Nil)	Nil	1(Nil)	30(Nil)	Nil			
13	General manufacturing	2(50)	35(57)	Nil	2(50)	220(9)	100	4(50)	255(16)	100			
14	Iron and steel	1(100)	200(100)	Nil	3(33)	42140(3)	7083	4(50)	42340(3)	7083			
15	Laundering and dry cleaning												
16	Leather tanning				3(Nil)	173(Nil)	Nil	3(Nil)	173(Nil)	Nil			
17	Metal smelting				1(Nil)	150(Nil)	Nil	1(Nil)	150(Nil)	Nil			
18	Paint making	1(100)	180(100)	Nil				1(100)	180(100)	Nil			
19	Paper and board making	1(Nil)	240(Nil)	Nil	2(Nil)	450(Nil)	Nil	3(Nil)	690(Nil)	Nil			
20	Petroleum refining				4(100)	5265(100)	Nil	4(100)	5265(100)	Nil			
21	Plastics manufacture												
22	Plating and metal finishing												
23	Pottery making												
24	Printing ink etc												
25	Quarrying and mining	11(54)	904(55)	Nil				11(54)	904(55)	Nil			
26	Rubber processing												
27	Soap and detergent												
28	Textile, cotton and man-made				1(Nil)	17400(Nil)	870	1(Nil)	17400(Nil)	870			
29	Textile, wool												
30	General farming												
31	Atomic energy establishments												
50	Water treatment	9(33)	1175(73)	Nil				9(33)	1175(73)	Nil			
51	Disposal tip drainage	1(Nil)	5(Nil)	Nil				1(Nil)	5(Nil)	Nil			
	TOTAL FOR CODES 1–51	42(43)	7731(54)	Nil	40(30)	281101(8·8)	194817	82(37)	288832(10)	194817			
52	Derelict coal mines	11	1590	Nil	1	1000	Nil	12	2590	Nil			
53	Other derelict mines	1	1000	Nil				1	1000	Nil			
54	Active coal mines	65	66471	Nil	6	2937	Nil	71	69408	Nil			
55	Other active mines	5	750	Nil				5	750	Nil			
	TOTAL FOR CODES 52–55	82	69811	Nil	7	3937	Nil	89	73748	Nil			
	TOTAL	124	77542	Nil	47	285038	194817	171	362580	194817			

* For uniformity of presentation a standard table layout has been adopted for each of the Tables 188-217, whether there are canals in the area or not. The reader should refer to Table 130 of Volume 2 for the list of river authorities in whose areas there are canals.

	rivers and canals		Discharges of solely cooling water to										Total Cooling Water Discharged 1000s galls	Total Process Water Discharged 1000s galls	Total Discharge 1000s galls	CBI Code
			Non-tidal rivers		Tidal rivers		All rivers		Canals		All rivers and canals					
	Total Volume 1000s galls (% satis-factory)	Volume of Cooling Water 1000s galls	Total Number (% satis-factory)	Total Volume 1000s galls (% satis-factory)	Total Number (% satis-factory)	Total Volume 1000s galls (% satis-factory)	Total Number (% satis-factory)	Total Volume 1000s galls (% satis-factory)	Total Number (% satis-factory)	Total Volume 1000s galls (% satis-factory)	Total Number (% satis-factory)	Total Volume 1000s galls (% satis-factory)				
																01
																02
(Nil)	6(Nil)	Nil	1(100)	240(100)			1(100)	240(100)			1(100)	240(100)	240	6	246	03
(23)	216035(1)	186764											186764	29271	216035	04
(33)	577(60)	Nil												577	577	05
																06
(100)	300(100)	Nil			4(100)	504530(100)	4(100)	504530(100)			4(100)	504530(100)	504530	300	504830	07
(40)	1175(82)	Nil	1(100)	29(100)	1(100)	11(100)	2(100)	40(100)			2(100)	40(100)	40	1175	1215	08
(25)	311(55)	Nil												311	311	09
(Nil)	1861(Nil)	Nil	1(100)	10(100)			1(100)	10(100)			1(100)	10(100)	10	1861	1871	10
																11
(Nil)	30(Nil)	Nil			1(100)	480(100)	1(100)	480(100)			1(100)	480(100)	480	30	510	12
(50)	255(16)	100											100	155	255	13
(50)	42340(3)	7083	1(100)	3500(100)			1(100)	3500(100)			1(100)	3500(100)	10583	35257	45840	14
																15
(Nil)	173(Nil)	Nil												173	173	16
(Nil)	150(Nil)	Nil	1(100)	100(100)	1(100)	40(100)	2(100)	140(100)			2(100)	140(100)	140	150	290	17
(100)	180(100)	Nil												180	180	18
(Nil)	690(Nil)	Nil												690	690	19
(100)	5265(100)	Nil												5265	5265	20
																21
																22
																23
																24
(54)	904(55)	Nil												904	904	25
																26
																27
(Nil)	17400(Nil)	870											870	16530	17400	28
																29
																30
																31
(33)	1175(73)	Nil												1175	1175	50
(Nil)	5(Nil)	Nil												5	5	51
(37)	288832(10)	194817	5(100)	3879(100)	7(100)	505061(100)	12(100)	508940(100)			12(100)	508940(100)	703757	94015	797772	
	2590	Nil												2590	2590	52
	1000	Nil												1000	1000	53
	69408	Nil												69408	69408	54
	750	Nil												750	750	55
	73748	Nil												73748	73748	
	362580	194817	5	3879	7	505061	12	508940			12	508940	703757	167763	871520	

YORKSHIRE RIVER AUTHORITY

TABLE 189 Main survey Numbers and volumes of discharges of industrial effluent to rivers and canals in England and Wales and percentage considered satisfactory

Industry		Discharges of total effluent (which may contain some cooling water) to											
		Non-tidal rivers			Tidal rivers			All rivers			Canals*		
CBI Code	Description	Total Number (% satisfactory)	Total Volume 1000s galls (% satisfactory)	Volume of Cooling Water 1000s galls	Total Number (% satisfactory)	Total Volume 1000s galls (% satisfactory)	Volume of Cooling Water 1000s galls	Total Number (% satisfactory)	Total Volume 1000s galls (% satisfactory)	Volume of Cooling Water 1000s galls	Total Number (% satisfactory)	Total Volume 1000s galls (% satisfactory)	Volume Cooling Water 1000s galls
01	Brewing	6(33)	165(50)	30				6(33)	165(50)	30			
02	Brickmaking	1(Nil)	5(Nil)	Nil				1(Nil)	5(Nil)	Nil			
03	Cement making	12(75)	68(78)	Nil				12(75)	68(78)	Nil	1(Nil)	3(Nil)	Nil
04	Chemical and allied industries	19(10)	1958(2)	477	1(Nil)	63(Nil)	Nil	20(10)	2021(2)	477	2(50)	342(9)	156
05	Coal mining	57(89)	3568(88)	Nil				57(89)	3568(88)	Nil	1(100)	113(100)	Nil
06	Distillation of ethanol												
07	Electricity generation	18(89)	43650(100)	735				18(89)	43650(100)	735			
08	Engineering	78(19)	8865(9)	243	1(Nil)	20(Nil)	Nil	79(19)	8885(9)	243	13(8)	393(8)	Nil
09	Food processing and manufacture	9(55)	20674(100)	9	4(Nil)	5939(Nil)	2385	13(38)	26613(78)	2394			
10	Gas and coke	20(30)	5850(4)	2992				20(30)	5850(4)	2992			
11	Glass making	1(Nil)	1000(Nil)	Nil				1(Nil)	1000(Nil)	Nil			
12	Glue and gelatine												
13	General manufacturing	9(33)	1502(68)	10	2(50)	125(4)	Nil	11(36)	1627(63)	10			
14	Iron and steel	53(47)	20626(22)	5033				53(47)	20626(22)	5033			
15	Laundering and dry cleaning	1(Nil)	40(Nil)	Nil				1(Nil)	40(Nil)	Nil			
16	Leather tanning	1(Nil)	3(Nil)	Nil				1(Nil)	3(Nil)	Nil			
17	Metal smelting	1(Nil)	17(Nil)	Nil				1(Nil)	17(Nil)	Nil			
18	Paint making												
19	Paper and board making	13(23)	7769(4)	Nil	1(Nil)	480(Nil)	Nil	14(21)	8249(4)	Nil			
20	Petroleum refining												
21	Plastics manufacture												
22	Plating and metal finishing	8(13)	335(30)	Nil				8(13)	335(30)	Nil			
23	Pottery making	1(100)	5(100)	Nil				1(100)	5(100)	Nil			
24	Printing ink etc												
25	Quarrying and mining	15(93)	6605(100)	Nil	1(100)	10(100)	Nil	16(94)	6615(100)	Nil			
26	Rubber processing	1(Nil)	1(Nil)	Nil				1(Nil)	1(Nil)	Nil			
27	Soap and detergent												
28	Textile, cotton and man-made	71(11)	3918(5)	28				71(11)	3918(5)	28	1(Nil)	1(Nil)	Nil
29	Textile, wool	78(12)	4118(18)	29				78(12)	4118(18)	29			
30	General farming												
31	Atomic energy establishments												
50	Water treatment	45(55)	3489(65)	Nil	3(100)	2000(100)	Nil	48(58)	5489(78)	Nil			
51	Disposal tip drainage	19(26)	485(5)	Nil				19(26)	485(5)	Nil			
	TOTAL FOR CODES 1-51	537(37)	134716(63)	9586	13(38)	8637(23)	2385	550(37)	143353(60)	11971	18(17)	852(20)	156
52	Derelict coal mines	45	10916					45	10916				
53	Other derelict mines	11	2335					11	2335				
54	Active coal mines	103	30643					103	30643		6	2668	
55	Other active mines	6	1158					6	1158		2	15	
	TOTAL FOR CODES 52-55	165	45052					165	45052		8	2683	
	TOTAL	702	179768	9586	13	8637	2385	715	188405	11971	26	3535	156

			Discharges of solely cooling water to										Total Cooling Water Discharged 1000s galls	Total Process Water Discharged 1000s galls	Total Discharge 1000s galls	
vers and canals			Non-tidal rivers		Tidal rivers		All rivers		Canals		All rivers and canals					
ber atis- ry)	Total Volume 1000s galls (% satis- factory)	Volume of Cooling Water 1000s galls	Total Number (% satis- factory)	Total Volume 1000s galls (% satis- factory)	Total Number (% satis- factory)	Total Volume 1000s galls (% satis- factory)	Total Number (% satis- factory)	Total Volume 1000s galls (% satis- factory)	Total Number (% satis- factory)	Total Volume 1000s galls (% satis- factory)	Total Number (% satis- factory)	Total Volume 1000s galls (% satis- factory)				CBI Code
33)	165(50)	30	3(100)	121(100)			3(100)	121(100)	1(100)	25(100)	4(100)	146(100)	176	135	311	01
Nil)	5(Nil)	Nil	2(100)	11(100)			2(100)	11(100)			2(100)	11(100)	11	5	16	02
69)	71(75)	Nil												71	71	03
14)	2363(3)	633	58(76)	25142(56)			58(76)	25142(56)	6(83)	2740(42)	64(77)	27882(55)	28515	1730	30245	04
90)	3681(88)	Nil	6(100)	891(100)			6(100)	891(100)	1(100)	5(100)	7(100)	896(100)	896	3681	4577	05
																06
89)	43650(100)	735	21(90)	1234900(93)	2(100)	42000(100)	23(91)	1276900(94)			23(91)	1276900(94)	1277635	42915	1320550	07
17)	9278(9)	243	33(79)	2819(55)			33(79)	2819(55)	8(100)	112(100)	41(83)	2931(56)	3174	9035	12209	08
38)	21613(78)	2394	12(92)	5323(35)	1(Nil)	750(Nil)	13(85)	6073(31)			13(85)	6073(31)	8467	24219	32686	09
30)	5850(4)	2992	3(67)	600(50)			3(67)	600(50)	1(Nil)	192(Nil)	4(50)	792(38)	3784	2858	6642	10
Nil)	1000(Nil)	Nil	2(100)	1445(100)			2(100)	1445(100)	2(100)	92(100)	4(100)	1537(100)	1537	1000	2537	11
			1(100)	250(100)			1(100)	250(100)			1(100)	250(100)	250		250	12
36)	1627(63)	10	2(100)	94(100)			2(100)	94(100)	5(80)	71(89)	7(86)	165(95)	175	1617	1792	13
47)	20626(22)	5033	32(47)	1904(58)			32(47)	1904(58)	1(100)	5(100)	33(48)	1909(58)	6942	15593	22535	14
Nil)	40(Nil)	Nil	3(100)	285(100)			3(100)	285(100)			3(100)	285(100)	285	40	325	15
														3	3	16
Nil)	3(Nil)	Nil							1(100)	12(100)	1(100)	12(100)	12	17	29	17
Nil)	17(Nil)	Nil														18
21)	8249(4)	Nil	1(100)	500(100)			1(100)	500(100)			1(100)	500(100)	500	8249	8749	19
																20
			6(100)	767(100)			6(100)	767(100)	1(100)	150(100)	7(100)	917(100)	917		917	21
(13)	335(30)	Nil	3(100)	358(100)			3(100)	358(100)	4(100)	177(100)	7(100)	535(100)	535	335	870	22
(100)	5(100)	Nil												5	5	23
			1(100)	8(100)			1(100)	8(100)			1(100)	8(100)	8		8	24
(94)	6615(100)	Nil												6615	6615	25
(Nil)	1(Nil)	Nil	3(100)	605(100)			3(100)	605(100)			3(100)	605(100)	605	1	606	26
			1(100)	30(100)			1(100)	30(100)			1(100)	30(100)	30		30	27
(11)	3919(4)	28	9(100)	1353(100)			9(100)	1353(100)	1(100)	100(100)	10(100)	1453(100)	1481	3891	5372	28
(12)	4118(18)	29	14(100)	1403(100)			14(100)	1403(100)	1(100)	100(100)	15(100)	1503(100)	1532	4089	5621	29
			1(100)	30(100)			1(100)	30(100)			1(100)	30(100)	30		30	30
																31
(58)	5489(78)	Nil												5489	5489	50
(26)	485(5)	Nil												485	485	51
(36)	144205(60)	12127	217(81)	1278839(92)	3(67)	42750(98)	220(80)	1321589(93)	33(91)	3781(52)	253(81)	1325370(93)	1337497	132078	1469575	
	10916													10916	10916	52
	2335													2335	2335	53
	33311													33311	33311	54
	1173													1173	1173	55
	47735													47735	47735	
	191940	12127	217	1278839	3	42750	220	1321589	33	3781	253	1325370	1337497	179813	1517310	

TRENT RIVER AUTHORITY

TABLE 190 Main survey Numbers and volumes of discharges of industrial effluent to rivers and canals in England and Wales and percentage considered satisfactory

Industry		Discharges of total effluent (which may contain some cooling water) to											
		Non-tidal rivers			Tidal rivers			All rivers			Canals		
CBI Code	Description	Total Number (% satisfactory)	Total Volume 1000s galls (% satisfactory)	Volume of Cooling Water 1000s galls	Total Number (% satisfactory)	Total Volume 1000s galls (% satisfactory)	Volume of Cooling Water 1000s galls	Total Number (% satisfactory)	Total Volume 1000s galls (% satisfactory)	Volume of Cooling Water 1000s galls	Total Number (% satisfactory)	Total Volume 1000s galls (% satisfactory)	Volume Cooling Water 1000s galls
01	Brewing												
02	Brickmaking	15(93)	1769(98)	Nil				15(93)	1769(98)	Nil			
03	Cement making	2(50)	41(88)	Nil				2(50)	41(88)	Nil			
04	Chemical and allied industries	12(50)	4938(58)	1212	2(Nil)	396(Nil)	Nil	14(43)	5334(54)	1212	8(38)	2582(28)	182
05	Coal mining	87(71)	15799(61)	1040				87(71)	15799(61)	1040	2(50)	174(46)	Nil
06	Distillation of ethanol												
07	Electricity generation	21(95)	133817(100)	80300	5(60)	13400(97)	Nil	26(88)	147217(100)	80300	4(50)	1515(74)	44
08	Engineering	13(69)	900(49)	113				13(69)	900(49)	113	11(45)	5529(84)	67
09	Food processing and manufacture	11(36)	2173(70)	1196	1(Nil)	48(Nil)	Nil	12(33)	2221(69)	1196			
10	Gas and coke	9(78)	1505(93)	691				9(78)	1505(93)	691			
11	Glass making	2(100)	325(100)	Nil				2(100)	325(100)	Nil	2(50)	250(96)	Nil
12	Glue and gelatine												
13	General manufacturing	1(Nil)	25(Nil)	Nil				1(Nil)	25(Nil)	Nil			
14	Iron and steel	28(57)	9258(32)	1282	1(Nil)	530(Nil)	Nil	29(55)	9788(31)	1282	12(42)	5468(15)	589
15	Laundering and dry cleaning												
16	Leather tanning										1(Nil)	10(Nil)	Nil
17	Metal smelting	6(67)	2101(98)	8				6(67)	2101(98)	8	3(67)	2013(31)	692
18	Paint making												
19	Paper and board making	3(Nil)	5940(Nil)	Nil				3(Nil)	5940(Nil)	Nil			
20	Petroleum refining												
21	Plastics manufacture												
22	Plating and metal finishing	11(36)	2172(25)	744				11(36)	2172(25)	744			
23	Pottery making	3(33)	125(48)	Nil				3(33)	125(48)	Nil			
24	Printing ink etc												
25	Quarrying and mining	32(94)	8808(87)	14				32(94)	8808(87)	14			
26	Rubber processing	9(67)	4626(47)	1287				9(67)	4626(47)	1287			
27	Soap and detergent												
28	Textile, cotton and man-made	1(100)	80(100)	Nil				1(100)	80(100)	Nil			
29	Textile, wool												
30	General farming												
31	Atomic energy establishments												
50	Water treatment	10(70)	4482(91)	150				10(70)	4482(91)	150			
51	Disposal tip drainage	6(83)	1489(48)	Nil				6(83)	1489(48)	Nil			
	TOTAL FOR CODES 1-51	282(71)	200373(86)	88037	9(33)	14374(90)	Nil	291(69)	214747(86)	88037	43(44)	17541(47)	1574
52	Derelict coal mines	12	8541					12	8541				
53	Other derelict mines												
54	Active coal mines	12	5740					12	5740				
55	Other active mines												
	TOTAL FOR CODES 52-55	24	14281					24	14281				
	TOTAL	306	214654	88037	9	14374	Nil	315	229028	88037	43	17541	1574

			Discharges of solely cooling water to										Total Cooling Water Discharged 1000s galls	Total Process Water Discharged 1000s galls	Total Discharge 1000s galls	
rivers and canals			Non-tidal rivers		Tidal rivers		All rivers		Canals		All rivers and canals					
ber atis- ory)	Total Volume 1000s galls (% satis- factory)	Volume of Cooling Water 1000s galls	Total Number (% satis- factory)	Total Volume 1000s galls (% satis- factory)	Total Number (% satis- factory)	Total Volume 1000s galls (% satis- factory)	Total Number (% satis- factory)	Total Volume 1000s galls (% satis- factory)	Total Number (% satis- factory)	Total Volume 1000s galls (% satis- factory)	Total Number (% satis- factory)	Total Volume 1000s galls (% satis- factory)				CBI Code
(93)	1769(98)	Nil	2(100)	64(100)			2(100)	64(100)			2(100)	64(100)	64	Nil	64	01
50)	41(88)	Nil	2(100)	160(100)			2(100)	160(100)			2(100)	160(100)	160	1769	1929	02
			1(100)	312(100)			1(100)	312(100)			1(100)	312(100)	312	41	353	03
41)	7916(45)	1394	9(89)	67428(100)			9(89)	67428(100)	11(64)	9231(67)	20(75)	76659(96)	78053	6522	84575	04
71)	15973(61)	1040	2(100)	134(100)			2(100)	134(100)			2(100)	134(100)	1174	14933	16107	05
																06
83)	148732(99·7)	80344	15(87)	1108724(98)	4(100)	213510(100)	19(89)	1322234(98)	2(100)	4850(100)	21(90)	1327084(98)	1407428	68388	1475816	07
58)	6429(79)	180	9(89)	2533(98)			9(89)	2533(98)	26(81)	7296(94)	35(83)	9829(95)	10009	6249	16258	08
33)	2221(69)	1196	9(55)	7240(43)			9(55)	7240(43)			9(55)	7240(43)	8436	1025	9461	09
78)	1505(93)	691	8(100)	1404(100)			8(100)	1404(100)	4(50)	1176(28)	12(83)	2580(67)	3271	814	4085	10
75)	575(98)	Nil							3(100)	773(100)	3(100)	773(100)	773	575	1348	11
			1(100)	700(100)			1(100)	700(100)			1(100)	700(100)	700	Nil	700	12
(Nil)	25(Nil)	Nil	1(100)	50(100)			1(100)	50(100)	1(100)	530(100)	2(100)	580(100)	580	25	605	13
51)	15256(25)	1871	11(91)	10410(99)			11(91)	10410(99)	10(40)	14740(63)	21(67)	25150(78)	27021	13385	40406	14
			1(100)	492(100)			1(100)	492(100)			1(100)	492(100)	492	Nil	492	15
(Nil)	10(Nil)	Nil	1(100)	65(100)			1(100)	65(100)	1(100)	20(100)	2(100)	85(100)	85	10	95	16
67)	4114(65)	700	10(70)	22499(99)			10(70)	22499(99)	4(25)	442(7)	14(57)	22941(97)	23641	3414	27055	17
			1(Nil)	10(Nil)			1(Nil)	10(Nil)			1(Nil)	10(Nil)	10	Nil	10	18
(Nil)	5940(Nil)	Nil	2(100)	4000(100)			2(100)	4000(100)			2(100)	4000(100)	4000	5940	9940	19
																20
																21
(36)	2172(25)	744	9(78)	367(59)			9(78)	367(59)			9(78)	367(59)	1111	1428	2539	22
(33)	125(48)	Nil											Nil	125	125	23
																24
(94)	8808(87)	14	4(100)	40(100)			4(100)	40(100)			4(100)	40(100)	54	8794	8848	25
(67)	4626(47)	1287	13(100)	1938(100)			13(100)	1938(100)			13(100)	1938(100)	3225	3339	6564	26
			1(100)	1220(100)			1(100)	1220(100)			1(100)	1220(100)	1220	Nil	1220	27
(100)	80(100)	Nil	1(100)	20(100)			1(100)	20(100)			1(100)	20(100)	20	80	100	28
																29
																30
																31
(70)	4482(91)	150											150	4332	4482	50
(83)	1489(48)	Nil											Nil	1489	1489	51
(66)	232288(83)	89611	113(87)	1229810(98)	4(100)	213510(100)	117(87)	1443320(98)	62(68)	39058(74)	179(80)	1482378(97)	1571989	142677	1714666	
2	8541													8541	8541	52
																53
2	5740													5740	5740	54
																55
4	14281													14281	14281	
8	246569	89611	113	1229810	4	213510	117	1443320	62	39058	179	1482378	1571989	156958	1728947	

LINCOLNSHIRE RIVER AUTHORITY

TABLE 191 Main survey Numbers and volumes of discharges of industrial effluent to rivers and canals in England and Wales and percentage considered satisfactory

Industry		Discharges of total effluent (which may contain some cooling water) to											
		Non-tidal rivers			Tidal rivers			All rivers			Canals		
CBI Code	Description	Total Number (% satisfactory)	Total Volume 1000s galls (% satisfactory)	Volume of Cooling Water 1000s galls	Total Number (% satisfactory)	Total Volume 1000s galls (% satisfactory)	Volume of Cooling Water 1000s galls	Total Number (% satisfactory)	Total Volume 1000s galls (% satisfactory)	Volume of Cooling Water 1000s galls	Total Number (% satisfactory)	Total Volume 1000s galls (% satisfactory)	Volume Cooling Water 1000s galls
01	Brewing												
02	Brickmaking												
03	Cement making	1(100)	84(100)	70				1(100)	84(100)	70			
04	Chemical and allied industries	2(Nil)	114(Nil)	Nil				2(Nil)	114(Nil)	Nil			
05	Coal mining												
06	Distillation of ethanol												
07	Electricity generation	1(100)	175(100)	Nil				1(100)	175(100)	Nil			
08	Engineering												
09	Food processing and manufacture	5(40)	814(17)	27				5(40)	814(17)	27			
10	Gas and coke												
11	Glass making												
12	Glue and gelatine												
13	General manufacturing	1(Nil)	6(Nil)	Nil				1(Nil)	6(Nil)	Nil			
14	Iron and steel	1(Nil)	1475(Nil)	Nil				1(Nil)	1475(Nil)	Nil			
15	Laundering and dry cleaning												
16	Leather tanning												
17	Metal smelting												
18	Paint making												
19	Paper and board making												
20	Petroleum refining	2(Nil)	2830(Nil)	260				2(Nil)	2830(Nil)	260			
21	Plastics manufacture												
22	Plating and metal finishing												
23	Pottery making												
24	Printing ink etc												
25	Quarrying and mining	2(50)	37(51)	Nil				2(50)	37(51)	Nil			
26	Rubber processing												
27	Soap and detergent												
28	Textile, cotton and man-made				1(Nil)	55(Nil)	28	1(Nil)	55(Nil)	28			
29	Textile, wool												
30	General farming	1(100)	100(100)	Nil				1(100)	100(100)	Nil			
31	Atomic energy establishments												
50	Water treatment	4(100)	146(100)	Nil				4(100)	146(100)	Nil			
51	Disposal tip drainage												
	TOTAL FOR CODES 1–51	20(50)	5781(11)	357	1(Nil)	55(Nil)	28	21(48)	5836(11)	385			
52	Derelict coal mines												
53	Other derelict mines												
54	Active coal mines												
55	Other active mines												
	TOTAL FOR CODES 52–55												
	TOTAL	20	5781	357	1	55	28	21	5836	385			

191

...rivers and canals			Discharges of solely cooling water to											Total Cooling Water Discharged 1000s galls	Total Process Water Discharged 1000s galls	Total Discharge 1000s galls	CBI Code
...ber (% satis- ...ry)	Total Volume 1000s galls (% satis- factory)	Volume of Cooling Water 1000s galls	Non-tidal rivers		Tidal rivers		All rivers		Canals		All rivers and canals						
			Total Number (% satis- factory)	Total Volume 1000s galls (% satis- factory)	Total Number (% satis- factory)	Total Volume 1000s galls (% satis- factory)	Total Number (% satis- factory)	Total Volume 1000s galls (% satis- factory)	Total Number (% satis- factory)	Total Volume 1000s galls (% satis- factory)	Total Number (% satis- factory)	Total Volume 1000s galls (% satis- factory)					
																01
																02
00)	84(100)	70											70	14	84	03
																04
il)	114(Nil)	Nil											Nil	114	114	05
																06
00)	175(100)	Nil											Nil	175	175	07
			4(100)	603(100)			4(100)	603(100)			4(100)	603(100)	603	603	603	08
0)	814(17)	27											27	787	814	09
																10
																11
																12
Nil)	6(Nil)	Nil											Nil	6	6	13
Nil)	1475(Nil)	Nil	3(Nil)	2421(Nil)			3(Nil)	2421(Nil)			3(Nil)	2421(Nil)	2421	1475	3896	14
																15
			1(100)	192(100)			1(100)	192(100)			1(100)	192(100)	192	Nil	192	16
																17
																18
Nil)	2830(Nil)	260											260	2570	2830	19
																20
																21
																22
																23
																24
50)	37(51)	Nil											Nil	37	37	25
																26
																27
Nil)	55(Nil)	28											28	27	55	28
																29
100)	100(100)	Nil											Nil	100	100	30
																31
100)	146(100)	Nil											Nil	146	146	50
																51
48)	5836(11)	385	8(63)	3216(25)			8(63)	3216(25)			8(63)	3216(25)	3601	5451	9052	
																52
																53
																54
																55
	5836	385	8	3216			8	3216			8	3216	3601	5451	9052	

WELLAND AND NENE RIVER AUTHORITY

TABLE 192 Main survey Numbers and volumes of discharges of industrial effluent to rivers and canals in England and Wales and percentage considered satisfactory

Industry		Discharges of total effluent (which may contain some cooling water) to											
		Non-tidal rivers			Tidal rivers			All rivers			Canals		
CBI Code	Description	Total Number (% satisfactory)	Total Volume 1000s galls (% satisfactory)	Volume of Cooling Water 1000s galls	Total Number (% satisfactory)	Total Volume 1000s galls (% satisfactory)	Volume of Cooling Water 1000s galls	Total Number (% satisfactory)	Total Volume 1000s galls (% satisfactory)	Volume of Cooling Water 1000s galls	Total Number (% satisfactory)	Total Volume 1000s galls (% satisfactory)	Volume Cooling Water 1000s galls
01	Brewing	1(100)	1407(100)	Nil	1(100)	10(100)	1	2(100)	1417(100)	1			
02	Brickmaking												
03	Cement making												
04	Chemical and allied industries	1(100)	19(100)	19				1(100)	19(100)	19			
05	Coal mining												
06	Distillation of ethanol												
07	Electricity generation												
08	Engineering	2(50)	23(56)	2				2(50)	23(56)	2			
09	Food processing and manufacture	3(33)	1075(1)	10	3(Nil)	2415(Nil)	516	6(70)	3490(Nil)	526			
10	Gas and coke	1(100)	19(100)	Nil				1(100)	19(100)	Nil			
11	Glass making												
12	Glue and gelatine												
13	General manufacturing												
14	Iron and steel	3(100)	6200(100)	4960				3(100)	6200(100)	4960			
15	Laundering and dry cleaning												
16	Leather tanning	5(20)	401(1)	3				5(20)	401(1)	3			
17	Metal smelting												
18	Paint making												
19	Paper and board making	1(Nil)	630(Nil)	Nil				1(Nil)	630(Nil)	Nil			
20	Petroleum refining												
21	Plastics manufacture												
22	Plating and metal finishing												
23	Pottery making												
24	Printing ink etc												
25	Quarrying and mining												
26	Rubber processing												
27	Soap and detergent												
28	Textile, cotton and man-made												
29	Textile, wool												
30	General farming												
31	Atomic energy establishments												
50	Water treatment												
51	Disposal tip drainage												
	TOTAL FOR CODES 1–51	17(53)	9774(78)	4994	4(25)	2425(Nil)	517	21(48)	12199(63)	5511			
52	Derelict coal mines												
53	Other derelict mines												
54	Active coal mines												
55	Other active mines												
	TOTAL FOR CODES 52–55												
	TOTAL	17	9774	4994	4	2425	517	21	12199	5511			

| | rivers and canals | | Discharges of solely cooling water to | | | | | | | | | | Total Cooling Water Discharged 1000s galls | Total Process Water Discharged 1000s galls | Total Discharge 1000s galls | |
| | | | Non-tidal rivers | | Tidal rivers | | All rivers | | Canals | | All rivers and canals | | | | | |
Total Volume 1000s galls (% satis-factory)	Total Volume 1000s galls (% satis-factory)	Volume of Cooling Water 1000s galls	Total Number (% satis-factory)	Total Volume 1000s galls (% satis-factory)	Total Number (% satis-factory)	Total Volume 1000s galls (% satis-factory)	Total Number (% satis-factory)	Total Volume 1000s galls (% satis-factory)	Total Number (% satis-factory)	Total Volume 1000s galls (% satis-factory)	Total Number (% satis-factory)	Total Volume 1000s galls (% satis-factory)				CBI Code
	1417(100)	1											1	1416	1417	01
																02
			1(100)	1107(100)			1(100)	1107(100)			1(100)	1107(100)	1107	Nil	1107	03
	19(100)	19	2(50)	1959(1)			2(50)	1959(1)			2(50)	1959(1)	1978	Nil	1978	04
																05
																06
	23(56)	2	2(100)	55196(100)			2(100)	55196(100)			2(100)	55196(100)	55196	Nil	55196	07
	3490(Nil)	526	1(100)	32(100)			1(100)	32(100)			1(100)	32(100)	34	21	55	08
			1(100)	132(100)			1(100)	132(100)			1(100)	132(100)	658	2964	3622	09
	19(100)	Nil											Nil	19	19	10
																11
									1(100)	42(100)	1(100)	42(100)	42	Nil	42	12
	6200(100)	4960	1(100)	35(100)			1(100)	35(100)			1(100)	35(100)	35	Nil	35	13
													4960	1240	6200	14
																15
	401(1)	3											3	398	401	16
																17
																18
	630(Nil)	Nil											Nil	630	630	19
																20
																21
																22
																23
																24
			1(100)	630(100)			1(100)	630(100)			1(100)	630(100)	630	Nil	630	25
																26
																27
																28
																29
																30
																31
																50
																51
(48)	12199(63)	5511	9(89)	59091(99·7)			9(89)	59091(99·7)	1(100)	42(100)	10(90)	59133(99·7)	64644	6688	71332	
																52
																53
																54
																55
	12199	5511	9	59091			9	59091	1	42	10	59133	64644	6688	71332	

GREAT OUSE RIVER AUTHORITY

TABLE 193 Main survey Numbers and volumes of discharges of industrial effluent to rivers and canals in England and Wales and percentage considered satisfactory

Industry		Discharges of total effluent (which may contain some cooling water) to											
		Non-tidal rivers			Tidal rivers			All rivers			Canals		
CBI Code	Description	Total Number (% satisfactory)	Total Volume 1000s galls (% satisfactory)	Volume of Cooling Water 1000s galls	Total Number (% satisfactory)	Total Volume 1000s galls (% satisfactory)	Volume of Cooling Water 1000s galls	Total Number (% satisfactory)	Total Volume 1000s galls (% satisfactory)	Volume of Cooling Water 1000s galls	Total Number (% satisfactory)	Total Volume 1000s galls (% satisfactory)	Volume Cooling Water 1000s galls
01	Brewing												
02	Brickmaking	1(Nil)	99(Nil)	Nil				1(Nil)	99(Nil)	Nil			
03	Cement making												
04	Chemical and allied industries	3(100)	4300(100)	3000	2(100)	580(100)	58	5(100)	4880(100)	3058			
05	Coal mining												
06	Distillation of ethanol												
07	Electricity generation	1(100)	25(100)	Nil				1(100)	25(100)	Nil			
08	Engineering	2(100)	104(100)	Nil				2(100)	104(100)	Nil			
09	Food processing and manufacture	9(78)	3205(90)	569	3(Nil)	5320(Nil)	32	12(58)	8525(36)	601			
10	Gas and coke	1(100)	120(100)	Nil				1(100)	120(100)	Nil			
11	Glass making												
12	Glue and gelatine												
13	General manufacturing												
14	Iron and steel												
15	Laundering and dry cleaning	1(Nil)	15(Nil)	Nil				1(Nil)	15(Nil)	Nil			
16	Leather tanning	1(Nil)	100(Nil)	Nil				1(Nil)	100(Nil)	Nil			
17	Metal smelting												
18	Paint making												
19	Paper and board making	1(100)	2000(100)	Nil				1(100)	2000(100)	Nil			
20	Petroleum refining												
21	Plastics manufacture												
22	Plating and metal finishing												
23	Pottery making												
24	Printing ink etc												
25	Quarrying and mining	2(100)	1010(100)	Nil				2(100)	1010(100)	Nil			
26	Rubber processing												
27	Soap and detergent												
28	Textile, cotton and man-made												
29	Textile, wool												
30	General farming	1(100)	70(100)	Nil				1(100)	70(100)	Nil			
31	Atomic energy establishments												
50	Water treatment												
51	Disposal tip drainage												
	TOTAL FOR CODES 1-51	23(78)	11048(95)	3569	5(40)	5900(10)	90	28(71)	16948(67)	3659			
52	Derelict coal mines												
53	Other derelict mines												
54	Active coal mines												
55	Other active mines												
	TOTAL FOR CODES 52–55												
	TOTAL	23	11048	3569	5	5900	90	28	16948	3659			

			Discharges of solely cooling water to										Total Cooling Water Discharged 1000s galls	Total Process Water Discharged 1000s galls	Total Discharge 1000s galls	
vers and canals			Non-tidal rivers		Tidal rivers		All rivers		Canals		All rivers and canals					
er atis- ry)	Total Volume 1000s galls (% satis- factory)	Volume of Cooling Water 1000s galls	Total Number (% satis- factory)	Total Volume 1000s galls (% satis- factory)	Total Number (% satis- factory)	Total Volume 1000s galls (% satis- factory)	Total Number (% satis- factory)	Total Volume 1000s galls (% satis- factory)	Total Number (% satis- factory)	Total Volume 1000s galls (% satis- factory)	Total Number (% satis- factory)	Total Volume 1000s galls (% satis- factory)				CBI Code
l)	99(Nil)	Nil	2(100) 1(100)	470(100) 10(100)			2(100) 1(100)	470(100) 10(100)			2(100) 1(100)	470(100) 10(100)	470 Nil 10	Nil 99 Nil	470 99 10	01 02 03
0)	4880(100)	3058											3058	1822	4880	04 05 06
0) 0) 3)	25(100) 104(100) 8525(36)	Nil Nil 601	2(100) 1(100) 4(100)	290000(100) 13(100) 283(100)			2(100) 1(100) 4(100)	290000(100) 13(100) 283(100)			2(100) 1(100) 4(100)	290000(100) 13(100) 283(100)	290000 13 884	25 104 7924	290025 117 8808	07 08 09
0)	120(100)	Nil											Nil	120	120	10 11 12
																13 14
il)	15(Nil)	Nil											Nil	15	15	15
il)	100(Nil)	Nil	1(100)	90(100)			1(100)	90(100)			1(100)	90(100)	90	100	190	16 17 18
00)	2000(100)	Nil											Nil	2000	2000	19 20 21
																22 23 24
00)	1010(100)	Nil											Nil	1010	1010	25 26 27
																28 29
00)	70(100)	Nil											Nil	70	70	30
																31 50 51
71)	16948(67)	3659	11(100)	290866(100)			11(100)	290866(100)			11(100)	290866(100)	294525	13289	307814	
																52 53 54 55
	16948	3659	11	290866			11	290866			11	290866	294525	13289	307814	

EAST SUFFOLK AND NORFOLK

TABLE 194 Main survey Numbers and volumes of discharges of industrial effluent to rivers and canals in England and Wales and percentage considered satisfactory

Industry		Discharges of total effluent (which may contain some cooling water) to											
		Non-tidal rivers			Tidal rivers			All rivers			Canals		
CBI Code	Description	Total Number (% satisfactory)	Total Volume 1000s galls (% satisfactory)	Volume of Cooling Water 1000s galls	Total Number (% satisfactory)	Total Volume 1000s galls (% satisfactory)	Volume of Cooling Water 1000s galls	Total Number (% satisfactory)	Total Volume 1000s galls (% satisfactory)	Volume of Cooling Water 1000s galls	Total Number (% satisfactory)	Total Volume 1000s galls (% satisfactory)	Volume Cooling Water 1000s galls
01	Brewing	5(80)	263(54)	Nil	2(Nil)	44(Nil)	Nil	7(57)	307(46)	Nil			
02	Brickmaking												
03	Cement making				1(100)	3(100)	Nil	1(100)	3(100)	Nil			
04	Chemical and allied industries				2(100)	675(100)	Nil	2(100)	675(100)	Nil			
05	Coal mining												
06	Distillation of ethanol												
07	Electricity generation												
08	Engineering	4(50)	64(20)	Nil	2(50)	8(13)	Nil	6(50)	72(19)	Nil			
09	Food processing and manufacture	9(55)	1812(52)	600	6(17)	305(1)	Nil	15(40)	4863(20)	600			
10	Gas and coke				1(Nil)	250(Nil)	Nil	1(Nil)	250(Nil)	Nil			
11	Glass making												
12	Glue and gelatine												
13	General manufacturing												
14	Iron and steel												
15	Laundering and dry cleaning												
16	Leather tanning	1(Nil)	10(Nil)	Nil	1(Nil)	20(Nil)	Nil	2(Nil)	30(Nil)	Nil			
17	Metal smelting												
18	Paint making												
19	Paper and board making												
20	Petroleum refining												
21	Plastics manufacture												
22	Plating and metal finishing												
23	Pottery making												
24	Printing ink etc												
25	Quarrying and mining	2(50)	1001(Nil)	Nil	1(Nil)	15(Nil)	Nil	3(33)	1016(Nil)	Nil			
26	Rubber processing												
27	Soap and detergent												
28	Textile, cotton and man-made												
29	Textile, wool												
30	General farming												
31	Atomic energy establishments												
50	Water treatment	2(100)	312(100)	Nil	1(100)	90(100)	Nil	3(100)	402(100)	Nil			
51	Disposal tip drainage												
	TOTAL FOR CODES 1-51	23(61)	3462(41)	600	17(35)	4156(19)	Nil	40(50)	7618(29)	600			
52	Derelict coal mines												
53	Other derelict mines												
54	Active coal mines												
55	Other active mines												
	TOTAL FOR CODES 52-55			Nil			Nil						
	TOTAL	23	3462	600	17	4156	Nil	40	7618	600			

			Discharges of solely cooling water to										Total Cooling Water Discharged 1000s galls	Total Process Water Discharged 1000s galls	Total Discharge 1000s galls	
ers and canals			Non-tidal rivers		Tidal rivers		All rivers		Canals		All rivers and canals					
satis-factory)	Total Volume 1000s galls (% satis-factory)	Volume of Cooling Water 1000s galls	Total Number (% satis-factory)	Total Volume 1000s galls (% satis-factory)	Total Number (% satis-factory)	Total Volume 1000s galls (% satis-factory)	Total Number (% satis-factory)	Total Volume 1000s galls (% satis-factory)	Total Number (% satis-factory)	Total Volume 1000s galls (% satis-factory)	Total Number (% satis-factory)	Total Volume 1000s galls (% satis-factory)				CBI Code
)	307(46)	Nil	1(100)	14(100)	2(100)	20(100)	3(100)	34(100)			3(100)	34(100)	34	307	341	01
																02
0)	3(100)	Nil											Nil	3	3	03
0)	675(100)	Nil	1(100)	200(100)			1(100)	200(100)			1(100)	200(100)	200	675	875	04
																05
																06
	72(19)	Nil			2(100)	335120(100)	2(100)	335120(100)			2(100)	335120(100)	335120	Nil	335120	07
)	4863(20)	600	4(50)	206(51)	1(100)	5(100)	1(100)	5(100)			1(100)	5(100)	5	72	77	08
)					3(100)	5300(100)	7(71)	5506(98)			7(71)	5506(98)	6106	4263	10369	09
il)	250(Nil)	Nil											Nil	250	250	10
																11
																12
			3(100)	89(100)	1(100)	130(100)	4(100)	219(100)			4(100)	219(100)	219	Nil	219	13
																14
																15
il)	30(Nil)	Nil											Nil	30	30	16
																17
			1(100)	150(100)			1(100)	150(100)			1(100)	150(100)	150	Nil	150	18
																19
																20
																21
																22
																23
																24
3)	1016(Nil)	Nil											Nil	1016	1016	25
																26
																27
																28
																29
																30
																31
100)	402(100)	Nil											Nil	402	402	50
																51
50)	7618(29)	600	10(80)	659(85)	9(100)	340575(100)	19(89)	341234(100)			19(89)	341234(100)	341834	7018	348852	
																52
																53
																54
																55
	7618	600	10	659	9	340575	19	341234			19	341234	341834	7018	348852	

ESSEX RIVER AUTHORITY

TABLE 195 Main survey Numbers and volumes of discharges of industrial effluent to rivers and canals in England and Wales and percentage considered satisfactory

Industry		Discharges of total effluent (which may contain some cooling water) to											
		Non-tidal rivers			Tidal rivers			All rivers			Canals		
CBI Code	Description	Total Number (% satisfactory)	Total Volume 1000s galls (% satisfactory)	Volume of Cooling Water 1000s galls	Total Number (% satisfactory)	Total Volume 1000s galls (% satisfactory)	Volume of Cooling Water 1000s galls	Total Number (% satisfactory)	Total Volume 1000s galls (% satisfactory)	Volume of Cooling Water 1000s galls	Total Number (% satisfactory)	Total Volume 1000s galls (% satisfactory)	Volume Cooling Water 1000s galls
01	Brewing	1(100)	27(100)	2	2(Nil)	15(Nil)	Nil	3(33)	42(64)	2			
02	Brickmaking												
03	Cement making												
04	Chemical and allied industries	4(Nil)	357(Nil)	Nil	1(Nil)	1300(Nil)	Nil	5(Nil)	1657(Nil)	Nil			
05	Coal mining												
06	Distillation of ethanol												
07	Electricity generation	2(50)	2200(59)	Nil				2(50)	2200(59)				
08	Engineering	2(Nil)	1250(Nil)	50				2(Nil)	1250(Nil)	50			
09	Food processing and manufacture	4(25)	412(56)	Nil	1(Nil)	50(Nil)	Nil	5(20)	462(50)				
10	Gas and coke	1(Nil)	245(Nil)	Nil	1(Nil)	80(Nil)	8	2(Nil)	325(Nil)	8			
11	Glass making												
12	Glue and gelatine												
13	General manufacturing	1(Nil)	480(Nil)	38				1(Nil)	480(Nil)	38			
14	Iron and steel												
15	Laundering and dry cleaning	1(100)	24(100)	Nil				1(100)	24(100)	Nil			
16	Leather tanning												
17	Metal smelting												
18	Paint making												
19	Paper and board making												
20	Petroleum refining												
21	Plastics manufacture												
22	Plating and metal finishing												
23	Pottery making												
24	Printing ink etc	1(100)	50(100)	Nil				1(100)	50(100)	Nil			
25	Quarrying and mining	6(67)	2475(93)	Nil	1(Nil)	40(Nil)	Nil	7(57)	2515(92)	Nil			
26	Rubber processing												
27	Soap and detergent												
28	Textile, cotton and man-made	1(Nil)	990(Nil)	624				1(Nil)	990(Nil)	624			
29	Textile, wool												
30	General farming												
31	Atomic energy establishments												
50	Water treatment	5(80)	1247(97)	Nil				5(80)	1247(97)	Nil			
51	Disposal tip drainage	1(100)	4(100)	Nil				1(100)	4(100)	Nil			
	TOTAL FOR CODES 1-51	30(40)	9761(53)	714	6(Nil)	1485(Nil)	8	36(39)	11246(46)	722			
52	Derelict coal mines												
53	Other derelict mines												
54	Active coal mines												
55	Other active mines												
	TOTAL FOR CODES 52-55												
	TOTAL	30	9761	714	6	1485	8	36	11246	722			

ers and canals		Discharges of solely cooling water to										Total Cooling Water Discharged 1000s galls	Total Process Water Discharged 1000s galls	Total Discharge 1000s galls	
		Non-tidal rivers		Tidal rivers		All rivers		Canals		All rivers and canals					
Total Volume 1000s galls (% satis- factory)	Volume of Cooling Water 1000s galls	Total Number (% satis- factory)	Total Volume 1000s galls (% satis- factory)	Total Number (% satis- factory)	Total Volume 1000s galls (% satis- factory)	Total Number (% satis- factory)	Total Volume 1000s galls (% satis- factory)	Total Number (% satis- factory)	Total Volume 1000s galls (% satis- factory)	Total Number (% satis- factory)	Total Volume 1000s galls (% satis- factory)				CBI Code
42(64)	2											2	40	42	01
															02
															03
1657(Nil)	Nil	1(100)	4500(100)			1(100)	4500(100)			1(100)	4500(100)	4500	1657	6157	04
															05
															06
2200(59)	Nil			1(100)	334000(100)	1(100)	334000(100)			1(100)	334000(100)	334000	2200	336200	07
1250(Nil)	50	3(33)	1112(95)			3(33)	1112(95)			3(33)	1112(95)	1162	1200	2362	08
462(50)	Nil											Nil	462	462	09
325(Nil)	8											8	317	325	10
															11
															12
480(Nil)	38	2(100)	190(100)			2(100)	190(100)			2(100)	190(100)	228	442	670	13
															14
24(100)	Nil											Nil	24	24	15
															16
															17
															18
															19
															20
															21
		1(100)	30(100)			1(100)	30(100)			1(100)	30(100)	30	Nil	30	22
															23
50(100)	Nil	1(100)	9(100)			1(100)	9(100)			1(100)	9(100)	9	50	59	24
2515(92)	Nil											Nil	2515	2515	25
															26
															27
990(Nil)	624											624	366	990	28
															29
															30
1247(97)	Nil											Nil	1247	1247	31
4(100)	Nil											Nil	4	4	50
															51
11246(46)	722	8(75)	5841(99)	1(100)	334000(100)	9(78)	339841(99·9)			9(78)	339841(99·9)	340563	10524	351087	
															52
															53
															54
															55
11246	722	8	5841	1	334000	9	339841			9	339841	340563	10524	351087	

KENT RIVER AUTHORITY

TABLE 196 Main survey Numbers and volumes of discharges of industrial effluent to rivers and canals in England and Wales and percentage considered satisfactory

Industry		Discharges of total effluent (which may contain some cooling water) to											
		Non-tidal rivers			Tidal rivers			All rivers			Canals		
CBI Code	Description	Total Number (% satis-factory)	Total Volume 1000s galls (% satis-factory)	Volume of Cooling Water 1000s galls	Total Number (% satis-factory)	Total Volume 1000s galls (% satis-factory)	Volume of Cooling Water 1000s galls	Total Number (% satis-factory)	Total Volume 1000s galls (% satis-factory)	Volume of Cooling Water 1000s galls	Total Number (% satis-factory)	Total Volume 1000s galls (% satis-factory)	Volume Cooling Water 1000s galls
01	Brewing												
02	Brickmaking												
03	Cement making				2(100)	550(100)	Nil	2(100)	550(100)	Nil			
04	Chemical and allied industries	1(Nil)	75(Nil)	4	4(25)	566(18)	Nil	5(20)	641(16)	4			
05	Coal mining												
06	Distillation of ethanol												
07	Electricity generation				1(100)	300(100)	Nil	1(100)	300(100)	Nil			
08	Engineering	1(100)	25(100)	24				1(100)	25(100)	24			
09	Food processing and manufacture	1(Nil)	7(Nil)	Nil				1(Nil)	7(Nil)	Nil			
10	Gas and coke				1(100)	250(100)	Nil	1(100)	250(100)	Nil			
11	Glass making												
12	Glue and gelatine												
13	General manufacturing				1(Nil)	450(Nil)	Nil	1(Nil)	450(Nil)	Nil			
14	Iron and steel												
15	Laundering and dry cleaning				1(100)	100(100)	Nil	1(100)	100(100)	Nil			
16	Leather tanning												
17	Metal smelting												
18	Paint making												
19	Paper and board making	6(67)	1045(73)	30	8(38)	22430(57)	Nil	14(50)	23475(58)	30			
20	Petroleum refining				2(100)	45033(100)	18000	2(100)	45033(100)	18000			
21	Plastics manufacture												
22	Plating and metal finishing												
23	Pottery making	1(Nil)	20(Nil)	Nil				1(Nil)	20(Nil)	Nil			
24	Printing ink etc												
25	Quarrying and mining	6(100)	1416(100)	Nil				6(100)	1416(100)	Nil			
26	Rubber processing												
27	Soap and detergent												
28	Textile, cotton and man-made												
29	Textile, wool												
30	General farming												
31	Atomic energy establishments												
50	Water treatment	1(100)	24(100)	Nil				1(100)	24(100)	Nil			
51	Disposal tip drainage												
	TOTAL FOR CODES 1-51	17(71)	2612(85)	58	20(65)	69679(85)	18000	37(62)	72291(85)	18058			
52	Derelict coal mines												
53	Other derelict mines												
54	Active coal mines	2	2000					2	2000				
55	Other active mines	2	230					2	230				
	TOTAL FOR CODES 52-55	4	2230					4	2230				
	TOTAL	21	4842	58	20	69679	18000	41	74521	18058			

rivers and canals			Discharges of solely cooling water to										Total Cooling Water Discharged 1000s galls	Total Process Water Discharged 1000s galls	Total Discharge 1000s galls	CBI Code
			Non-tidal rivers		Tidal rivers		All rivers		Canals		All rivers and canals					
al nber satis- ory)	Total Volume 1000s galls (% satis- factory)	Volume of Cooling Water 1000s galls	Total Number (% satis- factory)	Total Volume 1000s galls (% satis- factory)	Total Number (% satis- factory)	Total Volume 1000s galls (% satis- factory)	Total Number (% satis- factory)	Total Volume 1000s galls (% satis- factory)	Total Number (% satis- factory)	Total Volume 1000s galls (% satis- factory)	Total Number (% satis- factory)	Total Volume 1000s galls (% satis- factory)				
			2(100)	106(100)	1(100)	30(100)	3(100)	136(100)			3(100)	136(100)	136	Nil	136	01
																02
100)	550(100)	Nil			1(100)	50(100)	1(100)	50(100)			1(100)	50(100)	50	550	600	03
20)	641(16)	4	4(100)	158(100)			4(100)	158(100)			4(100)	158(100)	162	637	799	04
																05
																06
100)	300(100)	Nil			4(100)	312420(100)	4(100)	312420(100)			4(100)	312420(100)	312420	300	312720	07
100)	25(100)	24	6(100)	103(100)			6(100)	103(100)			6(100)	103(100)	127	1	128	08
Nil)	7(Nil)	Nil	2(100)	720(100)	1(100)	17(100)	3(100)	737(100)			3(100)	737(100)	737	7	744	09
100)	250(100)	Nil											Nil	250	250	10
																11
																12
Nil)	450(Nil)	Nil	1(100)	75(100)			1(100)	75(100)			1(100)	75(100)	75	450	525	13
																14
100)	100(100)	Nil											Nil	100	100	15
																16
																17
																18
50)	23475(58)	30	6(100)	4362(100)	4(100)	79000(100)	10(100)	83362(100)			10(100)	83362(100)	83392	23445	106837	19
100)	45033(100)	18000			1(100)	105000(100)	1(100)	105000(100)			1(100)	105000(100)	123000	27033	150033	20
																21
																22
(Nil)	20(Nil)	Nil											Nil	20	20	23
																24
(100)	1416(100)	Nil											Nil	1416	1416	25
																26
																27
																28
																29
																30
																31
(100)	24(100)	Nil											Nil	24	24	50
																51
(62)	72291(85)	18058	21(100)	5524(100)	12(92)	496517(79)	33(97)	502041(79)			33(97)	502041(79)	520099	54233	574332	
																52
																53
	2000												2000	2000		54
	230												230	230		55
	2230												2230	2230		
	74521	18058	21	5524	12	496517	33	502041			33	502041	520099	56463	576562	

SUSSEX RIVER AUTHORITY

TABLE 197 Main survey Numbers and volumes of discharges of industrial effluent to rivers and canals in England and Wales and percentage considered satisfactory

Industry		Discharges of total effluent (which may contain some cooling water) to											
		Non-tidal rivers			Tidal rivers			All rivers			Canals		
CBI Code	Description	Total Number (% satisfactory)	Total Volume 1000s galls (% satisfactory)	Volume of Cooling Water 1000s galls	Total Number (% satisfactory)	Total Volume 1000s galls (% satisfactory)	Volume of Cooling Water 1000s galls	Total Number (% satisfactory)	Total Volume 1000s galls (% satisfactory)	Volume of Cooling Water 1000s galls	Total Number (% satisfactory)	Total Volume 1000s galls (% satisfactory)	Volume Cooling Water 1000s galls
01	Brewing												
02	Brickmaking												
03	Cement making												
04	Chemical and allied industries												
05	Coal mining												
06	Distillation of ethanol												
07	Electricity generation												
08	Engineering												
09	Food processing and manufacture	1(Nil)	30(Nil)	Nil				1(Nil)	30(Nil)	Nil			
10	Gas and coke												
11	Glass making												
12	Glue and gelatine												
13	General manufacturing												
14	Iron and steel												
15	Laundering and dry cleaning												
16	Leather tanning												
17	Metal smelting												
18	Paint making												
19	Paper and board making												
20	Petroleum refining												
21	Plastics manufacture												
22	Plating and metal finishing	1(100)	29(100)	Nil	1(Nil)	46(Nil)		2(50)	75(39)	Nil			
23	Pottery making												
24	Printing ink etc												
25	Quarrying and mining												
26	Rubber processing												
27	Soap and detergent												
28	Textile, cotton and man-made												
29	Textile, wool												
30	General farming	4(Nil)	10(Nil)	Nil				4(Nil)	10(Nil)	Nil			
31	Atomic energy establishments												
50	Water treatment	3(100)	74(100)	Nil				3(100)	74(100)	Nil			
51	Disposal tip drainage	1(Nil)	5(Nil)	Nil				1(Nil)	5(Nil)	Nil			
	TOTAL FOR CODES 1-51	10(40)	148(70)	Nil	1(Nil)	46(Nil)		11(36)	194(53)	Nil			
52	Derelict coal mines												
53	Other derelict mines												
54	Active coal mines												
55	Other active mines												
	TOTAL FOR CODES 52–55												
	TOTAL	10	148	Nil	1	46		11	194	Nil			

	rivers and canals		Discharges of solely cooling water to										Total Cooling Water Discharged 1000s galls	Total Process Water Discharged 1000s galls	Total Discharge 1000s galls	
			Non-tidal rivers		Tidal rivers		All rivers		Canals		All rivers and canals					
al nber satis- ory)	Total Volume 1000s galls (% satis- factory)	Volume of Cooling Water 1000s galls	Total Number (% satis- factory)	Total Volume 1000s galls (% satis- factory)	Total Number (% satis- factory)	Total Volume 1000s galls (% satis- factory)	Total Number (% satis- factory)	Total Volume 1000s galls (% satis- factory)	Total Number (% satis- factory)	Total Volume 1000s galls (% satis- factory)	Total Number (% satis- factory)	Total Volume 1000s galls (% satis- factory)				CBI Code
																01
																02
																03
																04
																05
																06
																07
																08
(Nil)	30(Nil)	Nil											Nil	30	30	09
																10
																11
																12
																13
																14
																15
																16
																17
																18
																19
																20
																21
(50)	75(39)	Nil											Nil	75	75	22
																23
																24
																25
																26
																27
																28
																29
(Nil)	10(Nil)	Nil											Nil	10	10	30
																31
(100)	74(100)	Nil											Nil	74	74	50
(Nil)	5(Nil)	Nil											Nil	5	5	51
(36)	194(53)	Nil											Nil	194	194	
																52
																53
																54
																55
	194	Nil											Nil	194	194	

HAMPSHIRE RIVER AUTHORITY

TABLE 198 Main survey Numbers and volumes of discharges of industrial effluent to rivers and canals in England and Wales and percentage considered satisfactory

Industry		Discharges of total effluent (which may contain some cooling water) to											
		Non-tidal rivers			Tidal rivers			All rivers			Canals		
CBI Code	Description	Total Number (% satisfactory)	Total Volume 1000s galls (% satisfactory)	Volume of Cooling Water 1000s galls	Total Number (% satisfactory)	Total Volume 1000s galls (% satisfactory)	Volume of Cooling Water 1000s galls	Total Number (% satisfactory)	Total Volume 1000s galls (% satisfactory)	Volume of Cooling Water 1000s galls	Total Number (% satisfactory)	Total Volmme 1000s galls (% satisfactory)	Volume Cooling Water 1000s galls
01	Brewing												
02	Brickmaking												
03	Cement making												
04	Chemical and allied industries	1(100)	1(100)	Nil	3(67)	14517(100)	10180	4(75)	14518(100)	10180			
05	Coal mining												
06	Distillation of ethanol												
07	Electricity generation												
08	Engineering	3(67)	49(98)	24				3(67)	49(98)	24			
09	Food processing and manufacture				1(100)	30(100)	Nil	1(100)	30(100)	Nil			
10	Gas and coke												
11	Glass making												
12	Glue and gelatine				1(Nil)	108(Nil)	Nil	1(Nil)	108(Nil)	Nil			
13	General manufacturing												
14	Iron and steel												
15	Laundering and dry cleaning												
16	Leather tanning												
17	Metal smelting												
18	Paint making												
19	Paper and board making	1(100)	3000(100)	90				1(100)	3000(100)	90			
20	Petroleum refining				3(100)	151200(100)	137592	3(100)	151200(100)	137592			
21	Plastics manufacture												
22	Plating and metal finishing												
23	Pottery making												
24	Printing ink etc												
25	Quarrying and mining	2(Nil)	330(Nil)	Nil				2(Nil)	330(Nil)	Nil			
26	Rubber processing												
27	Soap and detergent												
28	Textile, cotton and man-made												
29	Textile, wool												
30	General farming	1(Nil)	1(Nil)	Nil				1(Nil)	1(Nil)	Nil			
31	Atomic energy establishments												
50	Water treatment												
51	Disposal tip drainage	2(Nil)	20(Nil)	Nil				2(Nil)	20(Nil)	Nil			
	TOTAL FOR CODES 1–51	10(40)	3401(90)	114	8(75)	165855(99.9)	147772	18(55)	169256(99.9)	147886			
52	Derelict coal mines												
53	Other derelict mines												
54	Active coal mines												
55	Other active mines												
	TOTAL FOR CODES 52–55												
	TOTAL	10	3401	114	8	165855	147772	18	169256	147886			

rivers and canals			Discharges of solely cooling water to										Total Cooling Water Discharged 1000s galls	Total Process Water Discharged 1000s galls	Total Discharge 1000s galls	CBI Code
			Non-tidal rivers		Tidal rivers		All rivers		Canals		All rivers and canals					
Total Number (% satisfactory)	Total Volume 1000s galls (% satisfactory)	Volume of Cooling Water 1000s galls	Total Number (% satisfactory)	Total Volume 1000s galls (% satisfactory)	Total Number (% satisfactory)	Total Volume 1000s galls (% satisfactory)	Total Number (% satisfactory)	Total Volume 1000s galls (% satisfactory)	Total Number (% satisfactory)	Total Volume 1000s galls (% satisfactory)	Total Number (% satisfactory)	Total Volume 1000s galls (% satisfactory)				
			3(100)	106(100)			3(100)	106(100)			3(100)	106(100)	106	Nil	106	01
																02
																03
75)	14518(100)	10180			1(100)	2(100)	1(100)	2(100)			1(100)	2(100)	10182	4338	14520	04
																05
																06
					2(100)	1622000(100)	2(100)	1622000(100)			2(100)	1622000(100)	1622000	Nil	1622000	07
67)	49(98)	24	1(100)	60(100)			1(100)	60(100)			1(100)	60(100)	84	25	109	08
100)	30(100)	Nil											Nil	30	30	09
																10
																11
Nil)	108(Nil)	Nil											Nil	108	108	12
																13
																14
																15
																16
																17
																18
(100)	3000(100)	90											90	2910	3000	19
(100)	151200(100)	137592											137592	13608	151200	20
			1(Nil)	100(Nil)			1(Nil)	100(Nil)			1(Nil)	100(Nil)	100	Nil	100	21
																22
																23
																24
(Nil)	330(Nil)	Nil											Nil	330	330	25
																26
																27
																28
																29
(Nil)	1(Nil)	Nil											Nil	1	1	30
																31
																50
2(Nil)	20(Nil)	Nil											Nil	20	20	51
8(55)	169256(99·9)	147886	5(80)	266(62)	3(100)	1622002(100)	8(88)	1622268(100)			8(88)	1622268(100)	1770154	21370	1791524	
																52
																53
																54
																55
8	169256	147886	5	266	3	1622002	8	1622268			8	1622268	1770154	21370	1791524	

ISLE OF WIGHT RIVER AND WATER AUTHORITY

TABLE 199 Main survey Numbers and volumes of discharges of industrial effluent to rivers and canals in England and Wales and percentage considered satisfactory

Industry		Discharges of total effluent (which may contain some cooling water) to											
		Non-tidal rivers			Tidal rivers			All rivers			Canals		
CBI Code	Description	Total Number (% satisfactory)	Total Volume 1000s galls (% satisfactory)	Volume of Cooling Water 1000s galls	Total Number (% satisfactory)	Total Volume 1000s galls (% satisfactory)	Volume of Cooling Water 1000s galls	Total Number (% satisfactory)	Total Volume 1000s galls (% satisfactory)	Volume of Cooling Water 1000s galls	Total Number (% satisfactory)	Total Volume 1000s galls (% satisfactory)	Volume Cooling Water 1000s galls
01	Brewing												
02	Brickmaking												
03	Cement making												
04	Chemical and allied industries												
05	Coal mining												
06	Distillation of ethanol												
07	Electricity generation												
08	Engineering												
09	Food processing and manufacture												
10	Gas and coke												
11	Glass making												
12	Glue and gelatine												
13	General manufacturing												
14	Iron and steel												
15	Laundering and dry cleaning	1(Nil)	10(Nil)	Nil				1(Nil)	10(Nil)	Nil			
16	Leather tanning												
17	Metal smelting												
18	Paint making												
19	Paper and board making												
20	Petroleum refining												
21	Plastics manufacture												
22	Plating and metal finishing	1(Nil)	15(Nil)	Nil				1(Nil)	15(Nil)	Nil			
23	Pottery making												
24	Printing ink etc												
25	Quarrying and mining	1(100)	135(100)	Nil	1(100)	20(100)	Nil	2(100)	155(100)	Nil			
26	Rubber processing												
27	Soap and detergent												
28	Textile, cotton and man-made												
29	Textile, wool												
30	General farming	2(Nil)	5(Nil)	1				2(Nil)	5(Nil)	1			
31	Atomic energy establishments												
50	Water treatment	1(100)	12(100)	Nil				1(100)	12(100)	Nil			
51	Disposal tip drainage												
	TOTAL FOR CODES 1-51	6(33)	177(83)	1	1(100)	20(100)	Nil	7(43)	197(85)	1			
52	Derelict coal mines												
53	Other derelict mines												
54	Active coal mines												
55	Other active mines												
	TOTAL FOR CODES 52-55												
	TOTAL	6	177	1	1	20	Nil	7	197	1			

| ...rivers and canals | | | Discharges of solely cooling water to | | | | | | | | All rivers and canals | | Total Cooling Water Discharged 1000s galls | Total Process Water Discharged 1000s galls | Total Discharge 1000s galls | CBI Code |
| | | | Non-tidal rivers | | Tidal rivers | | All rivers | | Canals | | | | | | | |
...ber atis- ...ry)	Total Volume 1000s galls (% satis- factory)	Volume of Cooling Water 1000s galls	Total Number (% satis- factory)	Total Volume 1000s galls (% satis- factory)	Total Number (% satis- factory)	Total Volume 1000s galls (% satis- factory)	Total Number (% satis- factory)	Total Volume 1000s galls (% satis- factory)	Total Number (% satis- factory)	Total Volume 1000s galls (% satis- factory)	Total Number (% satis- factory)	Total Volume 1000s galls (% satis- factory)				
																01
																02
																03
																04
																05
																06
				1(100)	30000(100)	1(100)	30000(100)			1(100)	30000(100)	30000	Nil	30000		07
																08
																09
																10
																11
																12
																13
																14
...il)	10(Nil)	Nil											Nil	10	10	15
																16
																17
																18
																19
																20
																21
...il)	15(Nil)	Nil											Nil	15	15	22
																23
																24
...00)	155(100)	Nil											Nil	155	155	25
																26
																27
																28
																29
...Nil)	5(Nil)	1											1	4	5	30
																31
...00)	12(100)	Nil											Nil	12	12	50
																51
...43)	197(85)	1			1(100)	30000(100)	1(100)	30000(100)			1(100)	30000(100)	30001	196	30197	
																52
																53
																54
																55
	197	1			1	30000	1	30000			1	30000	30001	196	30197	

AVON AND DORSET RIVER AUTHORITY

TABLE 200 Main survey Numbers and volumes of discharges of industrial effluent to rivers and canals in England and Wales and percentage considered satisfactory

Industry		Discharges of total effluent (which may contain some cooling water) to									Canals		
		Non-tidal rivers			Tidal rivers			All rivers					
CBI Code	Description	Total Number (% satisfactory)	Total Volume 1000s galls (% satisfactory)	Volume of Cooling Water 1000s galls	Total Number (% satisfactory)	Total Volume 1000s galls (% satisfactory)	Volume of Cooling Water 1000s galls	Total Number (% satisfactory)	Total Volume 1000s galls (% satisfactory)	Volume of Cooling Water 1000s galls	Total Number (% satisfactory)	Total Volume 1000s galls (% satisfactory)	Volume Cooling Water 1000s galls
01	Brewing												
02	Brickmaking												
03	Cement making												
04	Chemical and allied industries				1(Nil)	800(Nil)	160	1(Nil)	800(Nil)	160			
05	Coal mining												
06	Distillation of ethanol												
07	Electricity generation	1(100)	41000(100)	410				1(100)	41000(100)	410			
08	Engineering												
09	Food processing and manufacture	3(33)	292(89)	6				3(33)	292(89)	6			
10	Gas and coke												
11	Glass making												
12	Glue and gelatine												
13	General manufacturing												
14	Iron and steel												
15	Laundering and dry cleaning												
16	Leather tanning	1(Nil)	20(Nil)	Nil				1(Nil)	20(Nil)	Nil			
17	Metal smelting												
18	Paint making												
19	Paper and board making	1(Nil)	120(Nil)	12				1(Nil)	120(Nil)	12			
20	Petroleum refining												
21	Plastics manufacture												
22	Plating and metal finishing	1(100)	39(100)	29				1(100)	39(100)	29			
23	Pottery making												
24	Printing ink etc												
25	Quarrying and mining												
26	Rubber processing												
27	Soap and detergent												
28	Textile, cotton and man-made												
29	Textile, wool	1(Nil)	16(Nil)	Nil				1(Nil)	16(Nil)	Nil			
30	General farming												
31	Atomic energy establishments												
50	Water treatment	1(100)	200(100)	Nil				1(100)	200(100)	Nil			
51	Disposal tip drainage												
	TOTAL FOR CODES 1-51	9(44)	41687(99·8)	457	1(Nil)	800(Nil)	160	10(40)	42487(98)	617			
52	Derelict coal mines												
53	Other derelict mines												
54	Active coal mines												
55	Other active mines												
	TOTAL FOR CODES 52-55												
	TOTAL	9	41687	457	1	800	160	10	42487	617			

	ers and canals		Discharges of solely cooling water to										Total Cooling Water Discharged 1000s galls	Total Process Water Discharged 1000s galls	Total Discharge 1000s galls	
			Non-tidal rivers		Tidal rivers		All rivers		Canals		All rivers and canals					
er tis- y)	Total Volume 1000s galls (% satis- factory)	Volume of Cooling Water 1000s galls	Total Number (% satis- factory)	Total Volume 1000s galls (% satis- factory)	Total Number (% satis- factory)	Total Volume 1000s galls (% satis- factory)	Total Number (% satis- factory)	Total Volume 1000s galls (% satis- factory)	Total Number (% satis- factory)	Total Volume 1000s galls (% satis- factory)	Total Number (% satis- factory)	Total Volume 1000s galls (% satis- factory)				CBI Code
																01 02 03
l)	800(Nil)	160											160	640	800	04 05 06
0)	41000(100)	410											410	40590	41000	07 08
)	292(89)	6											6	286	292	09
																10 11 12
																13 14 15
il)	20(Nil)	Nil											Nil	20	20	16 17 18
il)	120(Nil)	12											12	108	120	19 20 21
00)	39(100)	29											29	10	39	22 23 24
																25 26 27
Nil)	16(Nil)	Nil											Nil	16	16	28 29 30
																31
100)	200(100)	Nil	1(100)	500(100)			1(100)	500(100)			1(100)	500(100)	500	200	700	50 51
40)	42487(98)	617	1(100)	500(100)			1(100)	500(100)			1(100)	500(100)	1117	41870	42987	
																52 53 54 55
	42487	617	1	500			1	500			1	500	1117	41870	42987	

DEVON RIVER AUTHORITY

TABLE 201 Main survey Numbers and volumes of discharges of industrial effluent to rivers and canals in England and Wales and percentage considered satisfactory

Industry		Discharges of total effluent (which may contain some cooling water) to											
		Non-tidal rivers			Tidal rivers			All rivers			Canals		
CBI Code	Description	Total Number (% satisfactory)	Total Volume 1000s galls (% satisfactory)	Volume of Cooling Water 1000s galls	Total Number (% satisfactory)	Total Volume 1000s galls (% satisfactory)	Volume of Cooling Water 1000s galls	Total Number (% satisfactory)	Total Volume 1000s galls (% satisfactory)	Volume of Cooling Water 1000s galls	Total Number (% satisfactory)	Total Volume 1000s galls (% satisfactory)	Volume Cooling Water 1000s galls
01	Brewing	1(Nil)	5(Nil)	Nil				1(Nil)	5(Nil)	Nil			
02	Brickmaking												
03	Cement making												
04	Chemical and allied industries												
05	Coal mining												
06	Distillation of ethanol												
07	Electricity generation	1(100)	7690(100)	Nil				1(100)	7690(100)	Nil			
08	Engineering												
09	Food processing and manufacture	9(44)	779(44)	Nil				9(44)	779(44)	Nil			
10	Gas and coke												
11	Glass making												
12	Glue and gelatine												
13	General manufacturing												
14	Iron and steel												
15	Laundering and dry cleaning	5(Nil)	149(Nil)	Nil				5(Nil)	149(Nil)	Nil			
16	Leather tanning	2(Nil)	70(Nil)	Nil				2(Nil)	70(Nil)	Nil			
17	Metal smelting												
18	Paint making												
19	Paper and board making	9(55)	3957(43)	594	1(Nil)	750(Nil)	Nil	10(50)	4707(36)	594			
20	Petroleum refining												
21	Plastics manufacture												
22	Plating and metal finishing	1(100)	100(100)	Nil				1(100)	100(100)	Nil			
23	Pottery making												
24	Printing ink etc												
25	Quarrying and mining	18(11)	2828(1)	Nil	2(Nil)	324(Nil)	Nil	20(10)	3152(Nil)	Nil			
26	Rubber processing												
27	Soap and detergent												
28	Textile, cotton and man-made	2(Nil)	510(Nil)	Nil	1(100)	200(100)	Nil	3(33)	710(28)	Nil			
29	Textile, wool												
30	General farming	1(Nil)	1(Nil)	Nil				1(Nil)	1(Nil)	Nil			
31	Atomic energy establishments												
50	Water treatment	5(60)	380(70)	Nil	1(Nil)	40(Nil)	Nil	6(50)	420(63)	Nil			
51	Disposal tip drainage												
	TOTAL FOR CODES 1-51	54(30)	16469(61)	594	5(20)	1314(15)	Nil	59(29)	17783(58)	594			
52	Derelict coal mines												
53	Other derelict mines												
54	Active coal mines												
55	Other active mines												
	TOTAL FOR CODES 52-55												
	TOTAL	54	16469	594	5	1314	Nil	59	17783	594			

| ...rs and canals | | Discharges of solely cooling water to | | | | | | | | | | Total Cooling Water Discharged 1000s galls | Total Process Water Discharged 1000s galls | Total Discharge 1000s galls | CBI Code |
| | | Non-tidal rivers | | Tidal rivers | | All rivers | | Canals | | All rivers and canals | | | | | |
Total Volume 1000s galls (% satis-factory)	Volume of Cooling Water 1000s galls	Total Number (% satis-factory)	Total Volume 1000s galls (% satis-factory)	Total Number (% satis-factory)	Total Volume 1000s galls (% satis-factory)	Total Number (% satis-factory)	Total Volume 1000s galls (% satis-factory)	Total Number (% satis-factory)	Total Volume 1000s galls (% satis-factory)	Total Number (% satis-factory)	Total Volume 1000s galls (% satis-factory)				
5(Nil)	Nil											Nil	5	5	01
															02
															03
															04
															05
															06
7690(100)	Nil	1(100)	165(100)	1(100)	160000(100)	2(100)	160165(100)			2(100)	160165(100)	160165	7690	167855	07
															08
779(44)	Nil	12(83)	4251(78)			12(83)	4251(78)			12(83)	4251(78)	4251	779	5030	09
															10
															11
															12
															13
															14
149(Nil)	Nil											Nil	149	149	15
70(Nil)	Nil	1(100)	40(100)			1(100)	40(100)			1(100)	40(100)	40	70	110	16
															17
															18
4707(36)	594	5(100)	700(100)			5(100)	700(100)			5(100)	700(100)	1294	4113	5407	19
															20
															21
100(100)	Nil											Nil	100	100	22
															23
															24
3152(Nil)	Nil											Nil	3152	3152	25
															26
															27
710(28)	Nil	1(100)	10(100)			1(100)	10(100)			1(100)	10(100)	10	710	720	28
															29
1(Nil)	Nil											Nil	1	1	30
															31
420(63)	Nil											Nil	420	420	50
															51
17783(58)	594	20(90)	5166(82)	1(100)	160000(100)	21(90)	165166(99)			21(90)	165166(99)	165760	17189	182949	
															52
															53
															54
															55
17783	594	20	5166	1	160000	21	165166			21	165166	165760	17189	182949	

CORNWALL RIVER AUTHORITY

TABLE 202 Main survey Numbers and volumes of discharges of industrial effluent to rivers and canals in England and Wales and percentage considered satisfactory

Industry		Discharges of total effluent (which may contain some cooling water) to											
		Non-tidal rivers			Tidal rivers			All rivers			Canals		
CBI Code	Description	Total Number (% satisfactory)	Total Volume 1000s galls (% satisfactory)	Volume of Cooling Water 1000s galls	Total Number (% satisfactory)	Total Volume 1000s galls (% satisfactory)	Volume of Cooling Water 1000s galls	Total Number (% satisfactory)	Total Volume 1000s galls (% satisfactory)	Volume of Cooling Water 1000s galls	Total Number (% satisfactory)	Total Volume 1000s galls (% satisfactory)	Volume Cooling Water 1000s galls
01	Brewing												
02	Brickmaking												
03	Cement making				1(Nil)	7(Nil)	Nil	1(Nil)	7(Nil)	Nil			
04	Chemical and allied industries				2(Nil)	27005(Nil)	Nil	2(Nil)	27005(Nil)	Nil			
05	Coal mining												
06	Distillation of ethanol												
07	Electricity generation	1(100)	48000(100)	Nil	1(100)	21000(100)	Nil	2(100)	69000(100)	Nil			
08	Engineering												
09	Food processing and manufacture	2(50)	1003(93)	802	2(50)	325(62)	80	4(50)	1328(85)	882			
10	Gas and coke				1(100)	600(100)	300	1(100)	600(100)	300			
11	Glass making												
12	Glue and gelatine												
13	General manufacturing												
14	Iron and steel												
15	Laundering and dry cleaning	1(Nil)	20(Nil)	Nil				1(Nil)	20(Nil)	Nil			
16	Leather tanning												
17	Metal smelting												
18	Paint making												
19	Paper and board making												
20	Petroleum refining												
21	Plastics manufacture												
22	Plating and metal finishing												
23	Pottery making												
24	Printing ink etc												
25	Quarrying and mining	74(35)	61977(33)	1333	1(Nil)	2984(Nil)	Nil	75(35)	64961(32)	1333			
26	Rubber processing												
27	Soap and detergent												
28	Textile, cotton and man-made												
29	Textile, wool												
30	General farming												
31	Atomic energy establishments												
50	Water treatment	15(53)	2601(89)	Nil				15(53)	2601(89)	Nil			
51	Disposal tip drainage												
	TOTAL FOR CODES 1-51	93(38)	113601(63)	2135	8(38)	51921(42)	380	101(37)	165522(57)	2515			
52	Derelict coal mines												
53	Other derelict mines	5	3793					5	3793				
54	Active coal mines												
55	Other active mines	5	14380					5	14380				
	TOTAL FOR CODES 52–55	10	18173					10	18173				
	TOTAL	103	131774	2135	8	51921	380	111	183695	2515			

			Discharges of solely cooling water to										Total Cooling Water Discharged 1000s galls	Total Process Water Discharged 1000s galls	Total Discharge 1000s galls	
...ers and canals			Non-tidal rivers		Tidal rivers		All rivers		Canals		All rivers and canals					
...er tis- y)	Total Volume 1000s galls (% satis- factory)	Volume of Cooling Water 1000s galls	Total Number (% satis- factory)	Total Volume 1000s galls (% satis- factory)	Total Number (% satis- factory)	Total Volume 1000s galls (% satis- factory)	Total Number (% satis- factory)	Total Volume 1000s galls (% satis- factory)	Total Number (% satis- factory)	Total Volume 1000s galls (% satis- factory)	Total Number (% satis- factory)	Total Volume 1000s galls (% satis- factory)				CBI Code
																01
																02
Nil)	7(Nil)	Nil											Nil	7	7	03
Nil)	27005(Nil)	Nil											Nil	27005	27005	04
																05
																06
00)	69000(100)	Nil			3(100)	22800(100)	3(100)	22800(100)			3(100)	22800(100)	22800	69000	91800	07
																08
50)	1328(85)	882	1(100)	12(100)			1(100)	12(100)			1(100)	12(100)	894	446	1340	09
00)	600(100)	300											300	300	600	10
																11
																12
																13
																14
Nil)	20(Nil)	Nil											Nil	20	20	15
																16
																17
																18
																19
																20
																21
																22
																23
																24
35)	6496(32)	1333	1(100)	432(100)			1(100)	432(100)			1(100)	432(100)	1765	63628	65393	25
																26
																27
																28
																29
																30
																31
(53)	2601(89)	Nil											Nil	2601	2601	50
																51
(37)	165522(57)	2515	2(100)	444(100)	3(100)	22800(100)	5(100)	23244(100)			5(100)	23244(100)	25759	163007	188766	
																52
	3793													3793	3793	53
																54
	14380													14380	14380	55
	18173													18173	18173	
	183695	2515	2	21444	3	22800	5	23244			5	23244	25759	181180	206939	

SOMERSET RIVER AUTHORITY

TABLE 203 Main survey Numbers and volumes of discharges of industrial effluent to rivers and canals in England and Wales and percentage considered satisfactory

Industry		Discharges of total effluent (which may contain some cooling water) to											
		Non-tidal rivers			Tidal rivers			All rivers			Canals		
CBI Code	Description	Total Number (% satisfactory)	Total Volume 1000s galls (% satisfactory)	Volume of Cooling Water 1000s galls	Total Number (% satisfactory)	Total Volume 1000s galls (% satisfactory)	Volume of Cooling Water 1000s galls	Total Number (% satisfactory)	Total Volume 1000s galls (% satisfactory)	Volume of Cooling Water 1000s galls	Total Number (% satisfactory)	Total Volume 1000s galls (% satisfactory)	Volume Cooling Water 1000s galls
01	Brewing	2(Nil)	35(Nil)	Nil				2(Nil)	35(Nil)	Nil			
02	Brickmaking												
03	Cement making												
04	Chemical and allied industries				1(100)	2500(100)	Nil	1(100)	2500(100)	Nil			
05	Coal mining												
06	Distillation of ethanol												
07	Electricity generation												
08	Engineering												
09	Food processing and manufacture	5(40)	191(49)	9				5(40)	191(49)	9			
10	Gas and coke												
11	Glass making												
12	Glue and gelatine												
13	General manufacturing												
14	Iron and steel												
15	Laundering and dry cleaning												
16	Leather tanning												
17	Metal smelting												
18	Paint making												
19	Paper and board making	4(Nil)	5365(Nil)	2991				4(Nil)	5365(Nil)	2991			
20	Petroleum refining												
21	Plastics manufacture												
22	Plating and metal finishing												
23	Pottery making												
24	Printing ink etc												
25	Quarrying and mining	1(Nil)	600(Nil)	Nil				1(Nil)	600(Nil)	Nil			
26	Rubber processing												
27	Soap and detergent												
28	Textile, cotton and man-made	1(Nil)	300(Nil)	Nil	1(Nil)	2200(Nil)	Nil	2(Nil)	2500(Nil)	Nil			
29	Textile, wool												
30	General farming	114(Nil)	1191(Nil)	Nil	5(Nil)	45(Nil)	Nil	119(Nil)	1236(Nil)	Nil			
31	Atomic energy establishments												
50	Water treatment												
51	Disposal tip drainage												
	TOTAL FOR CODES 1–51	127(2)	7682(1)	3000	7(14)	4745(53)	Nil	134(2)	12427(21)	3000			
52	Derelict coal mines												
53	Other derelict mines												
54	Active coal mines												
55	Other active mines												
	TOTAL FOR CODES 52–55												
	TOTAL	127	7682	3000	7	4745	Nil	134	12427	3000			

	ers and canals		Discharges of solely cooling water to											Total Cooling Water Discharged 1000s galls	Total Process Water Discharged 1000s galls	Total Discharge 1000s galls	
			Non-tidal rivers		Tidal rivers		All rivers		Canals		All rivers and canals					
er tis- y)	Total Volume 1000s galls (% satis- factory)	Volume of Cooling Water 1000s galls	Total Number (% satis- factory)	Total Volume 1000s galls (% satis- factory)	Total Number (% satis- factory)	Total Volume 1000s galls (% satis- factory)	Total Number (% satis- factory)	Total Volume 1000s galls (% satis- factory)	Total Number (% satis- factory)	Total Volume 1000s galls (% satis- factory)	Total Number (% satis- factory)	Total Volume 1000s galls (% satis- factory)				CBI Code
il)	35(Nil)	Nil											Nil	35	35	01 02 03
00)	2500(100)	Nil											Nil	2500	2500	04 05 06
0)	191(49)	9											9	182	191	07 08 09
																10 11 12
																13 14 15
																16 17 18
Nil)	5365(Nil)	2991											2991	2374	5365	19 20 21
																22 23 24
Nil)	600(Nil)	Nil											Nil	600	600	25 26 27
Nil)	2500(Nil)	Nil											Nil	2500	2500	28 29
Nil)	1236(Nil)	Nil											Nil	1236	1236	30
																31 50 51
(2)	12427(21)	3000											3000	9427	12427	
																52 53 54 55
	12427	3000											3000	9427	12427	

BRISTOL AVON RIVER AUTHORITY

TABLE 204 Main survey Numbers and volumes of discharges of industrial effluent to rivers and canals in England and Wales and percentage considered satisfactory

| Industry | | Discharges of total effluent (which may contain some cooling water) to | | | | | | | | | Canals | | |
| | | Non-tidal rivers | | | Tidal rivers | | | All rivers | | | | | |
CBI Code	Description	Total Number (% satisfactory)	Total Volume 1000s galls (% satisfactory)	Volume of Cooling Water 1000s galls	Total Number (% satisfactory)	Total Volume 1000s galls (% satisfactory)	Volume of Cooling Water 1000s galls	Total Number (% satisfactory)	Total Volume 1000s galls (% satisfactory)	Volume of Cooling Water 1000s galls	Total Number (% satisfactory)	Total Volume 1000s galls (% satisfactory)	Volume Cooling Water 1000s galls
01	Brewing				1(Nil)	1750(Nil)	875	1(Nil)	1750(Nil)	875			
02	Brickmaking												
03	Cement making												
04	Chemical and allied industries	1(Nil)	2000(Nil)	1260	1(Nil)	50(Nil)	45	2(Nil)	2050(Nil)	1305			
05	Coal mining												
06	Distillation of ethanol												
07	Electricity generation												
08	Engineering												
09	Food processing and manufacture	7(43)	4292(21)	3207	1(100)	1000(100)	Nil	8(50)	5292(36)	3207			
10	Gas and coke												
11	Glass making												
12	Glue and gelatine												
13	General manufacturing												
14	Iron and steel												
15	Laundering and dry cleaning												
16	Leather tanning	2(Nil)	90(Nil)	Nil	2(Nil)	55(Nil)	Nil	4(Nil)	145(Nil)	Nil			
17	Metal smelting												
18	Paint making												
19	Paper and board making	5(Nil)	42680(Nil)	24796	1(100)	70(100)	Nil	6(17)	42750(Nil)	24796			
20	Petroleum refining												
21	Plastics manufacture												
22	Plating and metal finishing												
23	Pottery making												
24	Printing ink etc												
25	Quarrying and mining	2(50)	112(89)	Nil				2(50)	112(89)	Nil			
26	Rubber processing												
27	Soap and detergent												
28	Textile, cotton and man-made												
29	Textile, wool												
30	General farming	2(50)	2008(100)	660				2(50)	2008(100)	660			
31	Atomic energy establishments												
50	Water treatment												
51	Disposal tip drainage												
	TOTAL FOR CODES 1-51	19(26)	51182(59)	29923	6(33)	2925(37)	920	25(28)	54107(57)	30843			
52	Derelict coal mines	1	100					1	100				
53	Other derelict mines												
54	Active coal mines	1	500					1	500				
55	Other active mines												
	TOTAL FOR CODES 52–55	2	600					2	600				
	TOTAL	21	51782	29923	6	2925	920	27	54707	30843			

| | ...rs and canals | | Discharges of solely cooling water to | | | | | | | | | | Total Cooling Water Discharged 1000s galls | Total Process Water Discharged 1000s galls | Total Discharge 1000s galls | |
| | | | Non-tidal rivers | | Tidal rivers | | All rivers | | Canals | | All rivers and canals | | | | | |
...r ...tis- ...y)	Total Volume 1000s galls (% satis-factory)	Volume of Cooling Water 1000s galls	Total Number (% satis-factory)	Total Volume 1000s galls (% satis-factory)	Total Number (% satis-factory)	Total Volume 1000s galls (% satis-factory)	Total Number (% satis-factory)	Total Volume 1000s galls (% satis-factory)	Total Number (% satis-factory)	Total Volume 1000s galls (% satis-factory)	Total Number (% satis-factory)	Total Volume 1000s galls (% satis-factory)				CBI Code
)	1750(Nil)	875											875	875	1750	01 02 03
)	2050(Nil)	1305	2(100)	881(100)	1(Nil)	300(Nil)	3(67)	1181(75)			3(67)	1181(75)	2486	745	3231	04 05 06
)	5292(36)	3207	5(100) 2(50)	340(100) 2500(60)	1(100)	10000(100)	5(100) 3(67)	340(100) 12500(92)			5(100) 3(67)	340(100) 12500(92)	340 15707	Nil 2085	340 17792	07 08 09
			2(100)	2400(100)			2(100)	2400(100)			2(100)	2400(100)	2400	Nil	2400	10 11 12
																13 14 15
il)	145(Nil)	Nil	1(100)	30(100)			1(100)	30(100)			1(100)	30(100)	Nil 30	145 Nil	145 30	16 17 18
)	42750(Nil)	24796											24796	17954	42750	19 20 21
																22 23 24
0)	112(89)	Nil	4(75)	9330(99)			4(75)	9330(99)			4(75)	9330(99)	Nil 9330	112 Nil	112 9330	25 26 27
0)	2008(100)	660											660	1348	2008	28 29 30
																31 50 51
8)	54107(57)	30843	16(88)	15481(93)	2(50)	10300(97)	18(83)	25781(95)			18(83)	25781(95)	56624	23264	79888	
	100													100	100	52 53
	500													500	500	54 55
	600													600	600	
	54707	30843	16	15481	2	10300	18	25781			18	25781	56624	23864	80488	

SEVERN RIVER AUTHORITY

TABLE 205 Main survey Numbers and volumes of discharges of industrial effluent to rivers and canals in England and Wales and percentage considered satisfactory

Industry		Discharges of total effluent (which may contain some cooling water) to											
		Non-tidal rivers			Tidal rivers			All rivers			Canals		
CBI Code	Description	Total Number (% satisfactory)	Total Volume 1000s galls (% satisfactory)	Volume of Cooling Water 1000s galls	Total Number (% satisfactory)	Total Volume 1000s galls (% satisfactory)	Volume of Cooling Water 1000s galls	Total Number (% satisfactory)	Total Volume 1000s galls (% satisfactory)	Volume of Cooling Water 1000s galls	Total Number (% satisfactory)	Total Volume 1000s galls (% satisfactory)	Volume Cooling Water 1000s galls
01	Brewing												
02	Brickmaking	1(100)	240(100)	Nil				1(100)	240(100)	Nil			
03	Cement making	1(Nil)	20(Nil)	Nil				1(Nil)	20(Nil)	Nil			
04	Chemical and allied industries	4(25)	384(3)	5	5(100)	7772(100)	Nil	9(67)	8156(95)	5			
05	Coal mining	1(Nil)	26(Nil)	Nil				1(Nil)	26(Nil)	Nil			
06	Distillation of ethanol												
07	Electricity generation	5(60)	16535(91)	14250	1(100)	3000(100)	1980	6(67)	19535(92)	16230			
08	Engineering	10(60)	1424(82)	64				10(60)	1424(82)	64	1(Nil)	372(Nil)	Nil
09	Food processing and manufacture	18(39)	4364(85)	8	1(Nil)	15(Nil)	Nil	19(37)	4379(85)	8	2(50)	15(67)	Nil
10	Gas and coke												
11	Glass making										2(Nil)	10(Nil)	Nil
12	Glue and gelatine												
13	General manufacturing												
14	Iron and steel										1(Nil)	4900(Nil)	833
15	Laundering and dry cleaning												
16	Leather tanning												
17	Metal smelting												
18	Paint making												
19	Paper and board making	4(50)	822(75)	Nil				4(50)	822(75)	Nil			
20	Petroleum refining												
21	Plastics manufacture	2(100)	109(100)	Nil				2(100)	109(100)	Nil			
22	Plating and metal finishing	8(38)	452(84)	Nil				8(38)	452(84)	Nil			
23	Pottery making												
24	Printing ink etc												
25	Quarrying and mining	14(86)	2354(79)					14(86)	2354(79)	Nil	2(100)	122(100)	Nil
26	Rubber processing	2(50)	28(18)	3				2(50)	28(18)	3			
27	Soap and detergent												
28	Textile, cotton and man-made	2(50)	164(54)	81				2(50)	164(54)	81			
29	Textile, wool	1(Nil)	50(Nil)					1(Nil)	50(Nil)				
30	General farming	1(100)	10(100)	3				1(100)	10(100)	3			
31	Atomic energy establishments												
50	Water treatment	10(40)	2542(24)	Nil	1(100)	70(100)	Nil	11(45)	2612(26)	Nil			
51	Disposal tip drainage	1(Nil)	10(Nil)	Nil				1(Nil)	10(Nil)	Nil			
	TOTAL FOR CODES 1-51	85(52)	29534(81)	14414	8(88)	10857(99·9)	1980	93(55)	40391(86)	16394	8(38)	5419(2)	833
52	Derelict coal mines	2	5030					2	5030				
53	Other derelict mines	1	50					1	50				
54	Active coal mines	2	355					2	355				
55	Other active mines	1	864					1	864				
	TOTAL FOR CODES 52-55	6	6299					6	6299				
	TOTAL	91	35833	14414	8	10857	1980	99	46690	16394	8	5419	833

...rivers and canals			Discharges of solely cooling water to										Total Cooling Water Discharged 1000s galls	Total Process Water Discharged 1000s galls	Total Discharge 1000s galls	CBI Code
			Non-tidal rivers		Tidal rivers		All rivers		Canals		All rivers and canals					
...er ...is-...y)	Total Volume 1000s galls (% satis-factory)	Volume of Cooling Water 1000s galls	Total Number (% satis-factory)	Total Volume 1000s galls (% satis-factory)	Total Number (% satis-factory)	Total Volume 1000s galls (% satis-factory)	Total Number (% satis-factory)	Total Volume 1000s galls (% satis-factory)	Total Number (% satis-factory)	Total Volume 1000s galls (% satis-factory)	Total Number (% satis-factory)	Total Volume 1000s galls (% satis-factory)				
00)	240(100)	Nil											Nil	240	240	01
il)	20(Nil)	Nil											Nil	20	20	02
																03
7)	8156(95)	5	2(50)	148(32)			2(50)	148(32)			2(50)	148(32)	153	8151	8304	04
il)	26(Nil)	Nil												26	26	05
			1(100)	20(100)			1(100)	20(100)			1(100)	20(100)	20	Nil	20	06
7)	19535(92)	16230	5(100)	467284(100)	2(100)	1004000(100)	7(100)	1471284(100)	1(100)	722(100)	8(100)	1472006(100)	1488236	3305	1491541	07
5)	1796(65)	64	18(94)	1673(97)			18(94)	1673(97)	3(100)	520(100)	21(95)	2193(98)	2257	1732	3989	08
8)	4394(85)	8	13(92)	1226(98)			13(92)	1226(98)	3(67)	2905(100)	16(88)	4131(99)	4139	4386	8525	09
	10(Nil)	Nil							1(100)	60(100)	1(100)	60(100)	60	Nil	60	10
il)													Nil	10	10	11
																12
il)	4900(Nil)	833	1(Nil)	7(Nil)			1(Nil)	7(Nil)			1(Nil)	7(Nil)	7	Nil	7	13
													833	4067	4900	14
																15
																16
																17
																18
0)	822(75)	Nil											Nil	822	822	19
00)	109(100)	Nil	1(100)	10(100)			1(100)	10(100)	1(100)	24(100)	2(100)	34(100)	34	109	143	20/21
8)	452(84)	Nil	4(75)	189(75)			4(75)	189(75)			4(75)	189(75)	189	452	641	22
																23
																24
38)	2476(80)	Nil											Nil	2476	2476	25
50)	28(18)	3	3(100)	59(100)			3(100)	59(100)			3(100)	59(100)	62	25	87	26/27
50)	164(54)	81							3(100)	40(100)	3(100)	40(100)	121	83	204	28
Nil)	50(Nil)	Nil											Nil	50	50	29
100)	10(100)	3											3	7	10	30
																31
45)	2612(26)	Nil											Nil	2612	2612	50
Nil)	10(Nil)	Nil											Nil	10	10	51
53)	45810(76)	17227	48(90)	470616(99·9)	2(100)	1004000(100)	50(90)	1474616(99·9)	12(92)	4271(100)	62(92)	1478887(99·9)	1496114	28583	1524697	
	5030													5030	5030	52
	50													50	50	53
	355													355	355	54
	864													864	864	55
	6299													6299	6299	
	52109	17227	48	470616	2	1004000	50	1474616	12	4271	62	1478887	1496114	34882	1530996	

WYE RIVER AUTHORITY

TABLE 206 Main survey Numbers and volumes of discharges of industrial effluent to rivers and canals in England and Wales and percentage considered satisfactory

Industry		Discharges of total effluent (which may contain some cooling water) to											
		Non-tidal rivers			Tidal rivers			All rivers			Canals		
CBI Code	Description	Total Number (% satis-factory)	Total Volume 1000s galls (% satis-factory)	Volume of Cooling Water 1000s galls	Total Number (% satis-factory)	Total Volume 1000s galls (% satis-factory)	Volume of Cooling Water 1000s galls	Total Number (% satis-factory)	Total Volume 1000s galls (% satis-factory)	Volume of Cooling Water 1000s galls	Total Number (% satis-factory)	Total Volume 1000s galls (% satis-factory)	Volume Cooling Water 1000s galls
01	Brewing												
02	Brickmaking												
03	Cement making												
04	Chemical and allied industries												
05	Coal mining												
06	Distillation of ethanol												
07	Electricity generation												
08	Engineering												
09	Food processing and manufacture	1(Nil)	100(Nil)	Nil				1(Nil)	100(Nil)	Nil			
10	Gas and coke												
11	Glass making												
12	Glue and gelatine												
13	General manufacturing												
14	Iron and steel												
15	Laundering and dry cleaning	1(100)	16(100)	Nil				1(100)	16(100)	Nil			
16	Leather tanning												
17	Metal smelting												
18	Paint making												
19	Paper and board making												
20	Petroleum refining												
21	Plastics manufacture												
22	Plating and metal finishing												
23	Pottery making	1(Nil)	15(Nil)	Nil				1(Nil)	15(Nil)	Nil			
24	Printing ink etc												
25	Quarrying and mining												
26	Rubber processing												
27	Soap and detergent												
28	Textile, cotton and man-made												
29	Textile, wool												
30	General farming	1(Nil)	5(Nil)	Nil				1(Nil)	5(Nil)	Nil			
31	Atomic energy establishments												
50	Water treatment	2(100)	1100(100)	Nil				2(100)	1100(100)	Nil			
51	Disposal tip drainage												
	TOTAL FOR CODES 1–51	6(50)	1236(90)	Nil				6(50)	1236(90)	Nil			
52	Derelict coal mines												
53	Other derelict mines												
54	Active coal mines												
55	Other active mines												
	TOTAL FOR CODES 52–55												
	TOTAL	6	1236	Nil				6	1236	Nil			

	rivers and canals		Discharges of solely cooling water to										Total Cooling Water Discharged 1000s galls	Total Process Water Discharged 1000s galls	Total Discharge 1000s galls	
			Non-tidal rivers		Tidal rivers		All rivers		Canals		All rivers and canals					
...al ...ber ...satis-...ory)	Total Volume 1000s galls (% satis-factory)	Volume of Cooling Water 1000s galls	Total Number (% satis-factory)	Total Volume 1000s galls (% satis-factory)	Total Number (% satis-factory)	Total Volume 1000s galls (% satis-factory)	Total Number (% satis-factory)	Total Volume 1000s galls (% satis-factory)	Total Number (% satis-factory)	Total Volume 1000s galls (% satis-factory)	Total Number (% satis-factory)	Total Volume 1000s galls (% satis-factory)				CBI Code
			1(100)	500(100)			1(100)	500(100)			1(100)	500(100)	500	Nil	500	01
																02
																03
																04
																05
																06
																07
																08
...il)	100(Nil)	Nil	1(100)	2000(100)			1(100)	2000(100)			1(100)	2000(100)	2000	100	2100	09
																10
																11
																12
																13
																14
...00)	16(100)	Nil											Nil	16	16	15
																16
																17
																18
																19
																20
																21
																22
Nil)	15(Nil)	Nil											Nil	15	15	23
																24
																25
																26
																27
																28
Nil)	5(Nil)	Nil											Nil	5	5	29
																30
100)	1100(100)	Nil											Nil	1100	1100	31
																50
																51
...50)	1236(90)	Nil	2(100)	2500(100)			2(100)	2500(100)			2(100)	2500(100)	2500	1236	3736	
																52
																53
																54
																55
	1236	Nil	2	2500			2	2500			2	2500	2500	1236	3736	

USK RIVER AUTHORITY

TABLE 207 Main survey Numbers and volumes of discharges of industrial effluent to rivers and canals in England and Wales and percentage considered satisfactory

Industry		Discharges of total effluent (which may contain some cooling water) to										
		Non-tidal rivers			Tidal rivers			All rivers			Canals		
CBI Code	Description	Total Number (% satisfactory)	Total Volume 1000s galls (% satisfactory)	Volume of Cooling Water 1000s galls	Total Number (% satisfactory)	Total Volume 1000s galls (% satisfactory)	Volume of Cooling Water 1000s galls	Total Number (% satisfactory)	Total Volume 1000s galls (% satisfactory)	Volume of Cooling Water 1000s galls	Total Number (% satisfactory)	Total Volume 1000s galls (% satisfactory)	Volume of Cooling Water 1000s galls
01	Brewing												
02	Brickmaking												
03	Cement making												
04	Chemical and allied industries	1(100)	13(100)	Nil				1(100)	13(100)	Nil			
05	Coal mining	8(63)	1270(32)	Nil				8(63)	1270(32)	Nil			
06	Distillation of ethanol												
07	Electricity generation												
08	Engineering	2(50)	124(81)	95				2(50)	124(81)	95			
09	Food processing and manufacture	1(Nil)	136(Nil)	Nil				1(Nil)	136(Nil)	Nil			
10	Gas and coke												
11	Glass making												
12	Glue and gelatine												
13	General manufacturing												
14	Iron and steel	20(Nil)	11549(Nil)	360	1(100)	80(100)		21(5)	11629(1)	360			
15	Laundering and dry cleaning												
16	Leather tanning												
17	Metal smelting				1(100)	200(100)		1(100)	200(100)	Nil			
18	Paint making												
19	Paper and board making				3(Nil)	1345(Nil)		3(Nil)	1345(Nil)	Nil			
20	Petroleum refining												
21	Plastics manufacture												
22	Plating and metal finishing	4(75)	1232(51)	Nil				4(75)	1232(51)	Nil			
23	Pottery making												
24	Printing ink etc												
25	Quarrying and mining												
26	Rubber processing												
27	Soap and detergent												
28	Textile, cotton and man-made												
29	Textile, wool												
30	General farming												
31	Atomic energy establishments												
50	Water treatment												
51	Disposal tip drainage												
	TOTAL FOR CODES 1-51	36(30)	14324(8)	455	5(40)	1625(17)		41(29)	15949(9)	455			
52	Derelict coal mines	3	617					3	617				
53	Other derelict mines												
54	Active coal mines	9	8780					9	8780				
55	Other active mines												
	TOTAL FOR CODES 52-55	12	9397					12	9397				
	TOTAL	48	23721	455	5	1625		53	25346	455			

rivers and canals			Discharges of solely cooling water to										Total Cooling Water Discharged 1000s galls	Total Process Water Discharged 1000s galls	Total Discharge 1000s galls	
			Non-tidal rivers		Tidal rivers		All rivers		Canals		All rivers and canals					
al number satis-factory)	Total Volume 1000s galls (% satis-factory)	Volume of Cooling Water 1000s galls	Total Number (% satis-factory)	Total Volume 1000s galls (% satis-factory)	Total Number (% satis-factory)	Total Volume 1000s galls (% satis-factory)	Total Number (% satis-factory)	Total Volume 1000s galls (% satis-factory)	Total Number (% satis-factory)	Total Volume 1000s galls (% satis-factory)	Total Number (% satis-factory)	Total Volume 1000s galls (% satis-factory)				CBI Code
																01
																02
																03
(100)	13(100)	Nil											Nil	13	13	04
(63)	1270(32)	Nil											Nil	1270	1270	05
																06
(50)	124(81)	95	1(100)	1300(100)	2(100)	195999(100)	3(100)	197299(100)			3(100)	197299(100)	197299	Nil	197299	07
			1(100)	8(100)			1(100)	8(100)			1(100)	8(100)	103	29	132	08
(Nil)	136(Nil)	Nil	1(100)	64(100)			1(100)	64(100)			1(100)	64(100)	64	136	200	09
																10
																11
																12
																13
(5)	11629(1)	360											360	11269	11629	14
																15
																16
(100)	200(100)	Nil											Nil	200	200	17
																18
(Nil)	1345(Nil)	Nil											Nil	1345	1345	19
																20
			1(100)	1000(100)			1(100)	1000(100)			1(100)	1000(100)	1000	Nil	1000	21
(75)	1232(51)	Nil	1(Nil)	30(Nil)			1(Nil)	30(Nil)			1(Nil)	30(Nil)	30	1232	1262	22
																23
																24
																25
																26
																27
																28
																29
																30
																31
																50
																51
1(31)	15949(9)	455	5(80)	2402(99)	2(100)	195999(100)	7(86)	198401(99·9)			7(86)	198401(99·9)	198856	15494	214350	
3	617													617	617	52
																53
9	8780													8780	8780	54
																55
12	9397													9397	9397	
53	25346	455	5	2402	2	195999	7	198401			7	198401	198856	24891	223747	

GLAMORGAN RIVER AUTHORITY

TABLE 208 Main survey Numbers and volumes of discharges of industrial effluent to rivers and canals in England and Wales and percentage considered satisfactory

| Industry | | Discharges of total effluent (which may contain some cooling water) to | | | | | | | | | Canals | | |
| | | Non-tidal rivers | | | Tidal rivers | | | All rivers | | | | | |
CBI Code	Description	Total Number (% satisfactory)	Total Volume 1000s galls (% satisfactory)	Volume of Cooling Water 1000s galls	Total Number (% satisfactory)	Total Volume 1000s galls (% satisfactory)	Volume of Cooling Water 1000s galls	Total Number (% satisfactory)	Total Volume 1000s galls (% satisfactory)	Volume of Cooling Water 1000s galls	Total Number (% satisfactory)	Total Volume 1000s galls (% satisfactory)	Volume of Cooling Water 1000s galls
01	Brewing												
02	Brickmaking												
03	Cement making												
04	Chemical and allied industries	4(75)	2870(92)	Nil				4(75)	2870(92)	Nil			
05	Coal mining	28(54)	9692(83)	Nil				28(54)	9692(83)	Nil			
06	Distillation of ethanol												
07	Electricity generation	2(100)	1500(100)	Nil				2(100)	1500(100)	Nil			
08	Engineering	2(50)	907(1)	Nil				2(50)	907(1)	Nil			
09	Food processing and manufacture												
10	Gas and coke	6(83)	7748(93)	Nil	1(Nil)	1(Nil)	Nil	7(71)	7749(93)	Nil			
11	Glass making												
12	Glue and gelatine	1(Nil)	3500(Nil)	Nil				1(Nil)	3500(Nil)	Nil			
13	General manufacturing	3(67)	866(47)	Nil				3(67)	866(47)	Nil			
14	Iron and steel				1(Nil)	7000(Nil)	Nil	1(Nil)	7000(Nil)	Nil			
15	Laundering and dry cleaning												
16	Leather tanning												
17	Metal smelting												
18	Paint making	2(100)	1220(100)	Nil				2(100)	1220(100)	Nil			
19	Paper and board making	2(100)	3580(100)	Nil	2(50)	9000(53)	Nil	4(75)	12580(67)	Nil			
20	Petroleum refining												
21	Plastics manufacture												
22	Plating and metal finishing	2(50)	560(36)	Nil	2(50)	112(51)	Nil	4(50)	672(38)	Nil			
23	Pottery making												
24	Printing ink etc												
25	Quarrying and mining												
26	Rubber processing												
27	Soap and detergent												
28	Textile, cotton and man-made												
29	Textile, wool												
30	General farming												
31	Atomic energy establishments												
50	Water treatment	4(75)	1100(54)	Nil				4(75)	1100(54)	Nil			
51	Disposal tip drainage	3(67)	236(48)	Nil				3(67)	236(48)	Nil			
	TOTAL FOR CODES 1-51	59(64)	33779(76)	Nil	6(33)	16113(30)	Nil	65(62)	49892(61)	Nil			
52	Derelict coal mines	4	1838					4	1838				
53	Other derelict mines												
54	Active coal mines	11	6833					11	6833				
55	Other active mines	1	300					1	300				
	TOTAL FOR CODES 52-55	16	8971					16	8971				
	TOTAL	75	42750	Nil	6	16113	Nil	81	58863	Nil			

| ...vers and canals | | | Discharges of solely cooling water to | | | | | | | | | | Total Cooling Water Discharged 1000s galls | Total Process Water Discharged 1000s galls | Total Discharge 1000s galls | CBI Code |
| | | | Non-tidal rivers | | Tidal rivers | | All rivers | | Canals | | All rivers and canals | | | | | |
	Total Volume 1000s galls (% satisfactory)	Volume of Cooling Water 1000s galls	Total Number (% satisfactory)	Total Volume 1000s galls (% satisfactory)	Total Number (% satisfactory)	Total Volume 1000s galls (% satisfactory)	Total Number (% satisfactory)	Total Volume 1000s galls (% satisfactory)	Total Number (% satisfactory)	Total Volume 1000s galls (% satisfactory)	Total Number (% satisfactory)	Total Volume 1000s galls (% satisfactory)				
																01
																02
																03
	2870(92)	Nil											Nil	2870	2870	04
	9692(83)	Nil	2(100)	2200(100)			2(100)	2200(100)			2(100)	2200(100)	2200	9692	11892	05
																06
	1500(100)	Nil	3(100)	98024(100)			3(100)	98024(100)			3(100)	98024(100)	98024	1500	99524	07
	807(1)	Nil											Nil	907	907	08
																09
	7749(93)	Nil			1(Nil)	460(Nil)	1(Nil)	460(Nil)			1(Nil)	460(Nil)	460	7749	8209	10
																11
	3500(Nil)	Nil											Nil	3500	3500	12
	866(47)	Nil											Nil	866	866	13
	7000(Nil)	Nil			1(100)	18600(100)	1(100)	18600(100)			1(100)	18600(100)	18600	7000	25600	14
																15
																16
																17
	1220(100)	Nil											Nil	1220	1220	18
	12580(67)	Nil											Nil	12580	12580	19
																20
																21
	672(38)	Nil											Nil	672	672	22
																23
																24
																25
																26
																27
																28
																29
																30
																31
	1100(54)	Nil											Nil	1100	1100	50
	236(48)	Nil											Nil	236	236	51
	49892(61)	Nil	5(100)	100224(100)	2(50)	19060(98)	7(86)	119284(99·9)			7(86)	119284(99·9)	119284	49892	169176	
	1838													1838	1838	52
																53
	6833													6833	6833	54
	300													300	300	55
	8971													8971	8971	
	58863	Nil	5	100224	2	19060	7	119284			7	119284	119284	58863	178147	

SOUTH WEST WALES RIVER AUTHORITY

TABLE 209 Main survey Numbers and volumes of discharges of industrial effluent to rivers and canals in England and Wales and percentage considered satisfactory

Industry		Discharges of total effluent (which may contain some cooling water) to											
		Non-tidal rivers			Tidal rivers			All rivers			Canals		
CBI Code	Description	Total Number (% satisfactory)	Total Volume 1000s galls (% satisfactory)	Volume of Cooling Water 1000s galls	Total Number (% satisfactory)	Total Volume 1000s galls (% satisfactory)	Volume of Cooling Water 1000s galls	Total Number (% satisfactory)	Total Volume 1000s galls (% satisfactory)	Volume of Cooling Water 1000s galls	Total Number (% satisfactory)	Total Volume 1000s galls (% satisfactory)	Volume Cooling Water 1000s galls
01	Brewing	2(Nil)	42(Nil)	Nil				2(Nil)	42(Nil)	Nil			
02	Brickmaking												
03	Cement making												
04	Chemical and allied industries				2(50)	420(29)	Nil	2(50)	420(29)	Nil			
05	Coal mining	9(67)	970(51)	Nil				9(67)	970(51)	Nil			
06	Distillation of ethanol												
07	Electricity generation	1(100)	83000(100)	Nil	1(100)	4000(100)	Nil	2(100)	87000(100)	Nil			
08	Engineering	1(Nil)	180(Nil)	Nil				1(Nil)	180(Nil)	Nil	1(Nil)	30(Nil)	Nil
09	Food processing and manufacture	6(67)	898(87)	Nil	3(Nil)	516(Nil)	150	9(44)	1414(55)	150			
10	Gas and coke												
11	Glass making												
12	Glue and gelatine												
13	General manufacturing	5(40)	274(22)	Nil				5(40)	274(22)	Nil			
14	Iron and steel	1(100)	3500(100)	Nil	2(100)	3311(100)	Nil	3(100)	6811(100)	Nil			
15	Laundering and dry cleaning				1(100)	20(100)	Nil	1(100)	20(100)	Nil			
16	Leather tanning												
17	Metal smelting	9(78)	8502(51)	Nil				9(78)	8502(51)	Nil			
18	Paint making												
19	Paper and board making												
20	Petroleum refining				3(100)	6364(100)	Nil	3(100)	6364(100)	Nil			
21	Plastics manufacture												
22	Plating and metal finishing	2(50)	148(73)	Nil	1(100)	7(100)	Nil	3(67)	155(74)	Nil			
23	Pottery making												
24	Printing ink etc												
25	Quarrying and mining	2(100)	26(100)	Nil				2(100)	26(100)	Nil			
26	Rubber processing												
27	Soap and detergent												
28	Textile, cotton and man-made												
29	Textile, wool												
30	General farming	1(Nil)	1(Nil)	Nil				1(Nil)	1(Nil)	Nil			
31	Atomic energy establishments												
50	Water treatment	1(100)	5(100)	Nil				1(100)	5(100)	Nil			
51	Disposal tip drainage	3(Nil)	15(Nil)	Nil				3(Nil)	15(Nil)	Nil			
	TOTAL FOR CODES 1–51	43(58)	97651(95)	Nil	13(69)	14638(94)	150	56(60)	112199(95)	150	1(Nil)	30(Nil)	Nil
52	Derelict coal mines												
53	Other derelict mines	26	13102					26	13102				
54	Active coal mines	1	20		1	1728		2	1748				
55	Other active mines												
	TOTAL FOR CODES 52–55	27	13122		1	1728		28	14850				
	TOTAL	70	110683	Nil	14	16366	150	84	127049	150	1	30	Nil

			Discharges of solely cooling water to										Total Cooling Water Discharged 1000s galls	Total Process Water Discharged 1000s galls	Total Discharge 1000s galls	
rivers and canals			Non-tidal rivers		Tidal rivers		All rivers		Canals		All rivers and canals					
er atis- ry)	Total Volume 1000s galls (% satis-factory)	Volume of Cooling Water 1000s galls	Total Number (% satis-factory)	Total Volume 1000s galls (% satis-factory)	Total Number (% satis-factory)	Total Volume 1000s galls (% satis-factory)	Total Number (% satis-factory)	Total Volume 1000s galls (% satis-factory)	Total Number (% satis-factory)	Total Volume 1000s galls (% satis-factory)	Total Number (% satis-factory)	Total Volume 1000s galls (% satis-factory)				CBI Code
il)	42(Nil)	Nil	1(100)	100(100)			1(100)	100(100)			1(100)	100(100)	100	42	142	01
) 7)	420(29) 970(51)	Nil Nil	1(100)	6(100)			1(100)	6(100)			1(100)	6(100)	Nil 6	420 970	420 976	04 05 06
00) il) 4)	87000(100) 210(Nil) 1414(55)	Nil Nil 150	1(100)	2650(100)	1(100)	18900(100)	1(100) 1(100)	18900(100) 2650(100)			1(100) 1(100)	18900(100) 2650(100)	18900 Nil 2800	87000 210 1264	105900 210 4064	07 08 09
																10 11 12
0) 100) 100)	274(22) 6811(100) 20(100)	Nil Nil Nil	1(100)	8(100)			1(100)	8(100)			1(100)	8(100)	8 Nil Nil	274 6811 20	282 6811 20	13 14 15
78)	8502(51)	Nil	2(100)	1224(100)			2(100)	1224(100)	1(100)	480(100)	3(100)	1704(100)	1704	8502	10206	16 17 18
100)	6364(100)	Nil							2(100)	834(100)	2(100)	834(100)	834	6364	7198	19 20 21
(67)	155(74)	Nil	2(100)	72(100)			2(100)	72(100)			2(100)	72(100)	72	155	227	22 23 24
(100)	26(100)	Nil											Nil	26	26	25 26 27
																28 29
(Nil)	1(Nil)	Nil											Nil	1	1	30
																31
(100) (Nil)	5(100) 15(Nil)	Nil Nil											Nil Nil	5 15	5 15	50 51
(60)	112229(95)	150	8(100)	4060(100)	1(100)	18900(100)	9(100)	22960(100)	3(100)	1314(100)	12(100)	24274(100)	24424	112079	136503	
																52 53
	13102 1748													13102 1748	13102 1748	54 55
	14850													14850	14850	
	127079	150	8	4060	1	18900	9	22960	3	1314	12	24274	24424	126929	151353	

RP—H*

GWYNEDD RIVER AUTHORITY

TABLE 210 Main survey Numbers and volumes of discharges of industrial effluent to rivers and canals in England and Wales and percentage considered satisfactory

Industry		Discharges of total effluent (which may contain some cooling water) to											
		Non-tidal rivers			Tidal rivers			All rivers			Canals		
CBI Code	Description	Total Number (% satisfactory)	Total Volume 1000s galls (% satisfactory)	Volume of Cooling Water 1000s galls	Total Number (% satisfactory)	Total Volume 1000s galls (% satisfactory)	Volume of Cooling Water 1000s galls	Total Number (% satisfactory)	Total Volume 1000s galls (% satisfactory)	Volume of Cooling Water 1000s galls	Total Number (% satisfactory)	Total Volume 1000s galls (% satisfactory)	Volume Cooling Water 1000s galls
01	Brewing												
02	Brickmaking												
03	Cement making												
04	Chemical and allied industries												
05	Coal mining												
06	Distillation of ethanol												
07	Electricity generation												
08	Engineering												
09	Food processing and manufacture												
10	Gas and coke												
11	Glass making												
12	Glue and gelatine												
13	General manufacturing												
14	Iron and steel												
15	Laundering and dry cleaning	1(Nil)	26(Nil)	Nil				1(Nil)	26(Nil)	Nil			
16	Leather tanning												
17	Metal smelting												
18	Paint making												
19	Paper and board making												
20	Petroleum refining												
21	Plastics manufacture												
22	Plating and metal finishing												
23	Pottery making												
24	Printing ink etc												
25	Quarrying and mining												
26	Rubber processing												
27	Soap and detergent												
28	Textile, cotton and man-made												
29	Textile, wool												
30	General farming												
31	Atomic energy establishments												
50	Water treatment												
51	Disposal tip drainage												
	TOTAL FOR CODES 1-51	1(Nil)	26(Nil)	Nil				1(Nil)	26(Nil)	Nil			
52	Derelict coal mines												
53	Other derelict mines												
54	Active coal mines												
55	Other active mines												
	TOTAL FOR CODES 52-55												
	TOTAL	1	26	Nil				1	26	Nil			

	rivers and canals		Discharges of solely cooling water to										Total Cooling Water Discharged 1000s galls	Total Process Water Discharged 1000s galls	Total Discharge 1000s galls	
			Non-tidal rivers		Tidal rivers		All rivers		Canals		All rivers and canals					
Total Number (% satis- factory)	Total Volume 1000s galls (% satis- factory)	Volume of Cooling Water 1000s galls	Total Number (% satis- factory)	Total Volume 1000s galls (% satis- factory)	Total Number (% satis- factory)	Total Volume 1000s galls (% satis- factory)	Total Number (% satis- factory)	Total Volume 1000s galls (% satis- factory)	Total Number (% satis- factory)	Total Volume 1000s galls (% satis- factory)	Total Number (% satis- factory)	Total Volume 1000s galls (% satis- factory)				CBI Code
																01
																02
																03
																04
																05
																06
																07
																08
																09
																10
																11
																12
																13
																14
1(Nil)	26(Nil)	Nil											Nil	26	26	15
																16
																17
																18
																19
																20
																21
																22
																23
																24
																25
																26
																27
																28
																29
																30
																31
																50
																51
1(Nil)	26(Nil)	Nil											Nil	26	26	
																52
																53
																54
																55
1	26	Nil											Nil	26	26	

DEE AND CLWYD RIVER AUTHORITY

TABLE 211 Main survey Numbers and volumes of discharges of industrial effluent to rivers and canals in England and Wales and percentage considered satisfactory

Industry		Discharges of total effluent (which may contain some cooling water) to											
		Non-tidal rivers			Tidal rivers			All rivers			Canals		
CBI Code	Description	Total Number (% satisfactory)	Total Volume 1000s galls (% satisfactory)	Volume of Cooling Water 1000s galls	Total Number (% satisfactory)	Total Volume 1000s galls (% satisfactory)	Volume of Cooling Water 1000s galls	Total Number (% satisfactory)	Total Volume 1000s galls (% satisfactory)	Volume of Cooling Water 1000s galls	Total Number (% satisfactory)	Total Volume 1000s galls (% satisfactory)	Volume of Cooling Water 1000s galls
01	Brewing												
02	Brickmaking												
03	Cement making												
04	Chemical and allied industries	5(80)	1819(92)	346	2(50)	11(91)	8	7(71)	1830(92)	354			
05	Coal mining	1(100)	100(100)	Nil				1(100)	100(100)	Nil			
06	Distillation of ethanol												
07	Electricity generation				1(100)	1473(100)	191	1(100)	1473(100)	191			
08	Engineering												
09	Food processing and manufacture	2(50)	60(83)	Nil				2(50)	60(83)	Nil			
10	Gas and coke	1(Nil)	3(Nil)	Nil				1(Nil)	3(Nil)	Nil			
11	Glass making	1(Nil)	39(Nil)	24				1(Nil)	39(Nil)	24			
12	Glue and gelatine												
13	General manufacturing												
14	Iron and steel	1(Nil)	604(Nil)	60	1(100)	37840(100)	1892	2(50)	38444(98)	1952			
15	Laundering and dry cleaning												
16	Leather tanning												
17	Metal smelting												
18	Paint making												
19	Paper and board making	3(Nil)	978(Nil)	Nil				3(Nil)	978(Nil)	Nil			
20	Petroleum refining												
21	Plastics manufacture												
22	Plating and metal finishing												
23	Pottery making												
24	Printing ink etc												
25	Quarrying and mining	6(67)	490(76)	Nil	1(Nil)	500(Nil)	Nil	7(57)	990(38)	Nil			
26	Rubber processing												
27	Soap and detergent												
28	Textile, cotton and man-made				2(Nil)	8820(Nil)	1202	2(Nil)	8820(Nil)	1202			
29	Textile, wool												
30	General farming	11(Nil)	67(Nil)	Nil				11(Nil)	67(Nil)	Nil			
31	Atomic energy establishments												
50	Water treatment	16(100)	905(100)	Nil				16(100)	905(100)	Nil			
51	Disposal tip drainage	1(Nil)	1(Nil)	Nil				1(Nil)	1(Nil)	Nil			
	TOTAL FOR CODES 1-51	48(54)	5066(61)	430	7(43)	48644(81)	3293	55(53)	53710(79)	3723			
52	Derelict coal mines												
53	Other derelict mines												
54	Active coal mines	1	125					1	125				
55	Other active mines	2	151					2	151				
	TOTAL FOR CODES 52-55	3	276					3	276				
	TOTAL	51	5342	430	7	48644	3293	58	53986	3723			

| rivers and canals | | | Discharges of solely cooling water to | | | | | | | | | | Total Cooling Water Discharged 1000s galls | Total Process Water Discharged 1000s galls | Total Discharge 1000s galls | CBI Code |
| | | | Non-tidal rivers | | Tidal rivers | | All rivers | | Canals | | All rivers and canals | | | | | |
Total Number (% satisfactory)	Total Volume 1000s galls (% satisfactory)	Volume of Cooling Water 1000s galls	Total Number (% satisfactory)	Total Volume 1000s galls (% satisfactory)	Total Number (% satisfactory)	Total Volume 1000s galls (% satisfactory)	Total Number (% satisfactory)	Total Volume 1000s galls (% satisfactory)	Total Number (% satisfactory)	Total Volume 1000s galls (% satisfactory)	Total Number (% satisfactory)	Total Volume 1000s galls (% satisfactory)				
																01
																02
			1(100)	112(100)			1(100)	112(100)			1(100)	112(100)	112	Nil	112	03
(71)	1830(92)	354											357	1476	1833	04
(100)	100(100)	Nil			1(Nil)	50(Nil)	1(Nil)	50(Nil)	1(100)	3(100)	1(100)	3(100)	50	100	150	05
											1(Nil)	50(Nil)				06
(100)	1473(100)	191											191	1282	1473	07
			1(100)	5(100)			1(100)	5(100)			1(100)	5(100)	5	Nil	5	08
2(50)	60(83)	Nil	2(100)	1258(100)			2(100)	1258(100)			2(100)	1258(100)	1258	60	1318	09
1(Nil)	3(Nil)	Nil											Nil	3	3	10
1(Nil)	39(Nil)	24											24	15	39	11
																12
2(50)	38444(98)	1952											1952	36492	38444	13
																14
																15
																16
																17
																18
3(Nil)	978(Nil)	Nil											Nil	978	978	19
																20
																21
																22
																23
																24
7(57)	990(38)	Nil	2(100)	22(100)			2(100)	22(100)			2(100)	22(100)	22	990	1012	25
																26
																27
2(Nil)	8820(Nil)	1202	1(100)	40(100)			1(100)	40(100)			1(100)	40(100)	1242	7618	8860	28
1(Nil)	67(Nil)	Nil											Nil	67	67	29
																30
16(100)	905(100)	Nil											Nil	905	905	31
1(Nil)	1(Nil)	Nil											Nil	1	1	50
																51
55(53)	53710(79)	3723	7(100)	1437(100)	1(Nil)	50(Nil)	8(88)	1487(97)	1(100)	3(100)	9(88)	1490(97)	5213	49987	55200	
																52
1	125													125	125	53
2	151													151	151	54
																55
3	276													276	276	
58	53986	3723	7	1437	1	50	8	1487	1	3	9	1490	5213	50263	55476	

MERSEY AND WEAVER RIVER AUTHORITY

TABLE 212 Main survey Numbers and volumes of discharges of industrial effluent to rivers and canals in England and Wales and percentage considered satisfactory

Industry		Discharges of total effluent (which may contain some cooling water) to											
		Non-tidal rivers			Tidal rivers			All rivers			Canals		
CBI Code	Description	Total Number (% satisfactory)	Total Volume 1000s galls (% satisfactory)	Volume of Cooling Water 1000s galls	Total Number (% satisfactory)	Total Volume 1000s galls (% satisfactory)	Volume of Cooling Water 1000s galls	Total Number (% satisfactory)	Total Volume 1000s galls (% satisfactory)	Volume of Cooling Water 1000s galls	Total Number (% satisfactory)	Total Volume 1000s galls (% satisfactory)	Volume of Cooling Water 1000s galls
01	Brewing												
02	Brickmaking	1(Nil)	58(Nil)	Nil				1(Nil)	58(Nil)	Nil			
03	Cement making	3(33)	211(3)	21				3(33)	211(3)	21			
04	Chemical and allied industries	45(27)	71313(9)	32650	35(29)	26912(28)	5322	80(27)	98225(14)	37972	6(50)	4548(13)	929
05	Coal mining	5(60)	268(55)	Nil				5(60)	268(55)	Nil			
06	Distillation of ethanol												
07	Electricity generation	13(92)	2049(98)	925	1(100)	250(100)	Nil	14(93)	2299(98)	925			
08	Engineering	4(75)	1465(88)	1024	2(Nil)	350(Nil)	34	6(50)	1815(71)	1058			
09	Food processing and manufacture	6(67)	8849(5)	2588	7(Nil)	4272(Nil)	164	13(31)	13121(3)	2752	2(Nil)	650(Nil)	Nil
10	Gas and coke	7(43)	4814(11)	Nil				7(43)	4814(11)	Nil			
11	Glass making	1(Nil)	100(Nil)	Nil				1(Nil)	100(Nil)	Nil			
12	Glue and gelatine	2(Nil)	62(Nil)	46				2(Nil)	62(Nil)	46			
13	General manufacturing	8(75)	8869(92)	634				8(75)	8869(92)	634			
14	Iron and steel	1(Nil)	150(Nil)	143				1(Nil)	150(Nil)	143			
15	Laundering and dry cleaning												
16	Leather tanning	1(Nil)	650(Nil)	Nil				1(Nil)	650(Nil)	Nil			
17	Metal smelting	8(38)	955(67)	378	2(50)	20(50)	2	10(40)	975(67)	380			
18	Paint making	1(Nil)	15(Nil)	Nil	1(Nil)	10(Nil)	Nil	2(Nil)	25(Nil)	Nil			
19	Paper and board making	27(33)	23907(32)	900	3(Nil)	11130(Nil)	Nil	30(30)	35037(22)	900	1(100)	270(100)	Nil
20	Petroleum refining	20(40)	5411(41)	1502	15(60)	159649(97)	139007	35(49)	165060(95)	140509	2(Nil)	170(Nil)	84
21	Plastics manufacture	2(Nil)	220(Nil)	146				2(Nil)	220(Nil)	146			
22	Plating and metal finishing	3(33)	286(6)	Nil	4(Nil)	540(Nil)	488	7(14)	826(2)	488			
23	Pottery making	2(50)	42(71)	Nil				2(50)	42(71)	Nil			
24	Printing Ink etc												
25	Quarrying and mining	5(60)	1314(31)	36				5(60)	1314(31)	36			
26	Rubber processing	1(100)	16(100)	Nil				1(100)	16(100)	Nil			
27	Soap and detergent	5(40)	3809(18)	2930	16(Nil)	5063(Nil)	602	21(9)	8872(8)	3532			
28	Textile, cotton and man-made	30(20)	19026(23)	656				30(20)	19026(23)	656			
29	Textile, wool	7(Nil)	450(Nil)	Nil				7(Nil)	450(Nil)	Nil			
30	General farming	17(Nil)	74(Nil)	Nil				17(Nil)	74(Nil)	Nil			
31	Atomic energy establishments												
50	Water treatment	11(100)	670(100)	Nil				11(100)	670(100)	Nil			
51	Disposal tip drainage	11(36)	794(45)	Nil				11(36)	794(45)	Nil			
	TOTAL FOR CODES 1-51	247(38)	155847(23)	44579	86(24)	208196(78)	145619	333(34)	164043(55)	190198	11(36)	5638(15)	1013
52	Derelict coal mines	2	1400					2	1400				
53	Other derelict mines												
54	Active coal mines	11	2825					11	2825		1	580	
55	Other active mines												
	TOTAL FOR CODES 52-55	13	4225					13	4225		1	580	
	TOTAL	260	160072	44579	86	208196	145619	346	368268	190198	12	6218	1013

rivers and canals			Discharges of solely cooling water to										Total Cooling Water Discharged 1000s galls	Total Process Water Discharged 1000s galls	Total Discharge 1000s galls	CBI Code
			Non-tidal rivers		Tidal rivers		All rivers		Canals		All rivers and canals					
l umber satis- ory)	Total Volume 1000s galls (% satis- factory)	Volume of Cooling Water 1000s galls	Total Number (% satis- factory)	Total Volume 1000s galls (% satis- factory)	Total Number (% satis- factory)	Total Volume 1000s galls (% satis- factory)	Total Number (% satis- factory)	Total Volume 1000s galls (% satis- factory)	Total Number (% satis- factory)	Total Volume 1000s galls (% satis- factory)	Total Number (% satis- factory)	Total Volume 1000s galls (% satis- factory)				
(Nil)	58(Nil)	Nil											Nil	58	58	01
(33)	211(3)	21											21	190	211	02
																03
(30)	102773(14)	38901	18(44)	69723(3)	5(40)	21770(95)	23(43)	91493(25)	5(80)	3968(77)	28(50)	95461(27)	134362	63872	198234	04
(60)	268(55)	Nil											Nil	268	268	05
																06
(93)	2299(98)	925	22(100)	612599(100)	5(100)	596000(100)	27(100)	1208599(100)	1(100)	98000(100)	28(100)	1306599(100)	1307524	1374	1308898	07
(50)	1815(71)	1058	3(100)	312(100)			3(100)	312(100)	1(100)	7000(100)	4(100)	7312(100)	8370	757	9127	08
(27)	13771(33)	2752							5(100)	2642(100)	5(100)	2642(100)	5394	11019	16413	09
(43)	4814(11)	Nil	2(50)	1400(71)			2(50)	1400(71)			2(50)	1400(71)	1400	4814	6214	10
(Nil)	100(Nil)	Nil	1(100)	380(100)			1(100)	380(100)	4(50)	850(41)	5(60)	1230(59)	1230	100	1330	11
(Nil)	62(Nil)	46	3(33)	2775(65)			3(33)	2775(65)			3(33)	2775(65)	2821	16	2837	12
(75)	8869(92)	634	9(89)	1229(97)			9(89)	1229(97)	2(100)	2033(100)	11(91)	3262(99)	3896	8235	12131	13
(Nil)	150(Nil)	143	1(Nil)	30000(Nil)	1(100)	688(100)	2(50)	30688(2)	3(100)	6300(100)	5(80)	36988(18)	37131	7	37138	14
																15
(Nil)	650(Nil)	Nil											Nil	650	650	16
(40)	975(67)	380			1(Nil)	226(Nil)	1(Nil)	226(Nil)			1(Nil)	226(Nil)	606	595	1201	17
(Nil)	25(Nil)	Nil							1(100)	13(100)	1(100)	13(100)	13	25	38	18
(32)	35307(23)	900	7(100)	8350(100)	1(100)	16800(100)	8(100)	25150(100)	1(100)	66(100)	9(100)	25216(100)	26116	34407	60523	19
(46)	165230(95)	140593	2(100)	7270(100)	2(50)	3500(20)	4(75)	10770(74)	1(100)	130(100)	5(80)	10900(74)	151493	24637	176130	20
(Nil)	220(Nil)	146	1(100)	100(100)			1(100)	100(100)	1(100)	210(100)	2(100)	310(100)	456	74	530	21
(14)	826(2)	488											488	338	826	22
(50)	42(71)	Nil											Nil	42	42	23
									1(100)	9(100)	1(100)	9(100)	9		9	24
(60)	1314(31)	36											36	1278	1314	25
(100)	16(100)	Nil	5(100)	3456(100)			5(100)	3456(100)	2(100)	70(100)	7(100)	3526(100)	3526	16	3542	26
(9)	8872(8)	3532	2(100)	780(100)			2(100)	780(100)			2(100)	780(100)	4312	5340	9652	27
(20)	19026(23)	656	9(100)	1814(100)			9(100)	1814(100)	1(100)	2500(100)	10(100)	4314(100)	4970	18370	23340	28
(Nil)	450(Nil)	Nil											Nil	450	450	29
(Nil)	74(Nil)	Nil											Nil	74	74	30
(100)	670(100)	Nil											Nil	670	670	31
(36)	794(45)	Nil											Nil	794	794	50
																51
(34)	369681(54)	191211	85(82)	740188(87)	15(67)	638984(99)	100(80)	1379172(92)	29(90)	123791(99)	129(82)	1502963(93)	1694174	178470	1872644	
2	1400													1400	1400	52
																53
12	3405													3405	3405	54
																55
14	4805													4805	4805	
358	374486	191211	85	740188	15	638984	100	1379172	29	123791	129	1502963	1694174	183275	1877449	

LANCASHIRE RIVER AUTHORITY

TABLE 213 Main survey Numbers and volumes of discharges of industrial effluent to rivers and canals in England and Wales and percentage considered satisfactory

Industry		Discharges of total effluent (which may contain some cooling water) to											
		Non-tidal rivers			Tidal rivers			All rivers			Canals		
CBI Code	Description	Total Number (% satis-factory)	Total Volume 1000s galls (% satis-factory)	Volume of Cooling Water 1000s galls	Total Number (% satis-factory)	Total Volume 1000s galls (% satis-factory)	Volume of Cooling Water 1000s galls	Total Number (% satis-factory)	Total Volume 1000s galls (% satis-factory)	Volume of Cooling Water 1000s galls	Total Number (% satis-factory)	Total Volume 1000s galls (% satis-factory)	Volume Cooling Water 1000s galls
01	Brewing												
02	Brickmaking												
03	Cement making	1(Nil)	200(Nil)	Nil				1(Nil)	200(Nil)				
04	Chemical and allied industries	4(Nil)	267(Nil)	Nil	7(43)	10910(38)	3605	11(27)	11177(37)	3605			
05	Coal mining												
06	Distillation of ethanol												
07	Electricity generation	2(100)	360(100)	Nil	1(100)	360(100)	Nil	3(100)	720(100)				
08	Engineering	3(Nil)	339(Nil)	163				3(Nil)	339(Nil)	163			
09	Food processing and manufacture	4(50)	275(60)	99	1(Nil)	350(Nil)	Nil	5(40)	625(26)	99			
10	Gas and coke	1(100)	20(Nil)	16				1(100)	20(100)	16			
11	Glass making												
12	Glue and gelatine	1(Nil)	75(Nil)	68				1(Nil)	75(Nil)	68			
13	General manufacturing												
14	Iron and steel												
15	Laundering and dry cleaning												
16	Leather tanning				1(Nil)	210(Nil)	Nil	1(Nil)	210(Nil)	Nil			
17	Metal smelting												
18	Paint making												
19	Paper and board making	15(7)	11100(4)	473	3(Nil)	4200(Nil)	Nil	18(5)	15300(3)	473			
20	Petroleum refining												
21	Plastics manufacture												
22	Plating and metal finishing												
23	Pottery making	2(Nil)	25(Nil)	Nil				2(Nil)	25(Nil)	Nil			
24	Printing ink etc												
25	Quarrying and mining	6(17)	941(1)	Nil				6(17)	941(1)	Nil			
26	Rubber processing												
27	Soap and detergent												
28	Textile, cotton and man-made	9(11)	3372(1)	139				9(11)	3372(1)	139			
29	Textile, wool												
30	General farming												
31	Atomic energy establishments												
50	Water treatment	20(75)	1978(76)	Nil				20(75)	1978(76)	Nil			
51	Disposal tip drainage	4(Nil)	1221(Nil)	Nil				4(Nil)	1221(Nil)	Nil			
	TOTAL FOR CODES 1-51	72(32)	20173(12)	958	13(31)	16030(28)	3605	85(32)	36203(22)	4563			
52	Derelict coal mines	6	1830					6	1830				
53	Other derelict mines												
54	Active coal mines	3	580					3	580		1	820	
55	Other active mines												
	TOTAL FOR CODES 52-55	9	2410					9	2410		1	820	
	TOTAL	81	22578	958	13	16030	3605	94	38608	4563	1	820	

vers and canals			Discharges of solely cooling water to										Total Cooling Water Discharged 1000s galls	Total Process Water Discharged 1000s galls	Total Discharge 1000s galls	CBI Code
			Non-tidal rivers		Tidal rivers		All rivers		Canals		All rivers and canals					
ber atis- ry)	Total Volume 1000s galls (% satis- factory)	Volume of Cooling Water 1000s galls	Total Number (% satis- factory)	Total Volume 1000s galls (% satis- factory)	Total Number (% satis- factory)	Total Volume 1000s galls (% satis- factory)	Total Number (% satis- factory)	Total Volume 1000s galls (% satis- factory)	Total Number (% satis- factory)	Total Volume 1000s galls (% satis- factory)	Total Number (% satis- factory)	Total Volume 1000s galls (% satis- factory)				
il)	200(Nil)	Nil											Nil	200	200	01 02 03
7)	11177(37)	3605	2(50)	1002(48)			2(50)	1002(48)			2(50)	1002(48)	4607	7572	12179	04 05 06
00) Nil) 0)	720(100) 339(Nil) 625(26)	Nil 163 99	3(100) 4(100) 5(100)	31315(100) 195(100) 3333(100)	4(100)	112100(100)	7(100) 4(100) 5(100)	143415(100) 195(100) 3333(100)	2(100) 2(100)	2020(100) 112(100)	9(100) 6(100) 5(100)	45435(100) 307(100) 3333(100)	145435 470 3432	720 176 526	146155 646 3958	07 08 09
00) Nil)	20(100) 75(Nil)	16 68	2(100) 1(100)	432(100) 150(100)			2(100) 1(100)	432(100) 150(100)	1(Nil)	1000(Nil)	2(100) 1(100) 1(Nil)	432(100) 150(100) 1000(Nil)	448 150 1068	4 Nil 7	452 150 1075	10 11 12
			1(100)	130(100)			1(100)	130(100)	1(100)	6(100)	2(100)	136(100)	136	Nil	136	13 14 15
Nil)	210(Nil)	Nil											Nil	210	210	16 17 18
5)	15300(3)	473	2(100)	2220(100)			2(100)	2220(100)			2(100)	2220(100)	2693	14827	17520	19 20
			2(100)	156(100)	1(100)	25(100)	3(100)	181(100)	1(Nil)	250(Nil)	4(75)	431(42)	431	Nil	431	21
Nil)	25(Nil)	Nil											Nil	25	25	22 23 24
(17)	941(1)	Nil											Nil	941	941	25
			1(100)	750(100)			1(100)	750(100)			1(100)	750(100)	750	Nil	750	26 27
(11)	3372(1)	139	2(100)	110(100)	4(100)	11150(100)	6(100)	11260(100)	4(75)	109(58)	10(90)	11369(99·6)	11508	3233	14741	28 29 30
(75) (Nil)	1978(76) 1221(Nil)	Nil Nil											Nil Nil	1978 1221	1978 1221	31 50 51
(32)	36203(22)	4563	25(96)	39793(99)	9(100)	123275(100)	34(97)	163068(99·7)	11(73)	3497(63)	45(91)	166565(99·7)	171128	31640	202768	
6	1830													1830	1830	52 53
4	1400													580 820	580 820	54 55
0	3230													3230	3230	
5	39433	4563	25	39793	9	123275	34	163068	11	3497	45	166565	171128	34870	205998	

CUMBERLAND RIVER AUTHORITY

TABLE 214 Main survey Numbers and volumes of discharges of industrial effluent to rivers and canals in England and Wales and percentage considered satisfactory

Industry		Discharges of total effluent (which may contain some cooling water) to											
		Non-tidal rivers			Tidal rivers			All rivers			Canals		
CBI Code	Description	Total Number (% satisfactory)	Total Volume 1000s galls (% satisfactory)	Volume of Cooling Water 1000s galls	Total Number (% satisfactory)	Total Volume 1000s galls (% satisfactory)	Volume of Cooling Water 1000s galls	Total Number (% satisfactory)	Total Volume 1000s galls (% satisfactory)	Volume of Cooling Water 1000s galls	Total Number (% satisfactory)	Total Volume 1000s galls (% satisfactory)	Volume Cooling Water 1000s galls
01	Brewing												
02	Brickmaking	2(Nil)	43(Nil)	Nil				2(Nil)	43(Nil)	Nil			
03	Cement making												
04	Chemical and allied industries				1(Nil)	720(Nil)	Nil	1(Nil)	720(Nil)	Nil			
05	Coal mining												
06	Distillation of ethanol												
07	Electricity generation	3(67)	516(99)	Nil				3(67)	516(99)	Nil			
08	Engineering	3(Nil)	45(Nil)	Nil				3(Nil)	45(Nil)	Nil			
09	Food processing and manufacture												
10	Gas and coke												
11	Glass making												
12	Glue and gelatine												
13	General manufacturing	1(100)	11(100)	Nil				1(100)	11(100)	Nil			
14	Iron and steel												
15	Laundering and dry cleaning												
16	Leather tanning	1(Nil)	8(Nil)	Nil				1(Nil)	8(Nil)	Nil			
17	Metal smelting	1(Nil)	41(Nil)	Nil				1(Nil)	41(Nil)	Nil			
18	Paint making												
19	Paper and board making	1(100)	50(100)	Nil				1(100)	50(100)	Nil			
20	Petroleum refining												
21	Plastics manufacture												
22	Plating and metal finishing												
23	Pottery making												
24	Printing ink etc												
25	Quarrying and mining	5(20)	527(12)	Nil				5(20)	527(12)	Nil			
26	Rubber processing												
27	Soap and detergent												
28	Textile, cotton and man-made	1(Nil)	500(Nil)	Nil				1(Nil)	500(Nil)	[Nil			
29	Textile, wool												
30	General farming												
31	Atomic energy establishments												
50	Water treatment	5(60)	541(95)	Nil				5(60)	541(95)	Nil			
51	Disposal tip drainage												
	TOTAL FOR CODES 1–51	23(35)	2282(50)	Nil	1(Nil)	720(Nil)	Nil	24(33)	3002(38)	Nil			
52	Derelict coal mines												
53	Other derelict mines												
54	Active coal mines												
55	Other active mines	7	2200					7	2200				
	TOTAL FOR CODES 52–55	7	2200					7	2200				
	TOTAL	30	4482	Nil	1	720	Nil	31	5202	Nil			

rs and canals		Discharges of solely cooling water to										Total Cooling Water Discharged 1000s galls	Total Process Water Discharged 1000s galls	Total Discharge 1000s galls	CBI Code
		Non-tidal rivers		Tidal rivers		All rivers		Canals		All rivers and canals					
Total Volume 1000s galls (% satis-factory)	Volume of Cooling Water 1000s galls	Total Number (% satis-factory)	Total Volume 1000s galls (% satis-factory)	Total Number (% satis-factory)	Total Volume 1000s galls (% satis-factory)	Total Number (% satis-factory)	Total Volume 1000s galls (% satis-factory)	Total Number (% satis-factory)	Total Volume 1000s galls (% satis-factory)	Total Number (% satis-factory)	Total Volume 1000s galls (% satis-factory)				
															01
43(Nil)	Nil											Nil	43	43	02
															03
720(Nil)	Nil	1(100)	960(100)			1(100)	960(100)			1(100)	960(100)	960	720	1680	04
															05
															06
516(99)	Nil	2(100)	77101(100)			2(100)	77101(100)			2(100)	77101(100)	77101	516	77617	07
45(Nil)	Nil											Nil	45	45	08
		3(100)	3596(100)			3(100)	3596(100)			3(100)	3596(100)	3596	Nil	3596	09
															10
															11
															12
11(100)	Nil											Nil	11	11	13
															14
															15
8(Nil)	Nil											Nil	8	8	16
41(Nil)	Nil											Nil	41	41	17
															18
50(100)	Nil											Nil	50	50	19
															20
															21
															22
															23
															24
527(12)	Nil											Nil	527	527	25
															26
															27
500(Nil)	Nil											Nil	500	500	28
															29
															30
															31
541(95)	Nil											Nil	541	541	50
															51
3002(38)	Nil	6(100)	81657(100)			6(100)	81657(100)			6(100)	81657(100)	81657	3002	84659	
															52
															53
															54
2200													2200	2200	55
2200													2200	2200	
5202	Nil	6	81657			6	81657			6	81657	81657	5202	86859	

THAMES CONSERVANCY

TABLE 215 Main survey Numbers and volumes of discharges of industrial effluent to rivers and canals in England and Wales and percentage considered satisfactory

Industry		Discharges of total effluent (which may contain some cooling water) to											
		Non-tidal rivers			Tidal rivers			All rivers			Canals		
CBI Code	Description	Total Number (% satisfactory)	Total Volume 1000s galls (% satisfactory)	Volume of Cooling Water 1000s galls	Total Number (% satisfactory)	Total Volume 1000s galls (% satisfactory)	Volume of Cooling Water 1000s galls	Total Number (% satisfactory)	Total Volume 1000s galls (% satisfactory)	Volume of Cooling Water 1000s galls	Total Number (% satisfactory)	Total Volume 1000s galls (% satisfactory)	Volume of Cooling Water 1000s galls
01	Brewing	1(100)	11(100)	10				1(100)	11(100)	10			
02	Brickmaking	2(100)	29(100)	Nil				2(100)	29(100)	Nil			
03	Cement making	1(100)	5500(100)	Nil				1(100)	5500(100)	Nil			
04	Chemical and allied industries	3(67)	394(97)	159				3(67)	394(97)	159			
05	Coal mining												
06	Distillation of ethanol												
07	Electricity generation												
08	Engineering	5(100)	768(100)	600				5(100)	768(100)	600	1(100)	45(100)	Nil
09	Food processing and manufacture	4(75)	84(88)	Nil				4(75)	84(88)	Nil			
10	Gas and coke	4(100)	3128(100)	2878				4(100)	3128(100)	2878			
11	Glass making												
12	Glue and gelatine												
13	General manufacturing	8(38)	1444(13)	851				8(38)	1444(13)	851			
14	Iron and steel												
15	Laundering and dry cleaning												
16	Leather tanning												
17	Metal smelting	1(100)	13(100)	Nil				1(100)	13(100)	Nil			
18	Paint making	1(Nil)	135(Nil)	27				1(Nil)	135(Nil)	27			
19	Paper and board making	8(75)	34910(16)	25590				8(75)	34910(16)	25590	1(100)	15(100)	Nil
20	Petroleum refining												
21	Plastics manufacture												
22	Plating and metal finishing	2(100)	308(100)	Nil				2(100)	308(100)	Nil			
23	Pottery making	1(Nil)	9(Nil)	Nil				1(Nil)	9(Nil)	Nil			
24	Printing ink etc	1(100)	327(100)	164				1(100)	327(100)	164			
25	Quarrying and mining	18(94)	22120(99·7)	Nil				18(94)	22120(99·7)	Nil	1(100)	635(100)	Nil
26	Rubber processing												
27	Soap and detergent												
28	Textile, cotton and man-made												
29	Textile, wool												
30	General farming	1(100)	10(100)	Nil				1(100)	10(100)	Nil			
31	Atomic energy establishments	4(100)	2016(100)	348				4(100)	2016(100)	348			
50	Water treatment	24(92)	14744(99)	15				24(92)	14744(99)	15			
51	Disposal tip drainage	1(100)	360(100)	Nil				1(100)	360(100)	Nil			
	TOTAL FOR CODES 1-51	90(84)	86310(64)	30642				90(84)	86310(64)	30642	3(100)	695(100)	Nil
52	Derelict coal mines												
53	Other derelict mines												
54	Active coal mines												
55	Other active mines												
	TOTAL FOR CODES 52-55												
	TOTAL	90	86310	30642				90	86310	30642	3	695	Nil

s and canals		Discharges of solely cooling water to										Total Cooling Water Discharged 1000s galls	Total Process Water Discharged 1000s galls	Total Discharge 1000s galls	
		Non-tidal rivers		Tidal rivers		All rivers		Canals		All rivers and canals					
Total Volume 1000s galls (% satis-factory)	Volume of Cooling Water 1000s galls	Total Number (% satis-factory)	Total Volume 1000s galls (% satis-factory)	Total Number (% satis-factory)	Total Volume 1000s galls (% satis-factory)	Total Number (% satis-factory)	Total Volume 1000s galls (% satis-factory)	Total Number (% satis-factory)	Total Volume 1000s galls (% satis-factory)	Total Number (% satis-factory)	Total Volume 1000s galls (% satis-factory)				CBI Code
11(100)	10	5(100)	928(100)			5(100)	928(100)			5(100)	928(100)	938	1	939	01
29(100)	Nil	1(100)	50(100)			1(100)	50(100)			1(100)	50(100)	50	29	79	02
5500(100)	Nil	1(100)	620(100)			1(100)	620(100)			1(100)	620(100)	620	5500	6120	03
394(97)	159	8(88)	427(98)			8(88)	427(98)	3(100)	112(100)	11(91)	539(98)	698	235	933	04
															05
															06
		3(100)	186240(100)			3(100)	186240(100)			3(100)	186240(100)	186240	Nil	186240	07
813(100)	600	15(93)	1943(100)			15(93)	1943(100)	2(100)	2740(100)	17(94)	4683(100)	5283	213	5496	08
84(88)	Nil	11(91)	6821(100)			11(91)	6821(100)	2(100)	175(100)	13(92)	6996(100)	6996	84	7080	09
3128(100)	2878											2878	250	3128	10
															11
															12
1444(13)	851	7(86)	1689(70)			7(86)	1689(70)	1(100)	29(100)	8(88)	1718(70)	2569	593	3162	13
															14
															15
															16
13(100)	Nil	1(100)	150(100)			1(100)	150(100)			1(100)	150(100)	150	13	163	17
135(Nil)	27	3(100)	155(100)			3(100)	155(100)			3(100)	155(100)	182	108	290	18
34925(16)	25590	6(100)	20275(100)			6(100)	20275(100)	1(100)	12(100)	7(100)	20287(100)	45877	9335	55212	19
		1(100)	164(100)			1(100)	164(100)			1(100)	164(100)	164	Nil	164	20
															21
308(100)	Nil	1(100)	140(100)			1(100)	140(100)			1(100)	140(100)	140	308	448	22
9(Nil)	Nil	2(100)	4(100)			2(100)	4(100)			2(100)	4(100)	4	9	13	23
327(100)	164	2(100)	174(100)			2(100)	174(100)			2(100)	174(100)	338	163	501	24
22755(99·7)	Nil											Nil	22755	22755	25
		3(100)	501(100)			3(100)	501(100)			3(100)	501(100)	501	Nil	501	26
															27
															28
															29
10(100)	Nil											Nil	10	10	30
2016(100)	348											348	1668	2016	31
14744(99)	15	5(100)	3034(100)			5(100)	3034(100)			5(100)	3034(100)	3049	14729	17778	51
360(100)	Nil											Nil	360	360	50
87005(64)	30642	75(95)	223315(99·8)			75(95)	223315(99·8)	9(100)	3068(100)	84(95)	226383(99·8)	257025	56363	313388	
															52
															53
															54
															55
87005	30642	75	223315			75	223315	9	3068	84	226383	257025	56363	313388	

LEE CONSERVANCY

TABLE 216 Main survey Numbers and volumes of discharges of industrial effluent to rivers and canals in England and Wales and percentage considered satisfactory

Industry		Discharges of total effluent (which may contain some cooling water) to											
		Non-tidal rivers			Tidal rivers			All rivers			Canals		
CBI Code	Description	Total Number (% satisfactory)	Total Volume 1000s galls (% satisfactory)	Volume of Cooling Water 1000s galls	Total Number (% satisfactory)	Total Volume 1000s galls (% satisfactory)	Volume of Cooling Water 1000s galls	Total Number (% satisfactory)	Total Volume 1000s galls (% satisfactory)	Volume of Cooling Water 1000s galls	Total Number (% satisfactory)	Total Volume 1000s galls (% satisfactory)	Vol Coo Wa 100 gall
01	Brewing												
02	Brickmaking												
03	Cement making												
04	Chemical and allied industries												
05	Coal mining												
06	Distillation of ethanol												
07	Electricity generation												
08	Engineering												
09	Food processing and manufacture												
10	Gas and coke	1(100)	500(100)	450				1(100)	500(100)	450			
11	Glass making												
12	Glue and gelatine												
13	General manufacturing												
14	Iron and steel												
15	Laundering and dry cleaning												
16	Leather tanning												
17	Metal smelting												
18	Paint making												
19	Paper and board making												
20	Petroleum refining												
21	Plastics manufacture												
22	Plating and metal finishing												
23	Pottery making												
24	Printing ink etc												
25	Quarrying and mining												
26	Rubber processing												
27	Soap and detergent												
28	Textile, cotton and man-made												
29	Textile, wool												
30	General farming												
31	Atomic energy establishments												
50	Water treatment												
51	Disposal tip drainage												
	TOTAL FOR CODES 1-51	1(100)	500(100)	450				1(100)	500(100)	450			
52	Derelict coal mines												
53	Other derelict mines												
54	Active coal mines												
55	Other active mines												
	TOTAL FOR CODES 52-55												
	TOTAL	1	500	450				1	500	450			

...rivers and canals		Discharges of solely cooling water to										Total Cooling Water Discharged 1000s galls	Total Process Water Discharged 1000s galls	Total Discharge 1000s galls	CBI Code
		Non-tidal rivers		Tidal rivers		All rivers		Canals		All rivers and canals					
Total Volume 1000s galls (% satisfactory)	Volume Cooling of Water 1000s galls	Total Number (% satisfactory)	Total Volume 1000s galls (% satisfactory)	Total Number (% satisfactory)	Total Volume 1000s galls (% satisfactory)	Total Number (% satisfactory)	Total Volume 1000s galls (% satisfactory)	Total Number (% satisfactory)	Total Volume 1000s galls (% satisfactory)	Total Number (% satisfactory)	Total Volume 1000s galls (% satisfactory)				
															01
															02
															03
		8(100)	3135(100)			8(100)	3135(100)			8(100)	3135(100)	3135	Nil	3135	04
															05
															06
		3(100)	112000(100)			3(100)	112000(100)			3(100)	112000(100)	112000	Nil	112000	07
		1(100)	50(100)			1(100)	50(100)			1(100)	50(100)	50	Nil	50	08
		4(100)	599(100)			4(100)	599(100)			4(100)	599(100)	599	Nil	599	09
500(100)	450											450	50	500	10
															11
															12
		3(100)	370(100)			3(100)	370(100)			3(100)	370(100)	370	Nil	370	13
															14
															15
		1(100)	150(100)			1(100)	150(100)			1(100)	150(100)	150	Nil	150	16
															17
															18
		1(100)	6000(100)			1(100)	6000(100)			1(100)	6000(100)	6000	Nil	6000	19
															20
															21
															22
															23
															24
															25
															26
															27
															28
															29
															30
															31
															50
															51
500(100)	450	21(100)	122304(100)			21(100)	122304(100)			21(100)	122304(100)	122754	50	122804	
															52
															53
															54
															55
500	450	21	122304			21	122304			21	122304	122754	50	122804	

PORT OF LONDON AUTHORITY (INCLUDING THE LONDON EXCLUDED AREA)

TABLE 217 Main survey Numbers and volumes of discharges of industrial effluent to rivers and canals in England and Wales and percentage considered satisfactory

Industry		Discharges of total effluent (which may contain some cooling water) to									Canals		
		Non-tidal rivers			Tidal rivers			All rivers					
CBI Code	Description	Total Number (% satisfactory)	Total Volume 1000s galls (% satisfactory)	Volume of Cooling Water 1000s galls	Total Number (% satisfactory)	Total Volume 1000s galls (% satisfactory)	Volume of Cooling Water 1000s galls	Total Number (% satisfactory)	Total Volume 1000s galls (% satisfactory)	Volume of Cooling Water 1000s galls	Total Number (% satisfactory)	Total Volume 1000s galls (% satisfactory)	Volume Cooling Water 1000s galls
01	Brewing				1(Nil)	40(Nil)	Nil	1(Nil)	40(Nil)	Nil			
02	Brickmaking												
03	Cement making												
04	Chemical and allied industries				7(43)	1512(76)	421	7(43)	1512(76)	421			
05	Coal mining												
06	Distillation of ethanol												
07	Electricity generation	1(100)	1250(100)	Nil	11(100)	204013(100)	172889	12(100)	205263(100)	172889			
08	Engineering												
09	Food processing and manufacture	1(100)	15(100)	4	1(Nil)	600(Nil)	Nil	2(50)	615(2)	4			
10	Gas and coke				1(100)	7000(100)	6790	1(100)	7000(100)	6790			
11	Glass making												
12	Glue and gelatine												
13	General manufacturing												
14	Iron and steel												
15	Laundering and dry cleaning												
16	Leather tanning												
17	Metal smelting												
18	Paint making												
19	Paper and board making				10(Nil)	32924(Nil)	11037	10(Nil)	32924(Nil)	11037			
20	Petroleum refining				2(50)	82317(100)	73800	2(50)	82317(100)	73800			
21	Plastics manufacture												
22	Plating and metal finishing				1(Nil)	172(Nil)	163	1(Nil)	172(Nil)	163			
23	Pottery making												
24	Printing ink etc												
25	Quarrying and mining				4(50)	13514(0·1)	Nil	4(50)	13514(0·1)	Nil	1(100)	10(100)	
26	Rubber processing												
27	Soap and detergent				1(100)	30(100)	Nil	1(100)	30(100)	Nil			
28	Textile, cotton and man-made												
29	Textile, wool												
30	General farming												
31	Atomic energy establishments												
50	Water treatment				1(Nil)	34(Nil)	Nil	1(Nil)	34(Nil)	Nil			
51	Disposal tip drainage				1(Nil)	50(Nil)	Nil	1(Nil)	50(Nil)	Nil			
	TOTAL FOR CODES 1–51	2(100)	1265(100)	4	41(46)	342206(86)	265100	43(49)	343471(86)	265104	1(100)	10(100)	
52	Derelict coal mines												
53	Other derelict mines												
54	Active coal mines												
55	Other active mines												
	TOTAL FOR CODES 52–55												
	TOTAL	2	1265	4	41	342206	265100	43	343471	265104	1	10	

ers and canals			Discharges of solely cooling water to										Total Cooling Water Discharged 1000s galls	Total Process Water Discharged 1000s galls	Total Discharge 1000s galls	
			Non-tidal rivers		Tidal rivers		All rivers		Canals		All rivers and canals					
er tis- y)	Total Volume 1000s galls (% satis-factory)	Volume of Cooling Water 1000s galls	Total Number (% satis-factory)	Total Volume 1000s galls (% satis-factory)	Total Number (% satis-factory)	Total Volume 1000s galls (% satis-factory)	Total Number (% satis-factory)	Total Volume 1000s galls (% satis-factory)	Total Number (% satis-factory)	Total Volume 1000s galls (% satis-factory)	Total Number (% satis-factory)	Total Volume 1000s galls (% satis-factory)				CBI Code
	40(Nil)	Nil			1(100) 1(100) 6(100)	250(100) 40(100) 471(100)	1(100) 1(100) 6(100)	250(100) 40(100) 471(100)			1(100) 1(100) 6(100)	250(100) 40(100) 471(100)	250 40 471	40 Nil Nil	290 40 471	01 02 03
	1512(76)	421	6(83)	2144(94)	7(86)	38003(100)	13(85)	40147(100)			13(85)	40147(100)	40568	1091	41659	04 05 06
0))	205263(100) 615(2)	172889 4	4(100) 1(100)	308(100) 120(100)	26(100) 3(100) 7(43)	2793972(100) 78722(100) 26641(59)	26(100) 7(100) 8(50)	2793972(100) 79030(100) 26761(59)	3(67) 5(100) 3(100)	7450(36) 179(100) 266(100)	29(97) 12(100) 11(47)	2801422(99·8) 79209(100) 27027(60)	2974311 79209 27031	32374 Nil 611	3006685 79209 27642	07 08 09
00)	7000(100)	6790			5(80)	24561(97)	5(80)	24561(97)			5(80)	24561(97)	31351	210	31561	10 11 12
																13 14 15
																16 17 18
il) 0)	32924(Nil) 82317(100)	11037 73800	2(100) 3(100)	6060(100) 17(100)	3(100) 4(100)	22516(100) 36337(100)	5(100) 4(100) 3(100)	28576(100) 36337(100) 17(100)			5(100) 4(100) 3(100)	28576(100) 36337(100) 17(100)	39613 110137 17	21887 8517 Nil	61500 118654 17	19 20 21
il)	172(Nil)	163											163	9	172	22 23 24
0) 100)	13524(0·1) 30(100)	Nil Nil	1(100)	40(100)	2(Nil)	2150(Nil)	3(33)	2190(2)	1(100)	2265(100)	1(100) 3(33)	2265(100) 2190(2)	Nil 2265 2190	13524 Nil 30	13524 2265 2220	25 26 27
																28 29 30
Nil) Nil)	34(Nil) 50(Nil)	Nil Nil			1(100)	131(100)	1(100)	131(100)			1(100)	131(100)	131 Nil	34 50	165 50	31 50 51
49)	343481(86)	265104	17(94)	8689(99)	66(88)	3023794(99·9)	83(89)	3032483(99·9)	12(92)	10160(53)	95(89)	3042643(99·8)	3307747	78377	3386124	
																52 53 54 55
	343481	265104	17	8689	66	3023794	83	3032483	12	10160	95	3042643	3307747	78377	3386124	

Table 218 Estimates of costs of remedial works on discharges of sewage effluent in £ millions

River Authority	Non-tidal rivers					Tidal rivers					All rivers
	Class 1	Class 2	Class 3	Class 4	Total non-tidal	Class 1	Class 2	Class 3	Class 4	Total tidal	Class 1
1 Northumbrian	3.090	0.542	0.591	1.3	5.523	0.7	0.1	0.75	0.864	2.414	3.79
2 Yorkshire	16.049	5.508	9.252	18.207	49.016	0.35	1.2	2.88	0.01	4.44	16.399
3 Trent	5.925	14.095	6.721	52.026	78.767	0	0	0	0.196	0.196	5.925
4 Lincolnshire	1.709	0.9	0.05	0	2.659	0.01	0	0.008	0	0.018	1.719
5 Welland and Nene	8.827	2.466	0.03	0	11.323	0	3.017	0	0	3.017	8.827
6 Great Ouse	7.452	12.291	1.471	0	21.214	0.635	0	0	0	0.635	8.087
7 East Suffolk and Norfolk	1.048	0.41	0	0	1.458	0.085	6.029	0	0	6.114	1.133
8 Essex	2.920	1.452	1.359	0	5.731	0.722	1.925	0.675	0	3.322	3.642
9 Kent	6.481	0	0	0	6.481	4.535	0.28	2.5	0.273	7.588	11.016
10 Sussex	3.231	1.039	0	0	4.27	0.615	0	0	0	0.615	3.846
11 Hampshire	1.53	0.467	0	0	1.997	1.8	1.815	0.37	1.3	5.285	3.33
12 Isle of Wight	0.24	0.04	0	0	0.28	0	0	0.3	0	0.3	0.24
13 Avon and Dorset	3.42	0.047	0	0	3.467	0.245	0	0	0	0.245	3.665
14 Devon	2.99	0.038	0	0	3.028	0.01	0.2	2.035	0	2.245	3.0
15 Cornwall	1.521	0.07	0.064	0.116	1.771	0.2	0	0	0	0.2	1.721
16 Somerset	4.094	1.175	0	0	5.269	1.25	0	0	0	1.25	5.344
17 Bristol Avon	2.978	4.166	0.09	0	7.234	0	0	0	0.06	0.06	2.978
18 Severn	15.253	3.759	3.437	5.955	28.404	2.622	0	0	0	2.622	17.875
19 Wye	1.175	0.063	1.6	0	2.838	0	0	0	0	0	1.175
20 Usk	0.79	0	0.04	0	0.83	0.7	2.5	0	0	3.2	1.49
21 Glamorgan	0.512	0.015	7.745	0	8.272	6.5	0	0	0	6.5	7.012
22 South West Wales	0.682	0.17	0.1	0	0.952	0.23	0	0.435	0.04	0.245	0.95	0.912
23 Gwynedd	0.839	0.12	0	0	0.959	0.175	0	0	0	0.175	1.014
24 Dee and Clwyd	1.179	1.455	0.05	0	2.684	0.21	0.75	0	0	0.96	1.389
25 Mersey and Weaver	2.914	6.261	5.403	15.1	29.678	0	0	2.055	1.965	4.02	2.914
26 Lancashire	1.327	3.315	0.884	10.215	15.741	1.86	0.097	1.85	0.02	3.827	3.187
27 Cumberland	1.326	0.531	0.02	0	1.877	0.092	0.03	0	0	0.122	1.418
28 Thames Conservancy	20.059	2.932	1.66	0	24.651	0	0	0	0	0	20.059
29 Lee Conservancy	5.414	0.24	0.17	0	5.824	0	0	0	0	0	5.414
30 Port of London Authority (incl. London Excluded Area)	0	0	0	0	0	0	1.25	30.146	0	31.396	0
Total England and Wales	124.975	63.567	40.737	102.919	332.198	23.546	19.628	43.609	4.933	91.716	148.521

			Canals					All rivers and canals					
Class 3	Class 4	Total all rivers	Class 1	Class 2	Class 3	Class 4	Total canals	Class 1	Class 2	Class 3	Class 4	Total all rivers and canals	
1.341	2.164	7.937	0	0	0	0	0	3.790	0.642	1.341	2.164	7.937	1
12.132	18.217	53.456	0	0	0	0.102	0.102	16.399	6.708	12.132	18.319	53.558	2
6.721	52.222	78.963	0	0.695	1.125	0	1.820	5.925	14.79	7.846	52.222	80.783	3
0.058	0	2.677	0	0	0	0	0	1.719	0.9	0.058	0	2.677	4
0.03	0	14.340	0.015	0	0	0	0.015	8.842	5.483	0.03	0	14.355	5
1.471	0	21.849	0	0	0	0	0	8.087	12.291	1.471	0	21.849	6
0	0	7.572	0	0	0	0	0	1.133	6.439	0	0	7.572	7
2.034	0	9.053	0	0	0	0	0	3.642	3.377	2.034	0	9.053	8
2.5	0.273	14.069	0	0	0	0	0	11.016	0.28	2.5	0.273	14.069	9
0	0	4.885	0	0	0	0	0	3.846	1.039	0	0	4.885	10
0.37	1.3	7.282	0	0	0	0	0	3.33	2.282	0.37	1.3	7.282	11
0.3	0	0.58	0	0	0	0	0	0.24	0.04	0.3	0	0.58	12
0	0	3.712	0	0	0	0	0	3.665	0.047	0	0	3.712	13
2.035	0	5.273	0	0	0	0	0	3.0	0.238	2.035	0	5.273	14
0.064	0.116	1.971	0	0	0	0	0	1.721	0.07	0.064	0.116	1.971	15
0	0	6.519	0	0	0	0	0	5.344	1.175	0	0	6.519	16
0.09	0.06	7.294	0	0	0	0	0	2.978	4.166	0.09	0.06	7.294	17
3.437	5.955	31.026	0	0.075	0	0	0.075	17.875	3.834	3.437	5.955	31.101	18
1.6	0	2.838	0	0	0	0	0	1.175	0.063	1.6	0	2.838	19
0.04	0	4.03	0	0	0	0	0	1.49	2.5	0.04	0	4.03	20
7.745	0	14.772	0	0	0	0	0	7.012	0.015	7.745	0	14.772	21
0.14	0.245	1.902	0	0	0	0	0	0.912	0.605	0.14	0.245	1.902	22
0	0	1.134	0	0	0	0	0	1.014	0.12	0	0	1.134	23
0.05	0	3.644	0	0	0	0	0	1.389	2.205	0.05	0	3.644	24
7.458	17.065	33.698	0	0	0	0	0	2.914	6.261	7.458	17.065	33.698	25
2.734	10.235	19.568	0	0	0	0	0	3.187	3.412	2.734	10.235	19.568	26
0.02	0	1.999	0	0	0	0	0	1.418	0.561	0.02	0	1.999	27
1.66	0	24.651	0.326	0	0	0	0.326	20.385	2.932	1.66	0	24.977	28
0.17	0	5.824	0	0	0	0	0	5.414	0.24	0.17	0	5.824	29
30.146	0	31.396	0	0	0	0	0	0	1.25	30.146	0	31.396	30
84.346	107.852	423.914	0.341	0.770	1.125	0.102	2.338	148.862	83.965	85.471	107.954	426.252	

Table 219 Estimates of costs of remedial works on discharges of crude sewage in £ millions

	River Authority	Non-tidal rivers					Tidal rivers				
		Class 1	Class 2	Class 3	Class 4	Total non-tidal	Class 1	Class 2	Class 3	Class 4	Total tidal
1	Northumbrian	0	0	0	0	0	0.895	1.526	4.89	60.3	67.61
2	Yorkshire	0.21	0	0	0	0.21	0.28	0	0.915	0	1.19
3	Trent	0	0	0	0	0	0	0	0	0	0
4	Lincolnshire	0	0	0	0	0	0	0	1.0	0	1.0
5	Welland and Nene	0	0	0	0	0	0	0	0	0	0
6	Great Ouse	0	0	0	0	0	0	0	0	0	0
7	East Suffolk and Norfolk	0	0	0	0	0	0.1	0.1	0	0	0.2
8	Essex	0	0	0	0	0	0.25	0	0.065	0.039	0.35
9	Kent	0	0	0	0	0	0.7	0.015	0.05	0	0.76
10	Sussex	0	0	0	0	0	1.5	0	0	0	1.5
11	Hampshire	0	0	0	0	0	0.1	0	0	0	0.1
12	Isle of Wight	0	0	0	0	0	0	1.2	0	0	1.2
13	Avon and Dorset	0	0	0	0	0	0	0	0	0	0
14	Devon	0.325	0.125	0	0	0.45	0.28	1.335	2.58	0	4.195
15	Cornwall	0	0	0	0	0	3.334	1.605	0	0	4.939
16	Somerset	0.33	0	0	0.03	0.36	0	0.51	0	3.5	4.01
17	Bristol Avon	0.03	0.763	0	0	0.793	0	0	0	1.715	1.715
18	Severn	0	0	0	0	0	0.85	0	0	0	0.85
19	Wye	0	0	0	0	0	0.11	0	0	0	0.11
20	Usk	0	0	0	0	0	0	0	0.762	0	0.762
21	Glamorgan	0	0	0	0	0	0	0	2.76	0.35	3.11
22	South West Wales	0.169	0	0	0	0.169	0.28	0.828	0.04	3.31	4.458
23	Gwynedd	0.577	0	0	0	0.577	1.174	0	0	0	1.174
24	Dee and Clwyd	0	0	0	0	0	0	0	0	0.3	0.3
25	Mersey and Weaver	0	0	0	0	0	0	27.425	7.165	3.57	38.16
26	Lancashire	0	0	0	0	0	0.7	1.595	0.16	1.515	3.97
27	Cumberland	0	0	0	0	0	0.13	0	0	0	0.13
28	Thames Conservancy	0	0	0	0	0	0	0	0	0	0
29	Lee Conservancy	0	0	0	0	0	0	0	0	0	0
30	Port of London Authority (incl. London Excluded Area)	0	0	0	0	0	0	0	0	0	0
	Total England and Wales	1.641	0.888	0	0.03	2.559	10.683	36.139	20.387	74.599	141.808

	Class 2	Class 3	Class 4	Total all rivers	
	1.526	4.89	60.3	67.611	1
	0	0.915	0	1.405	2
	0	0	0	0	3
	0	1.0	0	1.0	4
	0	0	0	0	5
	0	0	0	0	6
	0.1	0	0	0.2	7
	0	0.065	0.039	0.354	8
	0.015	0.05	0	0.765	9
	0	0	0	1.5	10
	0	0	0	0.1	11
	1.2	0	0	1.2	12
	0	0	0	0	13
	1.46	2.58	0	4.645	14
	1.605	0	0	4.939	15
	0.51	0	3.53	4.37	16
	0.763	0	1.715	2.508	17
	0	0	0	0.85	18
	0	0	0	0.11	19
	0	0.762	0	0.762	20
	0	2.76	0.35	3.11	21
	0.828	0.04	3.31	4.627	22
	0	0	0	1.751	23
	0	0	0.3	0.3	24
	27.425	7.165	3.57	38.16	25
	1.595	0.16	1.515	3.97	26
	0	0	0	0.13	27
	0	0	0	0	28
	0	0	0	0	29
	0	0	0	0	30
	37.027	20.387	74.629	144.367	

Table 220 Estimates of costs of remedial works on discharges of industrial effluent in £ millions

River Authority		Non-tidal rivers					Tidal rivers					All riv
		Class 1	Class 2	Class 3	Class 4	Total non-tidal	Class 1	Class 2	Class 3	Class 4	Total tidal	Class
1	Northumbrian	0.086	0.068	0.038	0	0.192	0	0	0.075	3.540	3.615	0.086
2	Yorkshire	0.724	0.268	0.734	2.678	4.404	0	0	0.272	0	0.272	0.724
3	Trent	0.118	0.153	0.272	0.531	1.074	0	0.100	0	1.000	1.100	0.118
4	Lincolnshire	0.019	0	0.080	0	0.099	0	0	0.002	0	0.002	0.019
5	Welland and Nene	0.033	0.005	0	0	0.038	0	0.003	0	0	0.003	0.033
6	Great Ouse	0.006	0.040	0	0	0.046	0.025	0	0	0	0.025	0.031
7	East Suffolk and Norfolk	0.035	0.010	0	0	0.045	0	0.079	0	0	0.079	0.035
8	Essex	0.110	0.003	0.003	0.027	0.143	0	0	0.005	0.011	0.016	0.110
9	Kent	0.008	0	0	0.002	0.010	0.406	0.750	0.102	0	1.258	0.414
10	Sussex	0.007	0.017	0	0	0.024	0	0.002	0	0	0.002	0.007
11	Hampshire	0.005	0.001	0.001	0.002	0.009	0	0	0	0.020	0.020	0.005
12	Isle of Wight	0.005	0	0	0	0.005	0	0	0	0	0	0.005
13	Avon and Dorset	0.013	0	0	0	0.013	0	0	0	0	0	0.013
14	Devon	0.070	0.010	0.063	0	0.143	0.020	0	0	0	0.020	0.090
15	Cornwall	0.080	0.165	0.682	2.731	3.658	0.002	0	0	0	0.002	0.082
16	Somerset	1.761	0.292	0.087	0	2.140	0	0.038	0	1.432	1.470	1.761
17	Bristol Avon	0.116	0.026	0.005	0.005	0.152	0	0	0	0	0	0.116
18	Severn	0.247	0.025	0	0.076	0.348	0.002	0	0	0	0.002	0.249
19	Wye	0.002	0	0	0	0.002	0	0	0	0	0	0.002
20	Usk	0.026	0.026	0.010	1.066	1.128	0	0	0	0	0	0.026
21	Glamorgan	0.206	0.020	0.143	0	0.369	0	0	0.003	0	0.003	0.206
22	South West Wales	0.475	0.863	0.050	0	1.388	0.150	0	0.200	0.115	0.465	0.625
23	Gwynedd	0.012	0	0	0	0.012	0	0	0	0	0	0.012
24	Dee and Clwyd	0.060	0.006	0.002	0.012	0.080	0.201	0.002	0	0	0.203	0.261
25	Mersey and Weaver	0.674	0.125	2.437	1.597	4.833	0	0.035	0.577	5.213	5.825	0.674
26	Lancashire	0.516	0.182	0.011	0.150	0.860	0	0.277	0.002	0.050	0.329	0.516
27	Cumberland	0.070	0.049	0	0	0.119	0	0.200	0	0	0.200	0.070
28	Thames Conservancy	0.208	0.052	0.020	0	0.280	0	0	0	0	0	0.208
29	Lee Conservancy	0	0	0	0	0	0	0	0	0	0	0
30	Port of London Authority (incl. London Excluded Area)	0	0	0	0	0	0	0	1.421	0	1.421	0
	Total England and Wales	5.692	2.406	4.638	8.886	21.622	0.806	1.486	2.659	11.381	16.332	6.498

See notes on Table 91 of Chapter 6.

				Canals					All rivers and canals					
s 2	Class 3	Class 4	Total all rivers	Class 1	Class 2	Class 3	Class 4	Total canals	Class 1	Class 2	Class 3	Class 4	Total all rivers and canals	
8	0.113	5.540	3.807	0	0	0	0	0	0.086	0.068	0.113	3.540	3.807	1
8	1.006	2.678	4.676	0.002	0	0.041	0.020	0.063	0.726	0.268	1.047	2.698	4.739	2
3	0.272	1.531	2.174	0	0.002	0.081	0.278	0.361	0.118	0.255	0.353	1.809	2.535	3
	0.082	0	0.101	0	0	0	0	0	0.019	0	0.082	0	0.101	4
08	0	0	0.041	0	0	0	0	0	0.033	0.008	0	0	0.041	5
40	0	0	0.071	0	0	0	0	0	0.031	0.040	0	0	0.071	6
89	0	0	0.124	0	0	0	0	0	0.035	0.089	0	0	0.124	7
03	0.008	0.038	0.159	0	0	0	0	0	0.110	0.003	0·008	0·038	0.159	8
50	0.102	0.002	1.268	0	0	0	0	0	0.414	0.750	0.102	0.002	1.268	9
19	0	0	0.026	0	0	0	0	0	0.007	0.019	0	0	0.026	10
01	0.001	0.022	0.029	0	0	0	0	0	0.005	0.001	0.001	0.022	0.029	11
	0	0	0.005	0	0	0	0	0	0.005	0	0	0	0.005	12
	0	0	0.013	0	0	0	0	0	0.013	0	0	0	0.013	13
10	0.063	0	0.163	0	0	0	0	0	0.090	0.010	0.063	0	0.163	14
65	0.682	2.731	3.660	0	0	0	0	0	0.082	0.165	0.682	2.731	3.660	15
30	0.087	1.432	3.610	0	0	0	0	0	1.761	0.330	0.087	1.432	3.610	16
26	0.005	0.005	0.152	0	0	0	0	0	0.116	0.026	0.005	0.005	0.152	17
25	0	0.076	0.350	0.022	0.104	0	0	0.126	0.271	0.129	0	0.076	0.476	18
	0	0	0.002	0	0	0	0	0	0.002	0	0	0	0.002	19
26	0.010	1.066	1.128	0	0	0	0	0	0.026	0.026	0.010	1.066	1.128	20
20	0.146	0	0.372	0	0	0	0	0	0.206	0.020	0.146	0	0.372	21
63	0.250	0.115	1.853	0.004	0	0	0	0.004	0.629	0.863	0.250	0.115	1.857	22
	0	0	0.012	0	0	0	0	0	0.012	0	0	0	0.012	23
08	0.002	0.012	0.283	0	0	0	0	0	0.261	0.008	0.002	0.012	0.283	24
60	3.014	6.810	10.658	0.020	0.060	0.120	0	0.200	0.694	0.220	3.134	6.810	10.858	25
459	0.013	0.209	1.197	0	0	0	0	0	0.516	0.459	0.013	0.209	1.197	26
249	0	0	0.319	0	0	0	0	0	0.070	0.249	0	0	0.319	27
052	0.020	0	0.280	0	0	0	0	0	0.208	0.052	0.020	0	0.280	28
	0	0	0	0	0	0	0	0	0	0	0	0	0	29
	1.421	0	1.421	0	0.025	0	0	0.025	0	0.025	1.421	0	1.446	30
892	7.297	20.267	37.954	0.048	0.191	0.242	0.298	0.779	6.546	4.083	7.539	20.565	38.733	

Printed in London for Her Majesty's Stationery Office by McCorquodale Printers Ltd.
HM4811 Dd. 505554 K34 12/72